KV-038-760

Surveillance of Environmental Pollution and

Resources by Electromagnetic Waves

NATO ADVANCED STUDY INSTITUTES SERIES

Proceedings of the Advanced Study Institute Programme, which aims at the dissemination of advanced knowledge and the formation of contacts among scientists from different countries

The series is published by an international board of publishers in conjunction with NATO Scientific Affairs Division

A Life Sciences B Physics	Plenum Publishing Corporation London and New York
C Mathematical and Physical Sciences	D. Reidel Publishing Company Dordrecht, Boston and London
D Behavioral and Social Sciences	Sijthoff International Publishing Company Leiden
E Applied Sciences	Noordhoff International Publishing Leiden

Series C – Mathematical and Physical Sciences

Volume 45 – Surveillance of Environmental Pollution and Resources by Electromagnetic Waves

Surveillance of Environmental Pollution and Resources by Electromagnetic Waves

Proceedings of the NATO Advanced Study Institute held in Spåtind, Norway, 9–19 April, 1978

edited by

TERJE LUND
Remote Sensing Technology Programme
Kjeller, Norway

D. Reidel Publishing Company

Dordrecht : Holland / Boston : U.S.A. / London : England

Published in cooperation with NATO Scientific Affairs Division

Library of Congress Cataloging in Publication Data

Nato Advanced Study Institute, Spåtind, Norway, 1978.
Surveillance of environmental pollution and resources by electromagnetic waves.

(NATO advanced study institutes series : Series C, Mathematical and physical sciences ; v. 45)
Bibliography: p.
1. Pollution–Remote sensing–Congresses. 2. Electromagnetic waves–Congresses. 3. Remote sensing–Congresses. I. Lund, Terje. II. Series.
TD177.N43 1978 628.5 78-12033
ISBN 90-277-0949-1

Published by D. Reidel Publishing Company
P.O. Box 17, Dordrecht, Holland

Sold and distributed in the U.S.A., Canada, and Mexico
by D. Reidel Publishing Company, Inc.
Lincoln Building, 160 Old Derby Street, Hingham, Mass. 02043, U.S.A.

Printed in The Netherlands

CONTENTS

PREFACE

These proceedings contain lectures, research papers and working group reports from the NATO Advanced Study Institute on "Surveillance of environmental pollution and resources by electromagnetic waves", held at Spåtind, Norway, April 9-19, 1978.

Remote sensing of the environment has developed into a very complex multidisciplinary field. It encompasses a huge range of different instrumental techniques and analytical methods, designed to provide information about a vast number of environmental parameters. Nevertheless, the approach to solve specific problems and the ways of handling the collected information are to a large extent the same or similar.

This commonality is the basis for the Advanced Study Institute. To provide the best possible background, both tutorially and for a fruitful exchange of research ideas and results, a number of outstanding scientists were invited to review some major fields.

The material presented in these proceedings is certainly not complete in the sense that it covers all aspects of the subject. The selection is deliberately due to the program committee and the editor.

The program committee would like to express their gratitude to Dr. Tilo Kester, head of the NATO Advanced Study Institute Program, NATO Scientific Affairs Division, for his support and encouragement during the organization of the Institute.

Oslo, June 1978.

Terje Lund
Proceedings editor

PARTICIPANTS

Allen, K.C.	1135 Lefthand Dr, Longmont, Colorado 80501, U.S.A
Alparslan, E.	Marmara Scientific and Industrial Research Inst., P.O.Box 141 Kadikøy, Istanbul, Turkey
Ataman, E.	Marmara Scientific and Industrial Research Inst., P.O.Box 141 Kadikøy, Istanbul, Turkey
Attema, E.P.W.	Microwave Laboratory, Delft Technical University, Mekelveg 2, Delft, Holland
Bartsch, N.	Institut für Flugfunk und Mikrowellen, D.F.V.L.R., Oberpfaffenhofen, 8031 Wessling, West Germany
Becker, G.	Deutsches Hydrographisches Institut, 4 Bernhard Nochtstrasse 78, Hamburg, West Germany
Berg, T.C.	NILU, Elvegt. 52, 2000 Lillestrøm, Norway
Bingen, E.	NDRE, P.O.Box 25, N-2007 Kjeller, Norway
Birks, A.R.	Appleton Laboratory, Ditton Park, Slough, Bucks. SL3 9JX, England
Bjor, H.	Elektrisk Bureau A/S, Bergerv. 12, 1360 Nesbru, Norway
Bratteng, O.	University of Tromsø, 9000 Tromsø, Norway
Burrows, W.G.	Head of Department, Marine Electronics, Brunel Technical College Bristol, Ashley Down, Bristol BS7 9BU, England
Capitini, M.R.	Centre d'Études Nucléaires de Saclay, Departement Recherche et Analyse, Boite Postale no 2, 91190 GIF-sur-YVETTE, France
Christensen, H.	NTNF, Gaustadalléen 30, Blindern, Oslo 3, Norway
Croom, D.	Appleton Laboratory, Ditton Park, Slough, Bucks. SL3 9JX, England
Dahl, H.	Christian Michelsens Institutt, Fantoftvegen 38, 5036 Fantoft, Norway
Ekengren, B.	MI-divisionen, L.M. Ericsson, Mölndal, Sweden

Eng, S.T. Institutionen för Mätteknik, Chalmers Tekniska Högskola, Gøteborg, Sweden
Fjeldly, T.A. University of Trondheim, 7034 Trondheim-NTH, Norway
Fredriksson, K. Department of Physics, Chalmers Tekniska Högskola, Fack, S-402 20 Gøteborg, Sweden
Gjessing, D.T. NTNF/PFF, P.O.Box 25, N-2007 Kjeller, Norway
Günneberg, F Bundesanstalt für Gewasserkunde, Postfach 309, 5400 Koblenz, West Germany
Gylling-Nielssen, O. Forsvarets forskningstjeneste, Østerbrogades Kaserne, København, Denmark
Haydn, R. Zentralstelle für Geophotogrammetrie und Fernerkundung (DFG), Luisenstrasse 37/IV, 800 München 2, West Germany
Heylen, R. Institut Royal Météorologique de Belgique, Avenye Circulaire 3, 1180 Bruxelles, Belgium
Holberg, K. NDRE, P.O.Box 25, N-2007 Kjeller, Norway
Holt, O. University of Tromsø, 9000 Tromsø, Norway
Jensen, A. Institute of Marine Biochemistry, University of Trondheim, N7034 Trondheim-NTH, Norway
Johannesen, O.M. Geofysisk Institutt, avd. A, University of Bergen, 5000 Bergen, Norway
Kjelaas, A.G. NTNF/PFF, P.O.Box 25, N-2007 Kjeller, Norway
Lund, T. NTNF/PFF, P.O.Box 25, N-2007 Kjeller, Norway
McKendrick, J.D. Office of Naval Research Branch Office, London 223/231 Old Marylebone Rd, London NW1 5TH, England
Mehlum, E. Sentralinstituttet for Industriell Forskning, Forskningsv. 1, Blindern, Oslo 3, Norway
Myrabø, H.K. NDRE, P.O.Box 25, N-2007 Kjeller, Norway
Offner, H. Institut für Physikalische Elektronik, Universität Stuttgart, Böblinger Str. 70, D-7000 Stuttgart 1, West Germany
Orhaug, T. National Defence Research Institute, Box 1165, S-581 11 Linkøping, Sweden
Ottar, B. NILU, Elvegt. 52, 2000 Lillestrøm, Norway
Parr, H. NTNF/PFF, P.O.Box 25, N-2007 Kjeller, Norway
Peckham, G. Dept. of Physics, Heriot-Watt University, Riccarton, Currie, Edinburgh EH14 4AS, U.K.
Rothe, K. Sektion Physik, Universität München, Am Coulombwall 1, 8046 Garching, West Germany

Røyset, H. — Geofysisk Inst., University of Bergen, 5000 Bergen, Norway

Schanda, E. — University of Bern, Inst. of Applied Physics, Sidlerstrasse 5, 3012 Berne, Switzerland

Schlesak, J. — Communications Research Centre, Box 11490 Station H, Ottawa K1N 8T5, Ontario, Canada

Schmalfuss, H. — Physikalisches Inst. der University of Erlangen-Nürnberg, Erwin-Rommel str. 1, D-8520 Erlangen, West Germany

Shemdin, O.H. — Jet Propulsion Lab., Caltech., 4800 Oak Grove Drive, Pasadena, Ca. 91103, U.S.A.

Steinmann, R. — D.F.V.L.R., Postfach 906058, 5000 Köln 90, West Germany

Stokseth, P.Aa. — NDRE, P.O.Box, N-2007 Kjeller, Norway

Svanberg, S. — Department of Physics, Chalmers Tekniska Högskola, Fack, S-402 20 Gøteborg, Sweden

Symeonides, C. — Filellinon 10, Kavala, Greece

Teillet, P.M. — Canada Centre for Remote Sensing, 2464 Sheffield Road, Ottawa K1A OY7, Canada

Thomas, J. — Ass. Director, Imperial College of Science and Technology, The Blackett Laboratory, Prince Consort Rd., London SW1 2BZ, England

Thompson, M.C. — Tropospheric Telecommunication Labs., Institute for Telecommunications Sciences, Boulder, Colorado 80302, U.S.A.

Tonning, A. — Department of Electrical Engineering, Institute for Electron Physcis, University of Trondheim, N-7034 Trondheim-NTH, Norway

Valenzuela, G. — Code 8344, Naval Research Laboratory, 4555 Overlook Ave SW, Washington D.C. 20375, U.S.A.

Vogel, M. — Deutsche Forschung und Versuchsanstalt für Luft- und Raumfahrt, D-8031 Oberpfaffenhofen, West Germany

Williams, P.D.L. — Decca Development Labs., Davis Td., Chessington, Surrey KT9 1TB, England

Wenstøp, Ø. — NDRE, P.O.Box 25, N-2007 Kjeller, Norway

Wiesemann, W. — Battelle-Institute e.V., 6000 Frankfurt am Main 90, Am Römerhof 35, Postfach 900160, West Germany

Wootton, N. — Blackett Lab., Imperial College, Prince Consort Rd., London SW7 2BZ, England

Yura, H.T. — The Aerospace Corporation, P.O.Box 92957, Los Angeles, California 90009, U.S.A.

PROGRAM COMMITTEE

Gjessing, D.T. director	Royal Norwegian Council for Scientific and Industrial Research, Remote Sensing Technology Programme, P.O.Box 25, N-2007 Kjeller, Norway
Thomas, J. co-director	Ass. Director, Imperial College of Science and Technology, The Blackett Laboratory, Prince Consort Rd., London SW1 2BZ, England
Thompson, M.C.	Tropospheric Telecommunication Labs., Institute for Telecommunications Sciences, Boulder, Colorado 80302, U.S.A.
Vogel, M.	Deutsche Forschung und Versuchsanstalt für Luft- und Raumfahrt, D-8031 Oberpfaffenhofen, West Germany
Lund, T. editor of Proceedings	Royal Norwegian Council for Scientific and Industrial Research, Remote Sensing Technology Programme, P.O.Box 25, N-2007 Kjeller, Norway

HOST NATION ADDRESS

Andreas Tonning

Department of Electrical Engineering, Institute for Electron Physics, University of Trondheim, N-7034 Trondheim - NTH, Norway

Dear Colleagues, gentlemen.

I have the great privilege of wishing you all warmly welcome to Norway as participants in this NATO Advanced Study Institute on "Surveillance of Environmental Pollution and Resources by Electromagnetic Waves". I am glad that everyone of you have successfully overcome the communication difficulties and found your way to this rather remote corner of Europe.

If you take a look at the surroundings, here at Spåtind, the snowclad mountains, the spruce trees, you might easily jump to the conclusion that pollution can hardly be the most serious problem for this part of the world. And it is indeed true that our country is thinly populated as compared to the average of Western Europe. There are about 10 Norwegians per square kilometer. In spite of this low population density a part of our industrial activities, our agriculture and the relatively high per capita consumption do cause pollution problems that must be taken seriously. Moreover, and perhaps more important the pollution in the atmosphere and in the ocean show a regrettably poor respect for national boundaries. They spread indiscriminately and rapidly over large distances and are therefore truly international. Hence they must be combated, if at all, by international collaboration.

Since NATO Science Committee was established about 20 years ago its primary aim has been to further international collaboration between scientist from the member countries of the Alliance. The development of general principles for the surveillance of the environmental pollution and resources constitute a logical first step in our common defence against pollution and in our efforts to preserve resources. It seems particularly appropriate, therefore, that this theme should be chosen for one of the NATO Advanced Study Institutes. I want to express the gratitude of the scientific community of Norway,to NATO and to the NATO Science Committee for financing this meeting. Likewise I want to thank the lecturers and participants for coming so far to take part in its work. I hope that you, in addition to the professional benefits, also will be able to form personal contacts and friendship, and that you will find the nature and the skiing facilities offered an enjoyable recreation.

The theme of the conference specifies electromagnetic waves as the means for carrying out the surveillance. The wide spectrum of electromagnetic waves we have at our disposal to-day is indeed the most natural, and most powerful, diagnostic tool for carrying out surveillance. As long as electromagnetic waves have been known they have been used as a means for exploring the macro-world around us all as well as the micro-world of atoms and molecules. The hypothesis of the existence of ionospheric layers around the earth is about as old as our century. It is worth remembering that its experimental basis was established during the first attempts to communicate by radio-waves across the Atlantic.

At mid-century, when microwave generators and receivers had become easily available, exploration of the atmosphere, of its layers, turbulences and other inhomogeneities became possible. This research field was then pursued with great vigour. At the outset the motivation was again predominantly due to radio communication interests. Scattering from the atmospheric inhomogeneities provided possibilities for long range, over-the-horizon communication. However, one soon realized that the radio waves were a tool of high potential for studying geophysical phenomena such as atmospheric winds, temperature and humidity variations in the atmosphere. The importance for the field of meteorology is obvious and, thanks to the methods of radio science, an improved understanding of the processes in the atmosphere has been achieved.

At the small wave-length end of the spectrum the identification of atoms and molecules by means of spectroscopy was made possible by the work of Kirchoff and Bunsen as early as around 1860, many years before it was definitely known that light is of electromagnetic nature.

The enormous importance of spectroscopy for chemistry and physics in general and for the identification of matter in particular, needs no comments here. During the years after the last war the availability of microwave generators for frequencies beyond 10 GHz made microwave spectroscopy possible, thus allowing for the first time the use of coherent electromagnetic sources for spectroscopic purposes. The advances of microwave techniques increased gradually the upper frequency limit of microwave generators into the millimeter wave region.

Then a most dramatic change occurred at the beginning of the nineteen-sixties with the invention of the laser. All of a sudden the possibility of having stable, coherent, monochromatic optical sources in the visible region as well as in the ultraviolet and infra-red seemed within reach. This has frequently been emphasized as one of the most important scientific and technological break-throughs in our generation. It is probably also true that the larger part of the possibilities offered by the laser still remains to be exploited.

From the point of view of radio science, faced with the problems of surveillance that concern us here, a number of possibilities appear to offer themselves. The possibility of laser heterodyne receivers creates visions of a dramatic increase in receiver sensitivity, similar to what happened with microwave receiver a generation ago. The advent of tunable lasers that can work dependably also outside of the laboratory might strongly increase the degrees of freedom for designers of systems for laser surveillance.

The use of advanced methods of signal processing has brought marked progress within the field of telecommunications and, still more, within radar techniques and radio astronomy. When the development of electro-optical components have reached a level of perfection that allows realization of similar principles in the optical region, great improvements can be expected. In this country developments in the direction just indicated have been advocated forcefully and eloquently by to-days chairman, the initiator of this meeting, Dag T. Gjessing. I expect that we shall have occasion to discuss them in some detail later in connection with some of his lectures.

Let me, in conclusion, point out that many of the resources we want to preserve are biological resources. Also, the pollution we want to fight very often have biological consequences of a serious nature. This leads to the need for a contact between radio science and the life sciences that up to this time have been rather unusual and that should prove beneficial and stimulating for all of us.

The theme of this conference abounds with problems and possibilities, I wish you all a pleasant stay here and thank you for your attention.

PART I : FUNDAMENTALS

FUNDAMENTAL BEHAVIOUR OF BIOSYSTEMS IN CONNECTION TO SURVEILLANCE OF ENVIRONMENTAL POLLUTION AND RESOURCES

Arne Jensen

Institute of Marine Biochemistry,
University of Trondheim,
N7034 Trondheim - NTH, Norway

1. INTRODUCTION - DEFINITIONS

The purpose of this meeting is to discuss surveillance of environmental pollution and resources by electromagnetic waves, and it seems that some sort of definition of environmental pollution and some agreement on what is ment by resources are needed as a fundament for the discussions to come. Electromagnetic waves are well defined. We do not want a captious discussion of what we mean by pollution. In our context we shall by pollution mean increased levels of materials or energy introduced by man into the environment. We are concerned mainly with inorganic compounds, some of which are plant nutrients (NO_3^-, PO_4^{2-}, NH_3, K^+, Mg^{2+}, Ca^{2+}, SO_2, etc.); non-toxic organic matter (organic wastes, sewage); toxic compounds (special chemicals, pesticides, insecticides, PCB, DDT, PAH); heavy metals, especially those that are only toxic (Pb, Cd, Hg), and pathogenic microorganisms (bacteria, vira). Also heat and suspended particles should be included.

Regarding resources we shall have to concentrate on organic resources and mainly on living ones. Food for man, raw materials for clothing, utilities and housing, feed for domesticated and wild animals are resources of obvious interest to us. The primary producers, higher plants on land and algae in the sea should be given special attention since they represent the only group which can turn inorganic material into organic matter.

T. Lund (ed.), Surveillance of Environmental Pollution and Resources by Electromagnetic Waves, 3–11.

2. EFFECTS OF POLLUTION

In general we are not interested in pollution as such. Only the type and degree of pollution which leads to unwanted effects on man himself or his environment is of interest in our context. This means that we are interested in the effects of pollution. Still nearly all monitoring and surveillance of pollution concentrate on the pollutant itself and on the occurrence and concentration of polluting chemicals. The effects, which are the decisive factor for the evaluation of the trouble caused by the pollutant are rarely used in quantitative monitoring. The reason for this is partly that determination of a well defined chemical is familiar to the analyst, and is fairly easy. It also seems to be an accepted rule that pollution studies are not complete unless they result in a chemical or physical method for monitoring the effective cause.

In some cases monitoring a pollutant may be easy, specific and can be carried out by reliable, automatic methods based on sound principles. Provided the effects of the pollutant on the environment are sufficiently known this is fine. The problem is, however, that no single pollutant works alone as the only variable in an otherwise constant system.

3. EFFECTIVE FORMS OF POLLUTANTS AND RELATION TO OTHER FACTORS

In most situations several pollutants occur simultaneously and their effects are always superimposed on the effects and interplay of effects that reign the non-polluted environment. To clarify the consept let us take an example of copper pollution in a Norwegian fjord. Our basic information regarding the pollutant will be its total concentration in the water, and the variations in space and time. I shall not go into all the problems this seemingly modest requirement introduces. We must, however, know what fraction of the copper occurs in the ionic state, what percentage is bound in complexes and chelates, how much is absorbed into inorganic particles and how much is tied up in living plankton; because all these forms of copper have different toxicities and act in different ways on the environment. Furthermore, it is necessary to know what other pollutants and chemicals are present; especially is information on other heavy metals needed since they may interfere in positive or negative ways (antagonistic or synergistic) with the copper ions. We often find that seasonal factors are important too. A certain copper level may have limited effects on the environment in late spring and early summer when the water is full of brown material from peat bogs and forest floors, while the same level may be significantly toxic in the winter water. In addition comes that high levels of nutrient salts reduce toxicity while stress factors, such as high

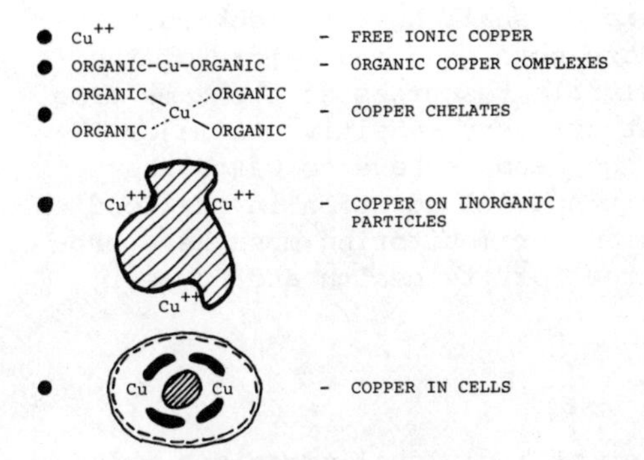

Table 1 Copper pollution in a fjord.

light intensity, lack of nutrients, low salinity a.s.o. increase the toxicity of copper to marine algae.

In other words copper monitoring in itself is pretty complex in sea water but can be done, while the effects on the environment are hard or nearly impossible to predict on the basis of copper monitoring data.

4. RATE OF REACTION OF ORGANISMS

There is also an additional problem which requires solution; namely the time factor involved or the reaction rate of the various organisms upon exposure to toxic chemicals. The concentration of a pollutant tends to vary in the environment, mainly because the influx is variable and geographically localized. Most pollutants are released in pulses and come from small source points, and we temporarily find relatively high concentrations in localized volumes of air, water and soil. The organisms which inhabit these biotopes may concequently be exposed to rather high doses for quite short periods, while near-by organisms may experience much lower concentrations over longer periods. The problem is that we have little knowledge in general of the dose-response relationship for most of the organism-pollutant pairs, and even less information about the time-scale involved.

5. MONITORING POLLUTION IS OFTEN BEST SOLVED BY MONITORING ORGANISMS

All this makes it rather obvious that monitoring pollution is best solved by monitoring the organisms of the environment, i.e., by monitoring resources. It is of course impossible to keep account on all organisms, and we shall have to look for specially suitable groups which are both good indicators of the health of the environment and valuable resources at the same time. We should also use organisms that are very sensitive to major pollutants to obtain early warnings, and we have to watch the primary producers because of their special function in the food web. Furthermore a useful organism for monitoring must have properties which allow easy measurement of its health and number.

6. BEHAVIOUR OF BIOSYSTEMS

In general biosystems consist of individual organisms which may be single cells or collection of cells, and which build up populations. We can therefore talk of behaviour on different levels: Subcellular level (atomic and molecular level); cellular and individual level; and population level.

What is going on inside the cell (sub-cellular level) is of course important to all organisms and can be used for living matter monitoring in general. Cellular behaviour will also involve all organisms but will have different complexity in unicellular compared to multicellular organisms. In general the cell will grow in size and divide describing both individual and population behaviour in unicellular and only individual behaviour in multicellular species. The development of populations is at a higher level of organization for multicellular organisms.

Another behavioural aspect of importance to surveillance and monitoring is mobility. In aquatic systems we separate between benthic, planktonic, and nectonic organisms. The benthic are tied to a substratum and do not move; the planktonic organisms are transported passively with the water they inhabit, and the necton swim positively against or along with the currents in the water.

On land most plants are benthic; they don't move around. They start from small units such as seeds and spores and grow to smaller and larger individuals of a more or less characteristic shape. Planktonic species are mainly lokalized to water, while migrating (self propelled organisms) are common both to water, land and air. Their migration complicates monitoring and evaluation.

7. SUBCELLULAR BEHAVIOUR

The subcellular level is the molecular level, and since all reactions of at least primitive organisms consist of chemical reactions any behaviour of a living resource can in principle be followed back to underlaying chemical reactions. The problem is to find crucial reactions which are suitable for measuring biomass, increments in biomass, and health of the biosystem, and to single out suitable reactions, products or intermediates which in a quantitative way can be used to characterize and measure loads of pollutants.

In biochemistry and enzymology the accumulated information on routes for biosynthesis, degradation and metabolism, including enzymes involved and to some extent the mechanisms operating is already enormous, and constantly increasing. Many key intermediates and regulating compounds have been identified. There is therefore a wealth of compounds which might serve as basis for monitoring purposes. Some are much more widespread than others.

All organisms, whether photosynthetic or heterotrofic (living on organic compounds) must make use of the same set of compounds for energy transfer , namely the ATP-ADP pair. Adenosinetriphosphate (ATP) stores chemical energy in its triphosphate group, and by splitting of this to give adenosinediphosphate (ADP) plus inorganic phosphate, 7.3 kcal are released per mole and used by the organism for biosynthesis, motion, work, pumping of ions, and other energy requiring essential reactions. The level of ATP and the proportion of ATP to ADP are regulating the rate of many decisive reactions in the cell, and this can be used both for biomass estimation and for evaluation of the physiological status of the cell.

There are many other systems the biologist could use for evaluation of health or as indicators of physiological state on the cellular level. One very promising set seems to be the enzymes for nitrate uptake and reduction which plants are dependent upon.

One may ask whether monitoring such systems would be remote sensing. From the point of view of the cell observations from some few centimeters away of the concentration of metabolites inside the cell, as is done by measurement of ATP by means of nuclear magnetic resonance (NMR), must seem quite remote. By this technique cells of a size of 10 μm in diameter held in a tube some 5 mm wide are placed in a strong and homogenous magnetic field and the proton magnetic resonance frequences determined. The characteristic frequences of ATP can be found and estimation of the cellular concentration of the compound carried out. I should like to look upon this as remote sensing on the subcellular level.

A related method, the electron paramagnetic resonance spectroscopy (EPR), should also be mentioned in the same context. This principle is specially suited for studying phenomena involving free radicals and paramagnetic ions in cells. Such phenomena are involved in photochemical reactions, for example photosynthesis.

I should like to make it clear that both the biochemist and the physiologist are looking to internal cell components for good indicators of pollution, and that rapid methods for non-destructive measurement of such compounds are in great demand and should form a first rate challenge to physisists.

8. BEHAVIOUR ON THE CELLULAR AND ORGANISM LEVEL

Both single cells and multicellular organisms have a phase of growth in size, which of course may be registered by analysis of size classes of populations of the organism in question. For unicellular organisms, division, which is another part of the life cycle, leads to a reduction in size, and an increase in number (usually a doubling).

An electronic particle counter which registers number and size of particles passing a narrow slit, by means of conductivity measurements, is some sort of a remote sensing instrument which gives information both on the growth in size of each organism and on the rate of division (growth of the population). It is to be expected that any significant changes in the environment, such as introduction of pollutants, will have some influence on growth and division.

In many cases photosynthesis in plants and respiration in animals are more sensitive to environmental changes than growth is. The photosynthetic activity is normally followed by monitoring incorporation of carbonate (^{14}C), production of oxygen or pH changes in the water, while respiration has to be measured by oxygen uptake or carbon dioxide production. Several of the chemical compounds mentioned have characteristic spectroscopic properties which allow monitoring by electromagnetic waves. This would open up for a very sensitive determination of pollution effects on important organisms and resources.

All photosynthetic plants contain chlorophyll *a* as an essential component of the light trapping system. When light is absorbed by living plants a small part of it is lost as fluorescence. Since fluorescence measurements usually are many times more sensitive than absorption measurements, this small fluorescence is used to determine chlorophyll contents in living cells. The fluorescence yield vary from species to species and is also dependent upon the physiological state of the cell. Safe estimates of biomass cannot be made today on the basis only on fluorescence measurements. Additional information on the species composition of the population and on the physiological state is required. When the species composition is known and the chlorophyll content has been determined by some other means, for example by absorption measurements, determination of chlorophyll a fluorescence will yield information on the physiological state of the cells.

Several recent investigations have shown that *in vivo* fluorescence of algal cells is very sensitive to environmental changes, and the biochemists are looking upon this parameter as a very promising indicator of photosynthetic efficiency and of physiological state of primary producers.

9. POPULATION BEHAVIOUR AND COMPOSITION

Again working with planktonic algae we expect to find single cells which grow and divide and give rise to population development. The growing population represents a certain production and is feed for zooplankton and in some cases for fish. Comprehensive experience clearly show that species composition and total production as well as total biomass of planktonic algae in an area are highly variable, making establishment of reliable baselines nearly impossible. This means that safe predictions cannot be made and influences of pollution will not be detected. It is, however, possible that long-term, detailed surveillance of total chlorophyll in the area will reveal systematic patterns which will allow detection of suspicious deviatons from normal behaviour, and which might indicate influence of pollution. To test this hypothesis one needs methods for rapid screening of large bodies of water for chlorophyll a content in the upper layers. The only principle which allows such surveys today involve electromagnetic waves.

Better than total chlorophyll a measurements may be determination of increments in chlorophyll a for defined species. The actual growth rate of an alga relative to its maximum growth rate obtained under optimum conditions forms a very good indicator of the quality of the environment as it is felt by the alga. This parameter looks therefore very promising for monitoring growth conditions of algae but does require a more complicated setup for its recording. Hitherto test algae in special cultures have been studied. The most promising method makes use of *in situ* cage cultures. These will, however, only give information about growth rate of the test algae at a single point.

Much more information could be obtained from remote measurement of chlorophyll if the signals could be assigned to the various algal species in the water. There is some hope for this. Different algal groups have different pigments and the pigment composition also varies between groups. In addition to chlorophyll *a*, the algae may have chlorophylls *b*, c_1 and c_2, biliproteins and carotenoids. There is accordingly a justified hope that each group of organisms may have a characteristic signature which can be recognized and allow determination of the growth rate of exactly the alga or algae involved.

In practise the observations of algal populations in the sea or a lake will consist of many components with different time frequencies.

The algeasensor which is being developed as part of NTNF Remote Sensing Technology Programme will hopefully give us profiles of pigment distribution both in space and time. There will be complicated variations which are built up of several components as indicated by Figure 1. Turbulence in the water will give microscale variations. The cells move around more or less stochastically in patterns of the mm to cm scale and in seconds and parts of minutes. Then there are water movements of larger size in space and time. Currents in the sea, rivers and lakes have a speed of several to many cm per second, and they may be periodic such as the tidal movement in the sea. They move the population under study away from the station and introduce new ones to confuse the observer. Then there is the development of the population itself, involving growth, division and removal. The time scale involved in population growth, the generation time, is in the order of 10 hours. Losses through grazing and some times sinking can be quite rapid.

It will be an important job to find ways to sort out the various components which make up the total information obtained through continuous measurement of pigment distribution in water volumes. Advanced data processing and clever use of redundant systems may really give the algologist a detailed wealth of information he has never seen before.

I do not want to end this challenge to the physisist and to the people concerned with the application of electromagnetic waves to environmental studies without giving a sort of warning to the biologists. There is no reason to sit down and wait for more sophisticated methods to come up. We do have many good means already which are not utilized or can be used much better. Let me give one example, namely echo sounding for fish. The method is extensively used in the sea and no modern fishing wessel can do without it. In fresh water, lakes and rivers, it is almost a revolution when the fish and wildlife research people bring echo sounding in for population studies.

It should be remembered that with the development of more and more sophisticated electromagnetic instruments, more and more effort to force the biologist to use it is required.

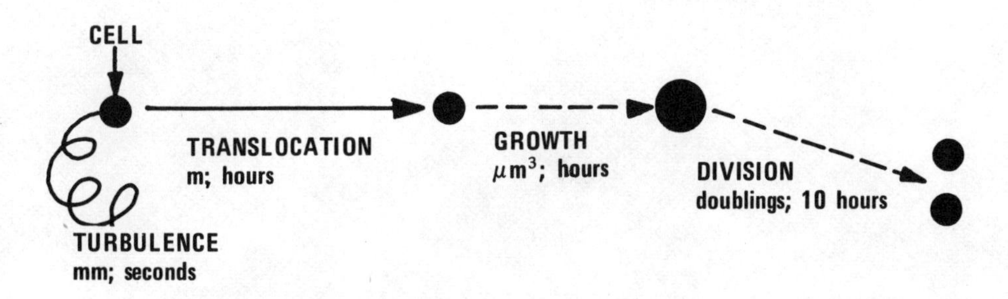

Figure 1 Scales of growth and motion for algea.

FUNDAMENTALS OF REMOTE SENSING METHODOLOGY

Dag T. Gjessing

Royal Norwegian Council for Scientific and Industrial Research,
Remote Sensing Technology Programme,
P.O.Box 25, N-2007 Kjeller, Norway

ABSTRACT. New components and methods in the field of microwave technique, electrooptics, computer technology, statistics and electromagnetics open new and very interesting possibilities with regard to detection and identification problems in relation to surveillance of environmental pollution and resources by electromagnetic waves. The current contribution introduces general remote sensing consepts and theories which conceivably may have an impact on several application areas (environmental surveillance, detection/identification of specific objects). The basic principle is the following: Most of the existing detection/idenfication systems do not make optimum use of all the a priori information that one generally is in possession of with regard to the object of interest. Knowing the geometrical shape of the object of interest and its molecular surface structure (e.g. structure of paint) an illumination function can be structured (matched filter concept) which gives optimum system sensitivity (minimum receiver bandwidth) with respect to the object of interest at the expense of the sensitivity for background objects (interferents). Theoretical results are given in the paper for a limited number of geometrical objects and for two different molecular surface compositions. It is shown that the system sensitivity and identification capability can be improved considerably using optimum methods which are based on fundamental principles from radio science and information theory.

1. INTRODUCTION

The last twenty years have been characterized by rapid progress in the field of radio science. At first the research was

T. Lund (ed.), Surveillance of Environmental Pollution and Resources by Electromagnetic Waves, 13–34.

stimulated predominantly by radio communication interests. In searching for better and more reliable communication channels one has been obliged to consider a wide spectrum of geophysical factors affecting the propagation of electromagnetic waves. Atmospheric waves and turbulence are responsible for long range over-the-horizon scatter communication in the troposphere, the earth topography and vegetation have a dominating influence on ground wave communication, and the sea surface irregularity structure influences the propagation of radio waves which rely on reflections between the ionosphere and the earth surface.

One soon realized that radio waves constitute a diagnostic tool of high potential in the study of these terrestrial geophysical phenomena. A new field of radio science emerged, that of radio physics, radio geophysics, radio meteorology. Radio techniques were developed with the objective of studying the motion pattern of atmospheric winds, of temperature and humidity in the troposphere. Through these radio science methods a very much improved understanding of the dynamic processes in our atmosphere was achieved.

Similarly, inaccessible regions in space were studied with electromagnetic waves. The moon, the planets and the sun were studied extensively and highly successfully by the aid of radio waves many years before modern space probes became available.

Looking back on the last two decades of research in the field of radio science, the potential of electromagnetic waves as a diagnostic tool in the study of a very wide spectrum of phenomena is evident.

In view of the historical background of radio science, it is not surprising that when the demand was accentuated for detailed information about the state of our environment as regards resources and pollution, powerful environmental surveillance methods rooted in the field of radio science emerged. Furthermore, the demand for such systems in the study of our environment is becoming increasingly urgent. The study of the earth's resources (minerals, vegetation, water, etc) calls for observation methods that will provide information about the relevant parameters with adequate resolution in time and space. The environmental pollution problem leads to a specific demand: an assessment of the pollution burden to which our environment is subjected. Also, environmental surveillance methods are sought for the purpose of establishing and verifying models on the basis of which assessments and predictions can be made. Finally, observation methods are needed for the study of the dynamic processes which control the state variables, including the motion patterns of the sea and of the atmosphere.

To meet the demands for efficient surveillance systems we need methods based on fundamental principles, and we need the technology - the components - to implement the methods.

During the last few years we have witnessed significant technological advances in many important fields. We now have tunable electromagnetic sources which enable us to illuminate the object under investigation - the scene - with electromagnetic waves ranging from microwaves, through infrared and optical waves, into the ultraviolet region. Microwave receiver elements coupled to micro-computers are being developed from which it is possible to determine the wavefront (amplitude and phase) over a large spatial region. The scattering of the illuminating wave from an object can thereby be investigated such that the object's geometrical shape can be determined. Electro-optical receiving systems, specifically superheterodyne receiver techniques, are about to give a dramatic improvement in receiver sensitivity (4-6 orders of magnitude) and frequency resolution, as was the case with microwaves some 30 years ago.

Low cost and high speed data processors (micro-computers) make it possible to perform the intricate and previously time-consuming computations necessary for detection/identification operations in real time.

In order to apply the techniques which have been developed, or techniques which are expected to become available in the near future, it will be necessary to develop the surveillance methods - architect the systems - to meet the specific demands.

To achieve this, we need to gain an understanding of the basic principles involved and of the general merits of electromagnetic waves in the study of environmental factors.

It is the endeavour of this introduction to the NATO Advanced Study Institute on Surveillance of Environmental Pollution and Resources by Electromagnetic Waves to provide a basis for such an understanding. The aim is to present a unified and general treatment of the subject of remote sensing, basing the work on general and well-established results from radio science, and our knowledge of the interaction of electromagnetic waves with gases, liquids and solids.

To reach this goal within a limited numbers of pages, we shall have to lean heavily on earlier published work. In particular, reference is made to a recent book by the author (1) (Remote Surveillance by Electromagnetic Waves for Air - Water - Land) and to a recent paper in Journal of Radio Science (2) where the various fundamental consepts are derived from first principles.

2. STATEMENT OF THE PROBLEM

In brief, the problem is the following:

A method of optimum sensitivity for the detection, identification and quantitative evaluation of an object or a chemical agent is sought. It is assumed that all the pertinent data pertaining to the target are known. These target "fingerprints" are typically:

- The shape, size and texture of the object, i.e. the distribution of scattering elements in height, width and depth (the macrostructure).
- The chemical composition of the surface.
- The chemical composition of the gases (or the liquids) which the object of interest may emit or leave behind it (scent from plants, odours from fermenting organic substances, gases from soil, exhaust gases from vehicles etc).
- The footprints, tracks or traces which the object leaves behind.

The object of interest will in general have fingerprints in various spectral ranges. Some of these are weak and some are pronounced. When it comes to determining which of these fingerprints the attention should be focused on, we shall also have to consider the fingerprints of the background, the noise, and we shall need information about to what extent the medium through which the illuminating waves must pass in order to illuminate the target is transparent to the various frequency bands of potential use. Thus, consideration will have to be given to the following:

- Characteristics (fingerprints) of the target (deterministic or statistical).
- Characteristics (statistics) of the background (additive noise).
- Characteristics (structure) of the propagation medium between the platform of observation and the target (distortion, attenuation, multiplicative noise source).

Since at least the two latter factors vary in time and space, it is not likely that there exists a unique frequency band within which optimum detection and identification capability is achieved at all times. One would therefore wish to have a redundant set of observations and leave it to a Kalman filtering method to sort out the best set of data under any given condition for the best assessment to be made.

A fundamental advantage of significant potential value in connection with the environmental surveillance problem is based on the knowledge we already have about the object (or environmental parameter) of interest. The merits of this situation can be summarized as follows:

- The search (detection identification, quantization) is very much simplified when we know what to look for. Knowing for example the detailed reflection or absorption spectrum of the chemical agent of interest (e.g. lead deposit on vegetation from the burning of high octane fuel), the illumination employed can be structured so as to give optimum detection capability for the particular target. In network theory this concept is known as "the matched filter".

- The identification can be based on several different properties of the object of interest. In this way a redundant set of information can be obtained from which the confidence level of the information is significantly enhanced. The following example may illustrate this point:

  The object is to detect minerals rich in e.g. copper on the earth's surface or subsurface by remote probing technology. The existence of such minerals gives rise to at least three mutually different parameters on the basis of which the mineral can be detected and identified:

  • The soil rich in copper leads to copper being present in the vegetation growing on the particular soil. This can be detected by emission or reflection spectroscopy remotely.

  • The particular soil influences the species that will grow on the soil. Identifying the particular vegetation being present (or absent) will give us information about the soil and its composition. The vegetation is identified by measurement of the geometrical shape of the vegetation using wavelength smaller than the geometrical scale (size) of the object of interest.

  • The particular mineral may under certain conditions influence the composition of the air above the ground. Measuring e.g. the amount of sulphur compounds in the air above the ground by laser spectroscopy methods, indirect information about the surface composition can be obtained.

  The degree to which a given parameter (signature) is dominant depends on the conditions prevailing. By a suitable weighting procedure (e.g. a Kalman filtering process) the redundant information set can be processed so as to reveal previously unidentifiable properties.

The following example may substantiate the arguments above. The objective is to detect a vehicle in a background of loose rocks (see Figure 1). You know the shape size and chemical surface structure of the target of interest. Hence you know what to look for. You are seeking two bits of information only: is the target there or is it no? From an information point of view only, an infinitesimally small bandwidth is required to make this assessment. For a given "signal power" (radio power scattered from the object of interest) the signal to noise ratio approaches infinity as the information bandwidth of the signal approaches zero. Any signal, however small, can in principle be extracted from a background of noise, however large, if the signal is a sine wave of known frequency and sufficient integration time is provided.

After this introduction we shall go on to elucidate the statements made above. We shall not go into detail in this introductory paper, but limit ourself to presenting the major concepts in a "semi-intuitive" manner referring to more rigorous treatments published elsewhere (1), (2).

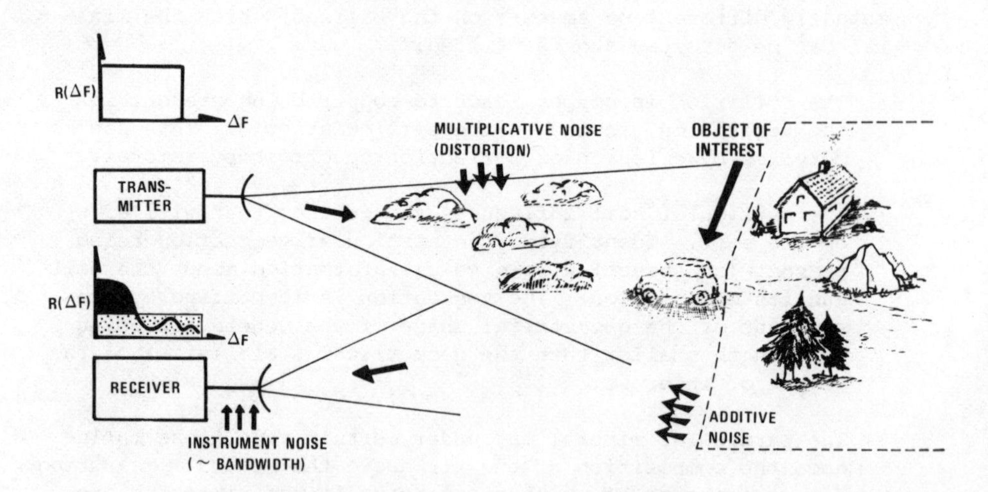

Figure 1 Symbolic presentation of the general detection/identification problem. Noise, additive and multiplicative, supress the signal from the object of interest. The task is to extract this signal from the noise so as to obtain an optimum signal to noise ratio.

We shall now proceed to discuss the two main aspects of the general complex of problems briefly mentioned above. These are:

a) Detection and identification of a given chemical agent, or a set of such, comprising the target surface.

b) Detection and identification of a given target shape.

3. OPTIMUM DETECTION AND IDENTIFICATION OF A CHEMICAL AGENT (CHEMICAL COMPOSITION OF TARGET SURFACE)

In this section of the presentation we shall discuss the basic principle for optimum, or selective, detection.

Probing our environment, e.g. the ground, with the aim of detecting or measuring the concentration of a chemical compound of known molecular structure, of known signature, how should we structure the illumination in order to peak the sensitivity of our detection system with regard to the compound of interest and suppress the influence of noise, i.e. contributions from all the uninteresting effects, the interferents? Most, if not all, the conventional spectroscopic methods do not make optimum use of the a priori information that is available about the agent to be measured qualitatively or quantitatively. The following example illustrates this statement. If we have no advance information about the surface molecular structure (the chemical composition) of a surface of interest, we illuminate this surface with electromagnetic waves, e.g. using a tunable laser source. We vary the wavelength of the illuminator in a linear (saw-tooth) manner as depicted in Figure 2 and we observe the signal which is scattered (reflected) back from the surface. If the surface consists of a higly conductive material with no absorptions lines or bands, we will upon backscattering obtain a signal which is virtually identical to that of the illuminator (disregarding the effect of surface roughness which can be compensated for by a special technique (1)).

If, however, the surface is such that waves of certain wavelengths are absorbed, then the scattered wave will give direct information about the absorption spectrum of the surface constituents. Figure 2 illustrates this.

We see that we shall require a broad-banded receiver in order to take care of the irregular "noisy" signal. The wider the receiver is the larger is, obviously, the contributions from the noise components.

Now let us use the classical results of information theory to structure the "colour composition" of our illuminator in such

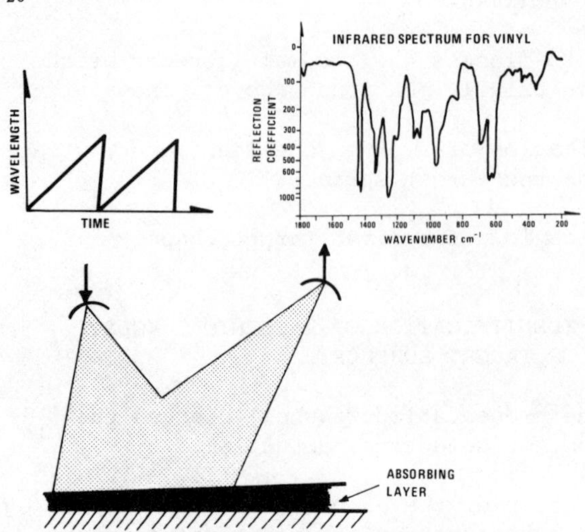

Figure 2 Reflection (absorption) spectrum of vinyl. When sweeping the frequency of the illuminator in a straight forward manner the irregular broad banded absorption spectrum appears directly from the receiver.

a way so as to obtimum coupling to the surface structure of interest and maximum noise suppression. In order to achieve this we shall need detailed information about the reflection spectrum of the material of interest. Having this information we know what to look for and should then on the basis of the fundamental concepts of Shannons information theory need only an infinitesimally small bandwidth to detect the object. It takes zero bandwidth (given sufficient integration time) to detect a sinusoidal waveform in an ocean of noise if you know the frequency of the sine-wave. We need two bits of information only, is the signal there, or is it not. The second question we may ask is; what is the amount present of the particular chemical compound of interest.

Hence, rather than using the crude method depicted in Figure 2 involving a linear frequency scan of the illuminator, we seek a "matched illumination" (analogous to the matched filter concept in information theory) so as to obtain optimum detection/identification capability.

To illustrate this optimum illumination concept let us consider Figure 3 and 4.

The illumination (microwaves, IR, visible light, UV) is amplitude and frequency modulated by particular waveforms (see Figure 4). These waveforms are the result of a detailed pro-

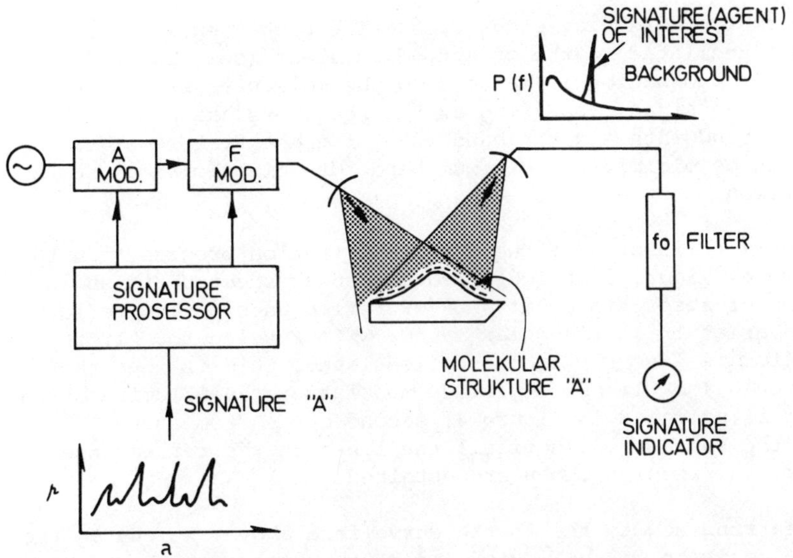

Figure 3 Schematic diagram of a practical system for optimum (matched filter) illumination.

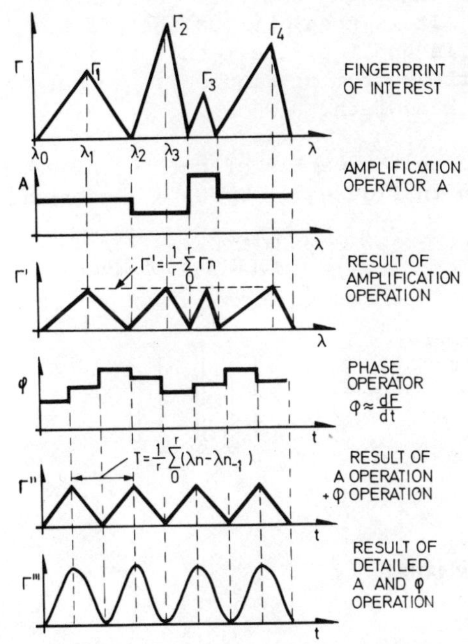

Figure 4 Example of an optimum structuring of the illumination function for the purpose of detecting/identifying a particular molecular structure.

cessing of the molecular signatures. The processed illumination is then transmitted to the object of interest (see Figure 3). Since the illumination is matched to the molecular structure of interest, the signal appearing at the receivers has minimum information bandwidth and the bandwidth of the entire receiving system can be minimized. Minimum bandwidth gives minimum noise contribution.

Figure 4 illustrates the A and ϕ operation process. In the top curve of Figure 4 an idealized molecular spectrum is shown (spectrum of absorption, emission or reflectance). The reflectance Γ varies in a triangular manner with wavelength. If the illuminator is linearly frequency modulated, this is what the signal would look like. Amplitude modulating the illuminator in a manner illustrated in Figure 4, second curve from above, seeking the same strength of all the lines in the reflectance spectrum, the results shown are obtained.

Referring now to the fourth curve from above, we change the relative position of the maxima and of the minima of the spectrum so as to get a periodic function shown in the fifth graphical representation, this is achieved by changing the rate dF/dt at which the frequency is changed. It is then obvious that if a more detailed A and ϕ operation is applied, the result is a sinusoidal variation, shown at the bottom of Figure 4. To detect this requires only a very small bandwidth.

For the purpose of emphasizing the merits of the technique, an illustrative example is given in Figure 5. Type "A" absorption

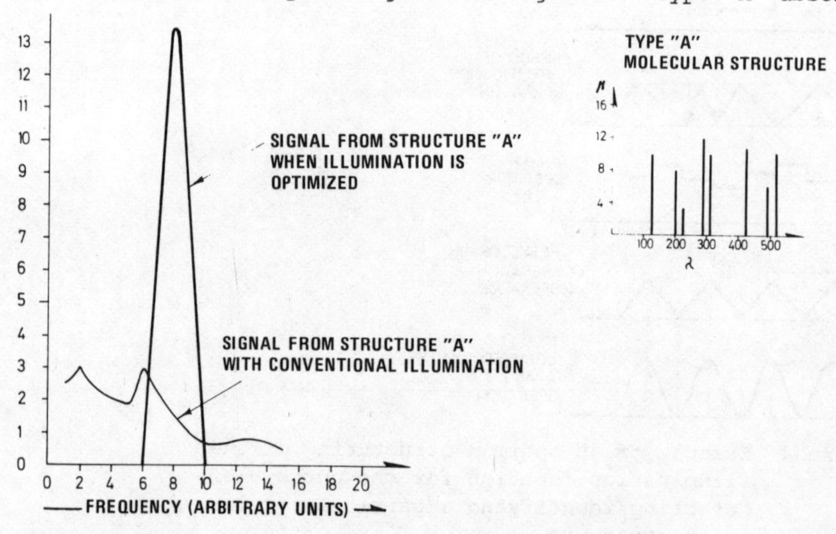

Figure 5 The effect of optimized illumination
Theoretical results

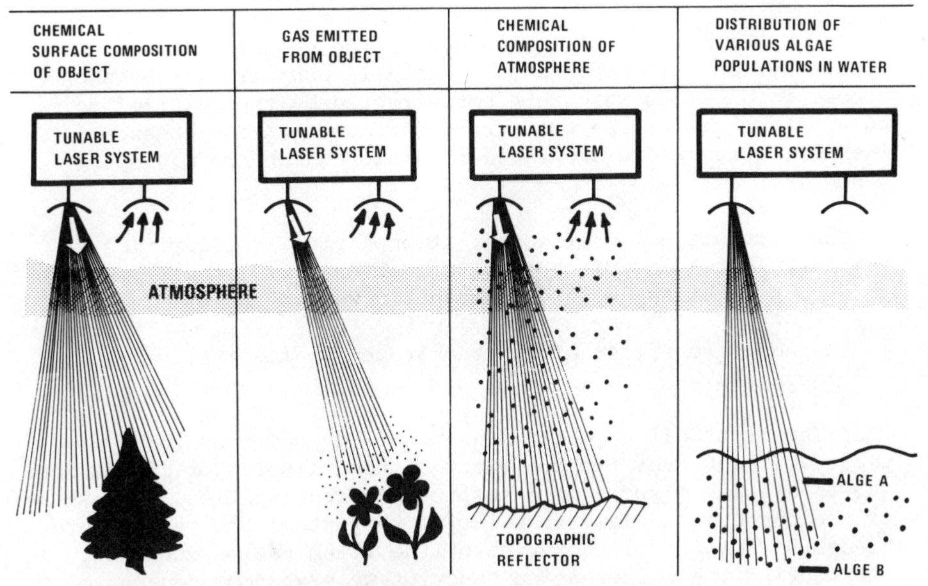

Figure 6 The matched illumination concept has many application areas.

spectrum is characterized by 8 absorption lines. Each absorption line is shaped by the characteristic Lorenz factor. In the interval between the lines the absorption/reflection is assumed to be zero.

Adopting the amplitude- and frequency-modulation scheme illustrated in Figure 4 above, the illumination is structured for type "A" molecules. Figure 5 shows the result of this optimization process. Note that the spectrum density function associated with optimized illumination is narrow, whereas the spectrum resulting from conventional illumination is wide. Note also that if an extensive structuring of the illumination had been accomplished, the resulting signal spectrum would have been a delta function.

In conclusion, it should be noted that the current matched illumination concept has a wide application area. Some of these are depicted in Figure 6.

We shall now consider the effect of the topography (the surface roughness) in relation to that of the surface chemistry.

4. OPTIMUM METHODS FOR DETECTION AND IDENTIFICATION ON THE BASIS OF GEOMETRICAL SHAPE OF OBJECT

For the purpose of making use of the geometrical shape of the object for detection purposes, we shall, contrary to the above, choose frequencies which are not affected by the detailed molecular structure of the scattering surface. In the absence of these effects, there are three remaining which are of importance (3), (4).

- The multiplicative effect of atmospheric irregularities (turbulence) and of particulate matter (rain, snow, dust etc) on our illuminating electromagnetic wave.

- The additive effect of the background scatterers (see Figure 1).

The first effect determines the frequency band to be employed. The higher the frequency of the illuminator is the more severe are the effects of atmospheric irregularities. The shape of the background elements relative to that of the target determines the detailed structuring of the illumination function. In structuring the illumination function we have four degrees of freedom, space, and time or frequency. We can structure the illuminating wave in the frequency domain and we can structure the wave in space (shape the wavefront) so as to be matched to the target of interest. First let us consider the frequency domain.

4.1 Optimum frequency distribution of an illuminating wave with regard to a target of known geometrical form

When studying the geometrical shape of an object, we have essentially 4 degrees of freedom at our disposal: 3 directions in space and time.

We shall now investigate what information we can obtain about the object by modulating the illuminating wave in time. There are, as we well know, several ways of doing this. In this survey we shall discuss one powerful method.

Knowing the distribution in depth of the scatterers, we can calculate the correlation properties in the frequency domain (bandwidth) of the signal reflected (scattered) back from the surface (5). We shall use the results of this reference for the purpose of establishing the optimum frequency distribution of the illuminating wave with a view to determining the distribution in depth (longitudinal) of the scatterers constituting the target of interest.

It has been shown (5) that the effect of a target having a finite depth is to decorrelate the frequency properties of the waves scattered back from the target.

Specifically, if a set of correlated wide band waves impinges on a target having a given distribution in depth of the scatterers P(z/c), then the correlation properties of the reflected waves are limited to some function of the Fourier transform of P(z/c). Accordingly, as depicted in Figure 7, the transmitted rectangular correlation function is reduced to a (sin x)/x function upon reception if the target delay spectrum is rectangular.

The identification potential of this multifrequency system is illustrated in Figure 8. Here the correlation properties of the scattered signal are calculated for a set of scattering objects having different shape and size. Note the very marked influence of object shape and the dramatic influence of size.

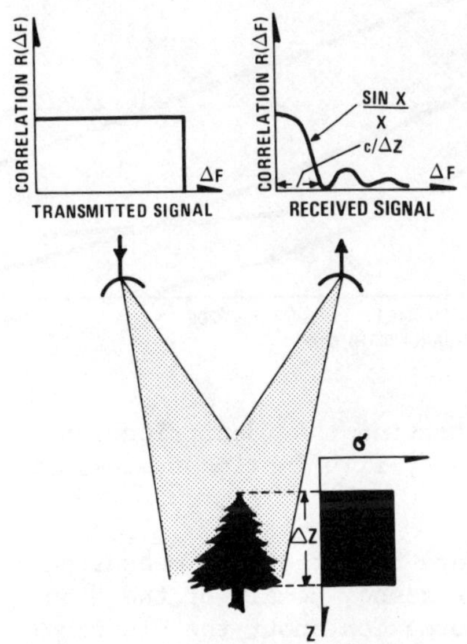

Figure 7 An object which is extended in depth (delay function is σ(z)) will cause decorrelation of the illuminating waves. If all the waves impinging on the target are correlated, the waves will, upon reflection (scattering) be decorrelated by a function which is the Fourier transform of the delay function (distribution in depth σ(z)) of the object.

In Figure 9 an object which is characterized by two independant parameters is shown. Here we have calculated the scattering/diffraction signature from an object in the shape of an exponentially damped sinusoid. The period of the sinusoidal curve (distance between branches of a tree) is Δz whereas the 1/e width of the exponential damping function (1/e height of the tree) is z_o.

Note that both Δz and z_o can be determined by the simple multifrequency experiment.

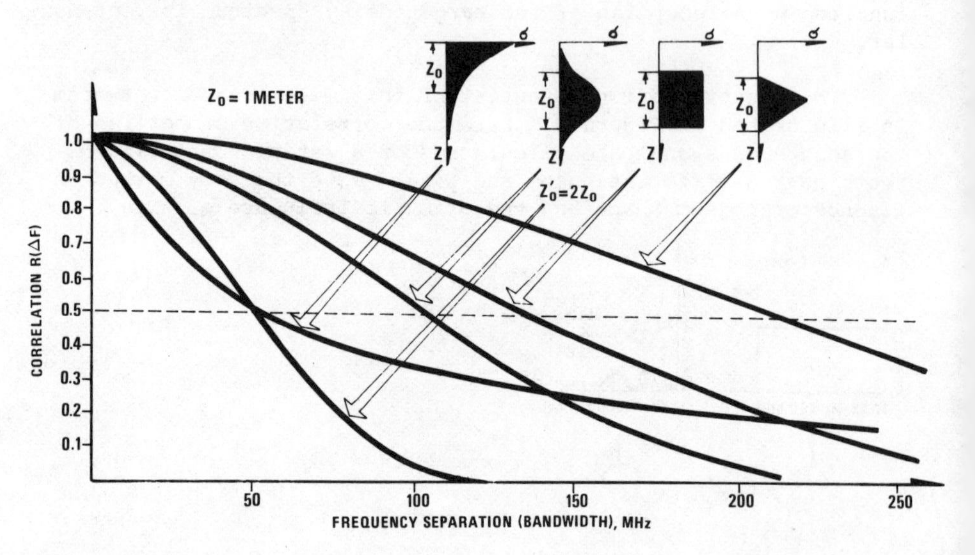

Figure 8 Theoretical results illustrating the identification potential of a multifrequency illumination system.

In conclusion it should therefore be noted that measuring the correlation properties in the frequency domain of the signal scattered back from the target, information about the distribution in depth of the target is obtained from a Fourier transformation of the received signal.

This simple, direct method would have been applicable if the target were suspended in the air so as to have no interfering background. In the case of an interfering background, one has to use a selective detection method. The procedure is analogous to that discussed in some detail in connection with the study of molecular structures under section above.

We should match the illuminating wave to the target of interest so as to get a narrow band signal of known frequency back

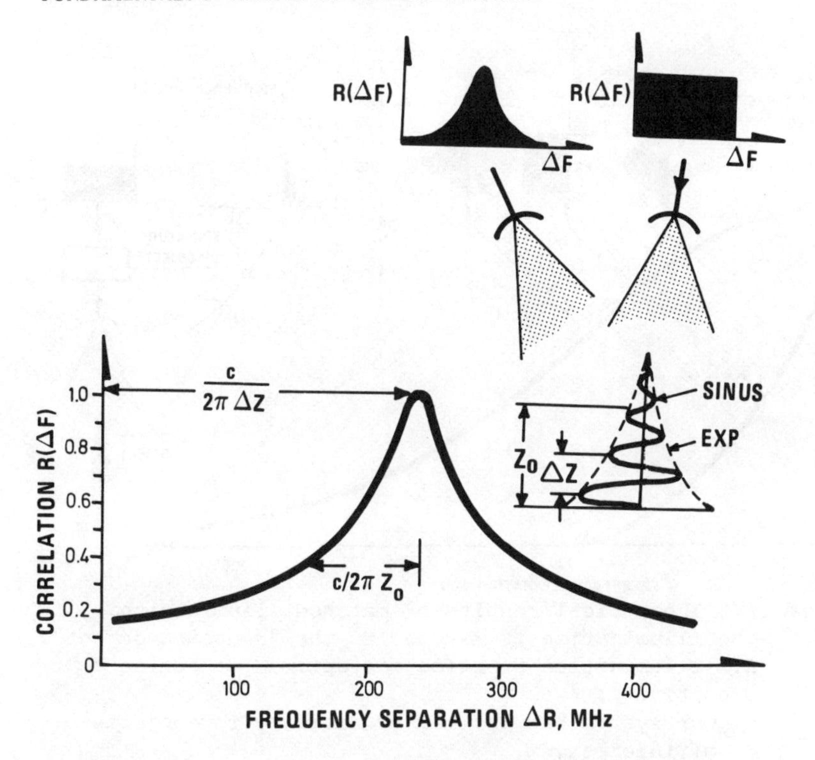

Figure 9 By the use of broad banded illumination, the footprint (signature) of an object characterized by several parameters can be determined. Here we have shown an idealized tree with height z_o and distance between branches Δz.

from the target of interest, the object with a gaussion delay function in the example, and a signal distributed in frequency from all the interferents, the background. By means of this selective detection method we have achieved the following:

- Maximized the system sensitivity with regard to the object of interest.

- Decreased the system sensitivity with respect to the background.

- Achieved an identification capability.

On the basis of the principles discussed above, we have structured the correlation properties of the illuminating waves so as to be matched to the object of interest (see Figure 10). In the current example, the illumination is matched to an object 1 meter deep having a gaussian distribution of the scattering

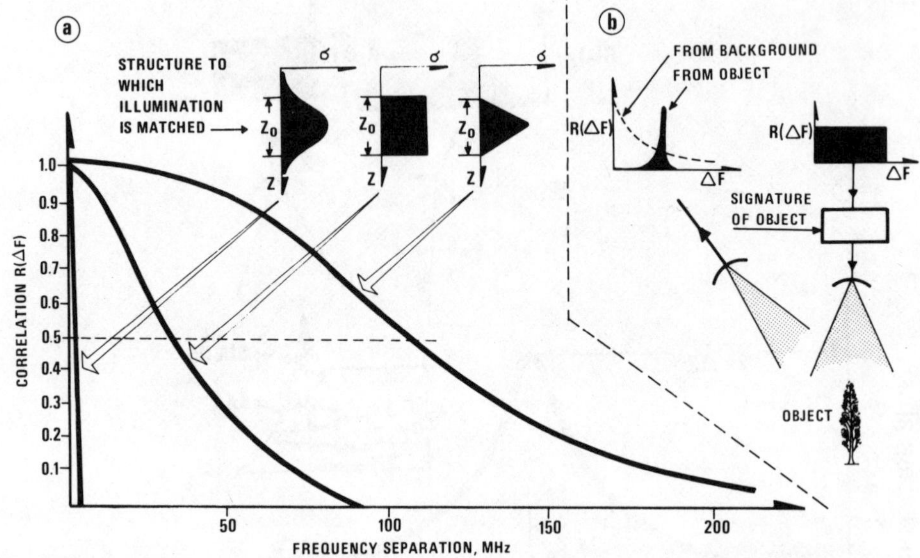

Figure 10 A. Theoretical results of matched illumination. The illumination is matched to the "gaussian object". B. Illumination function for optimum detection sensitivity.

- Our system has optimum sensitivity for objects of interest.
- Minimum sensitivity for other objects (background).

surface. We see that this gives rise to a very narrowbanded signal from the object of interest whereas the returns from targets to which the illumination is not optimized are relatively broadbanded. (For details, in particular regarding the question of matched illumination, the reader is referred to ref (1)).

4.2 Determination of target distribution in space by spatial correlation measurements of field strength

We have just completed a discussion about an experimental determination of the distribution in depth (in the direction of propagation) of the scattering elements constituting a scattering object. We have shown that if we illuminate the object by a set of EM waves the amplitude and phase of which are mutually correlated, then upon reflections by the object this set of correlated waves is subjected to a decorrelation process by the scattering object. The degree to which the waves are decorrelated depends on the distribution in depth of the scattering elements constituting the object. Specifically, the correlation function in the frequency domain (covariance function for waves having separate frequencies) is the Fourier transform of the distribution in depth of the scatterers (Fourier transform of delay function).

We shall now focus our attention on the spatial correlation properties of a scattered wave in relation to the spatial distribution of scatterers over the scattering object. In exactly the same way as for spaced frequencies, it can be shown (6) that the transverse (normal to direction of propagation) spatial correlation of field strength is the Fourier transform of the angle of arrival spectrum, i.e. the Fourier transform of some function which is determined by the transverse distribution of the scatterers constituting the target of interest. Thus, if X be a distance coordinate orthogonal to the Z-direction of propagation, then if $\sigma(X)$ be the distribution of scattering elements over the scattering object under investigation, the spatial correlation of field strength in the X-direction is given by

$$R_E(x) = \text{Fourier transform of } \sigma(X/R)$$

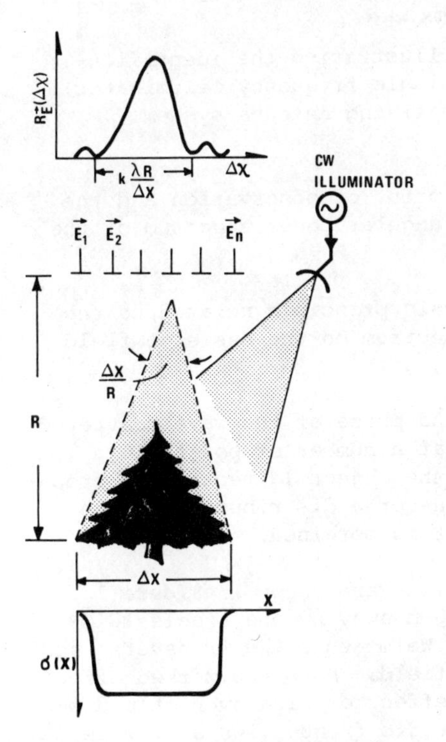

Figure 11 Wavefront analysis by array matrix. Spatial correlation of complex field $R_E(x)$ is the Fourier transform of the angular distribution of scattered power $P(\theta) \sim P(\frac{X}{R})$.

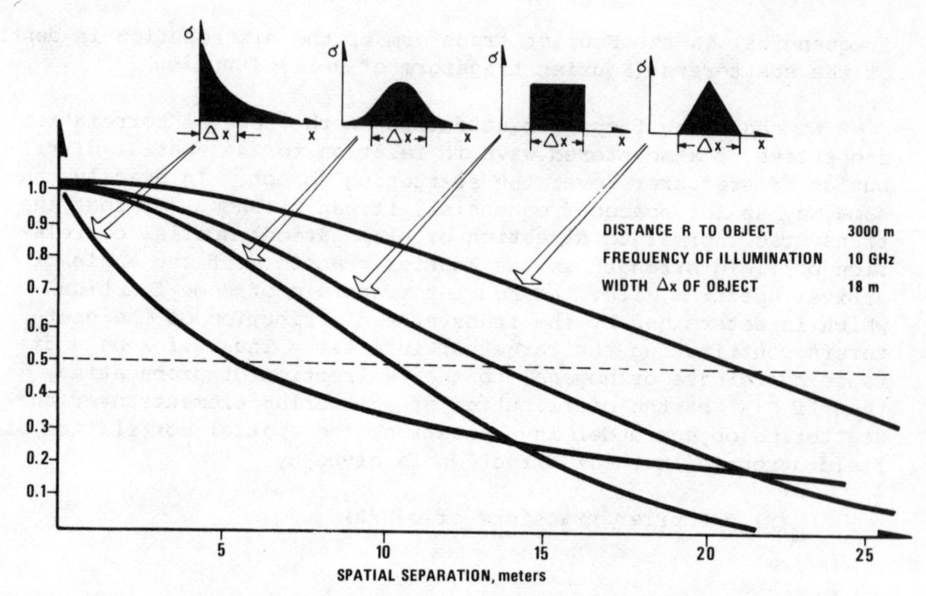

Figure 12. Theoretical results illustrating the identification potential of a single frequency illuminator and a multielement-receiving antenna system.

R being the distance between the point of observation and the target. Note that σ(X/R) is the angular power spectrum of the scattered wave.

Figure 11 illustrates the basic principle related to the determination of angular power spectrum on the basis of field strength correlation measurements.

By measuring the amplitude and phase of the wave scattered back from the object of interest at a number of points in a plane normal to the direction to the object (direction of propagation) information about the transverse distribution of the scatterers constituting the target is obtained.

The results of such calculations are shown in Figure 12. Here we illuminate the object 3000 m away by one single 10 GHz radio wave (wavelength is 3 cm). We measure the transverse spatial correlation of the received field. Note the marked effect of object shape and the dramatic effect of size even at a distance as great as 3000 m and for radio frequencies as low as 10 GHz.

Measuring in this manner the three projections of target surface, the aspect angle of the object relative to the point of observation is known if we have a priori information about the object size and shape. On the basis of this, a selective detection/identification system can be implemented in a manner very similar to that discussed in sections 3 and 4.1 above. Note that from simple reciprocity considerations, the same principle applies for a transmitting system. However, dynamic phase and amplitude control of a transmitting antenna matrix is very costly and only cost effective in few instances. The dynamic and adaptive receiving antenna array system sketched in Figure 11 could probably most efficiently be realized as follows:

Antenna elements consisting of an integrated assembly of spiral antenna, mixing diode, preamplifier feeding into a charge coupled storage device are positioned on a suitable surface. The position in space X, Y, Z of each element is measured and the information stored in the processor.

The target illuminator shines a small fraction of its power into the array matrix. (In practice it may be more practical to shift the frequency of the wave illuminating the array slightly relative to the frequency of the wave illuminating the target so as to obtain a suitable intermediate frequency for efficient amplification and subsequent quantization and storage in the receiving system). By this system the amplitude and relative phase are measured in each antenna element. An on-line computer processing system (parallel processing) reads the information about amplitude and phase stored at each antenna element and performs the appropriate computations.

Finally, the schematic diagram of Figure 13 sums up some essential application areas with regard to the technique based on scattering/diffraction of centimeter and millimeter waves. Note, that also when dealing with these relatively long wavelengths (long in comparison with the optical waves used for spectroscopic purposes and discussed in chapter 3) the atmospheric propagation medium plays an important role. The effect of the atmosphere on these wavelengths and for the current application is to perturb the phasefront (destruct the correlation properties of the wave in space and time). Under adverse conditions (see ref (7)) the atmosphere may constitute the limiting factor with regard to the detection/identification capability of a system.

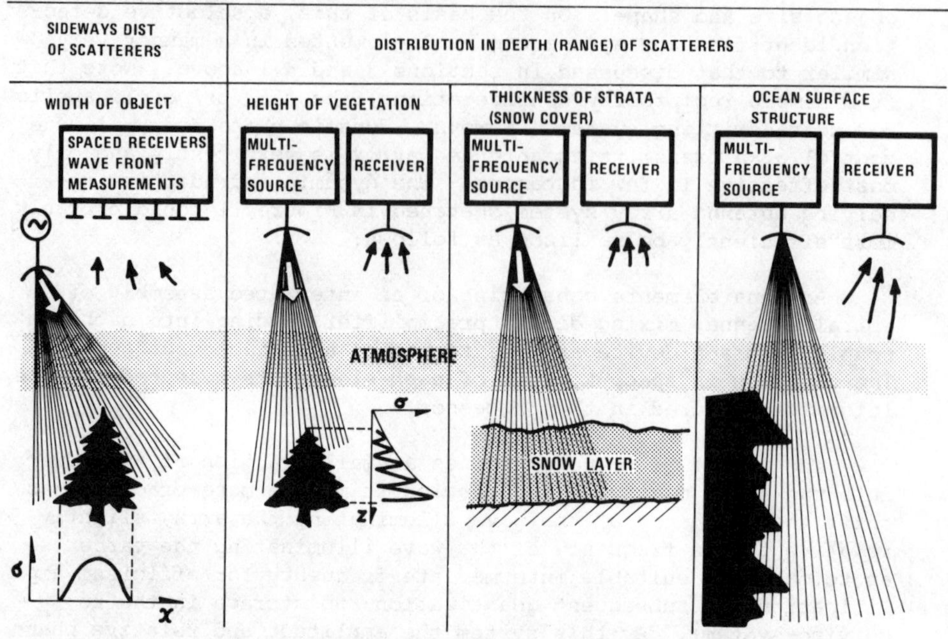

Figure 13 Basing the system performance on the scattering/diffraction capabilities of the object, detailed information about the size and shape of an object can be obtained.

5. CONCLUSIONS

New components and devices that open very interesting possibilities with regard to detection and identification problems are in the process of being developed (tunable laser and mm wave sources, superheterodyne electrooptical receivers, microwave receiving elements suitable for array matrix applications, charge coupled memory devices which can match the array to a computer parallel processing device, statistical signal processing methods, methods based on electromagnetic wave theory etc). The demand for powerful selective detection and identification methods is at the point of being articulated for environmental surveillance applications.

In the current introduction a set of unified detection methods is presented which conceivably may have a substantial impact on several application areas. The philosophy on which the methods are based is the following:

Most of the existing detection/identification systems do not make optimum use of all the a priori information that one generally is in possession of with regard to the target of interest. Knowing the geometrical shape of the target of interest, and its molecular surface structure (e.g. structure of paint) an illumination function can be structured (matched filter concept) which gives optimum system sensitivity (minimum receiver bandwidth) with respect to the object of interest at the expense of the sensitivity for background objects (interferents).

On the basis of these concepts and on the basis of electro-optical, microwave and computer devices which are expected to be available in the near future, it is believed that several areas of importance can be innovated. Examples of such are:

- Detection/identification of objects on the sea, in the field and in the air on the basis of object shape and chemical composition of surface.

- Detection/identification on the basis of traces of chemicals which the object emits (e.g. gasses emitted from minerals).

- Detection/identification on the basis of the trace (footprints) which the target of interest leaves behind (oil spills on ground,lead and sulphur compounds adsorbed on vegetation, belt-tracks on ground, upwelling of sea-water with different salinity,plankton content or temperature relative to the ambient conditions etc).

- Selective environmental surveillance (remote probing, pollution monitoring) of land sea, air with regard to minute concentrations of pollutants.

- Surveillance of land and sea with regard to vegetation and micro-organisms, to minerals, to water content in snow etc.

- Mapping of the wave structure of the sea surface.

REFERENCES

1. Gjessing, Dag T., Remote Surveillance by Electromagnetic Waves for Air-Water-Land. Ann Arbour Science Publishers Inc., Michigan, USA, 1978.
2. Gjessing, Dag T., A Generalized Method for Environmental Surveillance by Remote Probing. J. Radio Science, March/ April, 1978.
3. Gjessing, Dag T., On the influence of atmospheric refractive index irregularities on the resolution performance of a radar. NATO Advanced Institute Series. Atmospheric Effects on Radar Target Identification and Imaging, ed: H. Jeske, D. Reidel Publishing Company, 1975.
4. Lee, R.W. and Harp, J.C., Weak scattering in random media. IEE 57, 375, 1969.
5. Gjessing, Dag T., Atmospheric structure deduced from forward scatter wave propagation experiments, Radio Sci. 4, 12, 1195-1210, (1969)
6. Gjessing, Dag T. and J. Børresen, The influence of an irregular refractive index structure on the spatial field-strength correlation of a scattered radio wave, IEE Conf. Proc. 48, September, (1968)
7. Gjessing, Dag T., Target Detection and Identification Methods Based on Radio- and Optical Waves, AGARD Lecture Series No 93, page 12.1 to 12.23, May 1978.

PART II : THE ATMOSPHERE

FUNDAMENTALS OF ATMOSPHERIC SPECTROSCOPY

S. Svanberg

Department of Physics, Chalmers University of Technology
S-402 20 Göteborg, Sweden

ABSTRACT. Remote sensing of atmospheric pollution and general meteorological conditions is largely based on applied molecular spectroscopy. In the first part of this paper atomic and molecular energy-level structures and fundamental interactions between radiation and matter are reviewed. Based on this description different atmospheric-spectroscopy methods are then discussed. Passive techniques, where an analysis of the spectral distribution of the back-ground radiation is performed, can be used in both absorption and emission studies. Various kinds of measurements utilizing dispersive-, Fourier-, correlation- or heterodyne spectrometers are described. In active remote-sensing techniques, electro-magnetic radiation is transmitted into the atmosphere and the modification of this radiation due to interaction with the atmospheric constituents is studied. Long-path absorption and Lidar (Light detection and ranging) methods are discussed. Out of the various methods described, the Lidar technique is chosen for a more thorough presentation.

1. INTRODUCTION

Remote-sensing techniques are becoming increasingly important for monitoring the state of the earth's atmosphere. Weather conditions and atmospheric pollution are of great interest for both global and local monitoring. The remote-sensing observations can be performed from the ground or from air- or spaceborne platforms. Atmospheric monitoring is largely based on molecular spectroscopy, and in this paper we will discuss the basics of atmospheric spectroscopy. First the general energy-level structure of atoms and molecules and the basic interactions between radiation

T. Lund (ed.), Surveillance of Environmental Pollution and Resources by Electromagnetic Waves, 37-66.

and matter are reviewed as a back-ground for a description of different spectroscopic methods, useful for remote-sensing applications. A distinction is made between passive- and active techniques. In the former type of technique the spectral distribution of the back-ground radiation is utilized in absorption- emission- or reflection measurements. Although image-forming passive registrations in selected wavelength bands such as photography or area-scanning from weather- or land-resource satellites are very important we will not consider this kind of remote sensing here, but will rather concentrate on spectroscopic methods which employ dispersive, Fourier, correlation or heterodyne instruments in absorption or emission measurements.

In active techniques electro-magnetic radiation is transmitted into the probing volume from the measuring equipment, and the modification of this radiation due to interaction with atmospheric constituents is monitored. Active techniques can be used both in point measurements and for remote sensing over large distances. We will consider opto-acoustic measurements as an example of point monitoring. Long-path absorption- and Lidar (Light detection and ranging) techniques are powerful approaches to remote sensing of atmospheric gases and particles. In the final part of this paper, Lidar measurements of various kinds will be chosen as a topic for a somewhat more detailed discussion. A concluding paragraph provides an out-look for the future.

2. ENERGY LEVEL STRUCTURES

As a back-ground for atmospheric spectroscopy it is necessary to consider the possible energy states of the atmospheric molecules and the radiation processes in which they take part. This is the subject of the present and the following section. More detailed information can be found in Refs. [1-2]. The possible energy levels, E, of a quantum-mechanical system are given by the Schrödinger equation:

$$\underline{H}\psi = E\psi \quad , \tag{1}$$

where H is the Hamiltonian (energy) operator and ψ are the eigenfunctions of the system. Generally $\underline{H}$ has a complicated structure calling for suitable approximations to be applied. Before considering molecules we will discuss atomic energy-level structure.

2.1. Atomic energy levels

Exact solutions can be given for the case of the hydrogen atom. The energy levels of the alkali atoms, which have a single valence electron outside closed shells, can be expressed with the hydrogen-like formula

$$E = -hcR_y \frac{1}{n_{eff}^2} , \quad (2)$$

where h, c and R_y are the Planck constant, the light velocity and the Rydberg constant, respectively. The ionization energy has been chosen as the zero-point for energy. n_{eff} is, in contrast to the hydrogen case, not an integer, and depends on the shell structure as well as on how much the valence electron penetrates into the electronic case. The spin-orbit interaction gives rise to a fine-structure splitting of most alkali states.

General multi-electron atoms are more complicated to treat. Usually, a central-field approximation is used to obtain rough energy levels corresponding to the possible electron configurations. Non-central parts of the electrostatic repulsion between the electrons, and spin-orbit interactions are then treated as perturbations. Depending on whether the first or the second of these perturbations is the dominating one, the two basic coupling schemes, LS-coupling or jj-coupling, are obtained. The separation between low-lying electronic states, for both coupling cases characterized by a total angular-momentum quantum number J, is of the order of 1 eV.

2.2. Molecular energy levels

Naturally, molecules have a much more complicated energy-level structure than atoms. We will here primarily restrict ourselves to *diatomic* molecules. Such molecules have an axial symmetry, and the electric-dipole moment of the molecule, associated with the total orbital-momentum vector, $\underline{L}$, will precess around the strong electric field, directed along the symmetry axis. The projection $L_z = M_L\hbar$ is quantized with M_L ranging from L, L-1... down to -L. Due to the electric nature of the interaction, there is an energy degeneracy $M_L \leftrightarrow -M_L$. For the given electron configuration the energy levels are then designated according to their $|M_L|$ value:

$$|M_L| = \underset{\Sigma}{\overset{0}{\updownarrow}} , \underset{\Pi}{\overset{1}{\updownarrow}} , \underset{\Delta}{\overset{2}{\updownarrow}} \ldots \quad (3)$$

The total spin quantum number further specifies the state, and the multiplicity 2S+1 is added to the state $|M_L|$-symbol: $^1\Sigma$, $^3\Pi$ etc. Typical energy separations between the lower electronic energy levels are of the order of 1 eV.

In addition to the electronic energy, molecules also possess vibrational and rotational energy that we will now consider.

Vibrational motion. The vibrational motion of a diatomic molecule is governed by a vibrational potential curve often described by a Morse function:

$$V(r) = D\{1-e^{-\alpha(r-r_o)}\}^2 \qquad (4)$$

Here, D and r_o are the dissociation energy and the equilibrium nuclear separation, respectively. The Morse potential is illustrated in Fig. 1. Close to the bottom of the potential it can be approximated with a parabola, corresponding to the case of a harmonic oscillator. The frequency of the oscillation, ν_c, is classically

$$\nu_c = \frac{1}{2\pi}\sqrt{\frac{k}{\mu}} \qquad (5)$$

with the force constant k defined by $F = -k(r-r_o)$ and the reduced mass μ according to $\mu = m_1 m_2/(m_1+m_2)$. Quantum-mechanically the possible energy states of the harmonic oscillator are given by

$$E_v = (v + 1/2)h\nu_c \quad ; \quad v = 0, 1, 2, \ldots \qquad (6)$$

Typical vibrational level spacings are 0.1 eV. Higher up in the Morse potential the vibrational levels get closer and closer to-

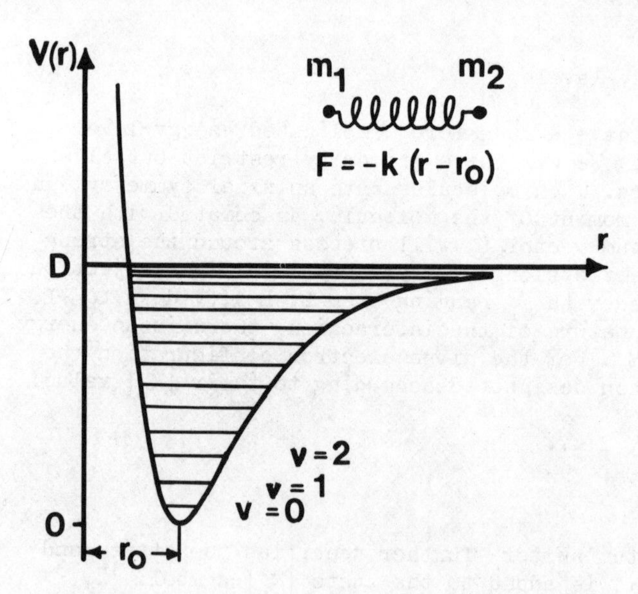

Fig. 1. The Morse potential function for a diatomic molecule. Vibrational energy levels are not equidistant for higher vibrational quantum numbers, v, due to anharmonicity of the potential.

gether, and the energy expression (6) must be augmented with higher-order terms.

Rotational motion. Classically, the energy E of a system of two masses m_1 and m_2, separated by a distance r and rotating around a centre of mass is

$$E = \frac{J^2}{2T} , \quad (7)$$

where $\underline{J}$ is the angular momentum and $T = \mu r^2$ is the moment of inertia of the system of reduced mass μ. The quantum-mechanically correct expression is obtained from (7) by inserting for $\underline{J}^2$ the value $J(J+1)\hbar^2$ ($\hbar = h/2\pi$):

$$E_J = \frac{J(J+1)\hbar^2}{2T} = BJ(J+1) \quad (8)$$

For high rotational states, r is slightly stretched, leading to an increase in T and a corresponding reduction in the E_J values as compared to Eq. 8. In Fig. 2 the rotation and energy levels for a rigid and an elastic rotator are shown. Rotational level spacings are of the order of 10^{-3} eV.

Poly-atomic molecules. Poly-atomic molecules possess much higher degrees of freedom with regard to vibrational and rotational motion. Such molecules are characterized by several equilibrium separations, force constants and dissociation energies. In a description of the basic properties, shapes and symmetries, discussed in the frame-work of group theory, play an important part. It can be shown, that a general n-atomic molecule has 3n-6 fundamental vibration frequencies, whereas a linear molecule has one further frequency. The vibrations are classified as stretch- or bending modes.

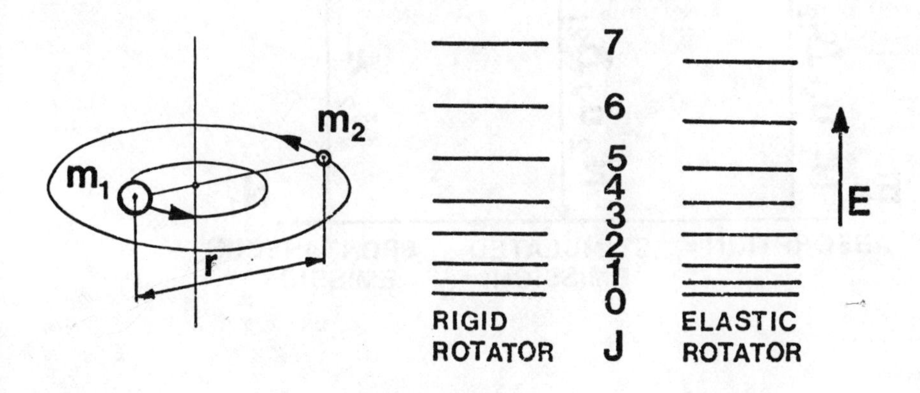

Fig. 2. Rotational energy-level diagrams for a diatomic molecule assuming a rigid and an elastic rotator, respectively.

3. RADIATIVE INTERACTIONS AND SPECTRA

In this section we will consider radiative interactions involving molecules or other particles in the atmosphere. As a result of these interactions, transitions between different states are induced according to certain selection rules and spectra, characterizing the particles, are obtained.

3.1. Electric dipole radiation

When an atom/molecule is irradiated with electromagnetic waves of a frequency ν, fulfilling the resonance condition between two energy levels E_1 and E_2 ($h\nu = E_2 - E_1$), an electric-dipole transition may be induced. The process is governed by the time-dependent Schrödinger equation

$$i\hbar \frac{\partial\Psi(t)}{\partial t} = (\underline{H}_o + \underline{H}'\cos 2\pi\nu t)\Psi(t) \quad , \tag{9}$$

where $\underline{H}_o$ is the Hamiltonian of the unperturbed system and $\underline{H}'\cos 2\pi\nu t$ is the time-dependent perturbation. The wavefunction $\Psi(t)$ is an expansion

$$\Psi(t) = \sum_i c_i(t)\psi_i \quad , \tag{10}$$

in eigenfunctions ψ_i of $\underline{H}_o$. The expressions $|c_i(t)|^2$ are the probabilities of finding the system in its i:th eigenstate under the influence of the perturbation, and these expressions can be readily calculated. However, we will give here a simpler, phenomenological treatment of the problem, originally due to Einstein.

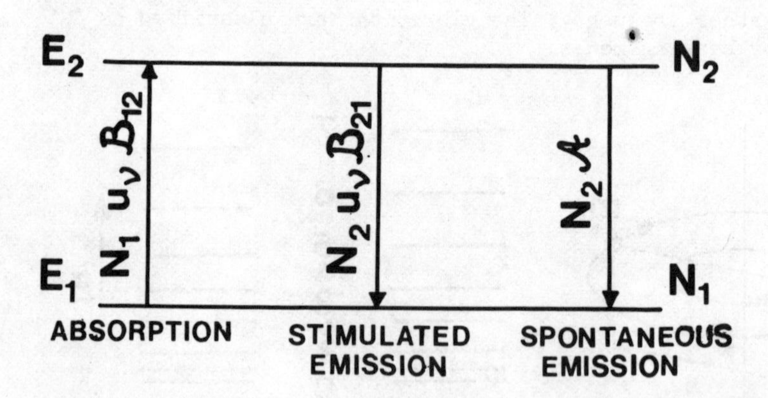

Fig. 3. Radiation processes for a two-level system in equilibrium with a radiation field.

We consider two-level atoms/molecules in equilibrium with a radiation field, u_ν. Under the influence of the three radiation processes absorption, stimulated- and spontaneous emission, the equilibrium populations of the two levels are N_1 and N_2 (see Fig. 3). We then have

$$N_1 u_\nu B_{12} = N_2 u_\nu B_{21} + N_2 A \tag{11}$$
$$\frac{N_2}{N_1} = \exp\left(-\frac{h\nu}{kT}\right)$$

B_{12}, B_{21} and A are called Einstein probability coefficients. By combining these equations we obtain

$$u_\nu = \frac{A}{B_{12}\exp(\frac{h\nu}{kT}) - B_{21}}\ , \tag{12}$$

which should be identified with the Planck radiation law, valid for such a system:

$$u_\nu = \frac{8\pi h\nu^3}{c^3}\ \frac{1}{\exp(\frac{h\nu}{kT}) - 1} \tag{13}$$

Thus we conclude the important relations

$$B_{12} = B_{21} = B \quad ; \quad A = \frac{8\pi h\nu^3}{c^3}\ B \tag{14}$$

From the quantum-mechanical treatment briefly outlined in the beginning of this section, the Einstein coefficients can be related to the wavefunctions, ψ_1 and ψ_2, of the two considered energy levels:

$$B \quad \propto \quad |\langle\psi_1|e\underline{r}|\psi_2\rangle|^2 \tag{15}$$

The transition probabilities are thus non-zero if the matrix element of the electric-dipole operator, $e\underline{r}$, between the two states does not vanish. From the expression (15) the selection rules for allowed transitions are then derived, reflecting the demand for conservation of angular momentum in the system, and certain symmetry considerations.

An excited state i will have a natural radiative lifetime τ_i, determined by the probabilities A_{ik} for allowed spontaneous transitions to lower-lying states, k:

$$\tau_i = \frac{1}{\sum_k A_{ik}} \qquad (16)$$

For excited electronic states we generally have $10^{-9} < \tau < 10^{-6}$ s.

For electronic transitions a number of selection rules determine which lines will occur in the absorption- and emission spectra. The specific rules depend on the electronic coupling schemes. E.g., for cases where the total spin quantum number is well defined, we have the rule $\Delta S = 0$, i.e.the multiplicity does not change. Of special interest in the present context is the consideration of vibrational-rotational transitions in molecules.

Vibrational-rotational transitions. The selection rule for electric-dipole transitions between harmonic-oscillator energy levels is

$$\Delta v = \pm 1 \quad ,$$

yielding the vibrational transition frequency, ν_c, according to (6). Such transitions generally occur in the wavelength region $\lambda_{vibr} \sim 10$ µm. Due to the anharmonicity of the true potential, weak transitions according to $\Delta v = \pm 2, \pm 3...$ can also occur (overtones).

A consideration of rotational wave-functions yield from (15) the selection rule for pure rotational transitions:

$$\Delta J = \pm 1$$

Thus pure rotational transition frequencies are obtained using (8):

$$\nu_{J+1 \leftrightarrow J} = \frac{2B}{h}(J+1) \quad ; \quad J = 0, 1, 2.. \qquad (17)$$

These transitions generally occur in the far infrared region ($\lambda_{rot} \sim 1000$ µm).

A vibrational transition is normally accompanied by a change in rotation. If the rotational constant, B, is the same in both vibrational levels (which means no interaction between rotational and vibrational motion) we obtain the vibrational-rotational transition frequencies

$$\nu = \nu_c \begin{cases} + \frac{2B}{h}(J+1) \quad ; \quad J \rightarrow J+1 \quad ; \quad J = 0, 1... \\ - \frac{2B}{h} J \quad ; \quad J \rightarrow J-1 \quad ; \quad J = 1, 2... \end{cases} \qquad (18)$$

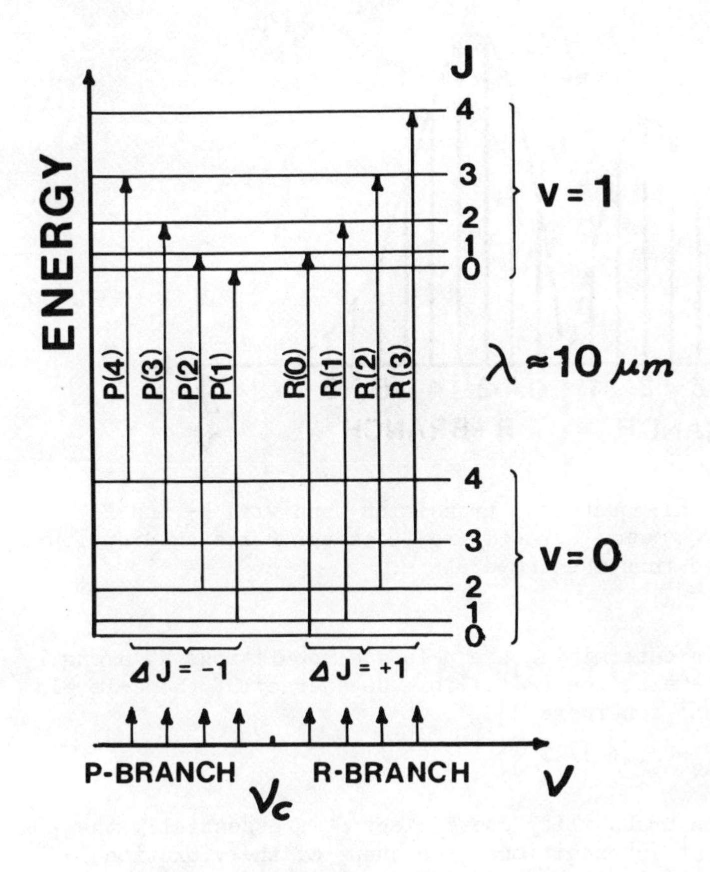

Fig. 4. Vibrational-rotational transitions between low-lying molecular levels. An R- and a P-branch are formed. Note, that no line occurs at the frequency, ν_c.

Two branches, called the R($\Delta J = +1$) and the P($\Delta J = -1$)-branches are obtained. In Fig. 4 an example of transitions of this kind is given. In the figure the usual designations of such transitions (e.g. R(3)) are also given. Higher-order effects may lead to a successive reduction of the frequencies in the R-branch leading to the formation of band-heads.

The strength of an absorption spectrum is, apart from transition probabilities B_{ki}, determined by the population of the lower absorbing levels. The population distribution is governed by the Boltzman factor $e^{-\Delta E/kT}$, which gives the relative population between two levels, separated by ΔE. At room temperature, $kT \sim 1/40$ eV and thus only the lowest vibrational level of a molecule is substantially populated. Consequently, the strongest transition is

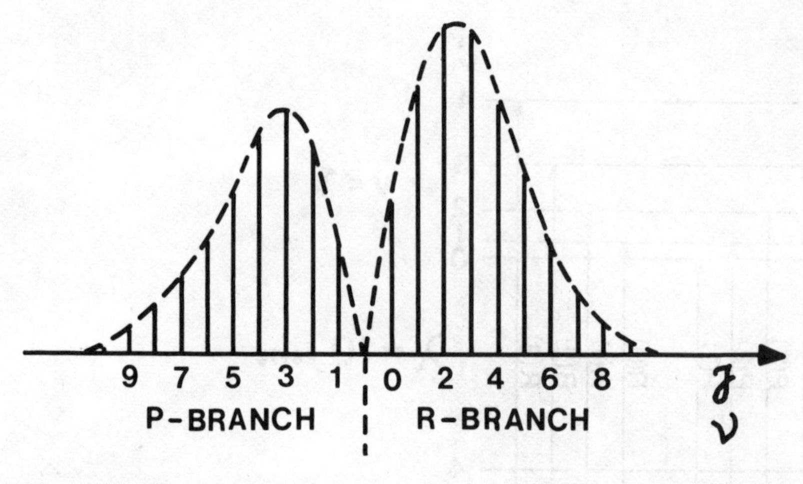

Fig. 5. Vibrational-rotational transition band with R- and P-branches. The asymmetry directly reflects the Boltzman distribution function and thus the temperature.

v = 0 → v = 1. In determining the relative populations N_J among the rotational levels, the (2J+1)-fold degeneracy of the J-levels must be taken into consideration.

$$N_J \propto (2J+1)e^{-BJ(J+1)/kT} \qquad (19)$$

As the absorption probability coefficient *B* is essentially the same for different J-transitions, the shape of the vibrational-rotational absorption band will reflect the populations and thus the temperature T according to (19). In Fig. 5 the asymmetry between the R and P-branches, is illustrated for a certain B/T value. This asymmetry is increasingly pronounced for higher B/T values and can be used for determining the temperature.

Atmospheric absorption. The sun radiates essentially as a Planck radiator of temperature 6000 K. However, this distribution is strongly modified by absorption in the earth's atmosphere. In Fig. 6 the absorption spectrum of the atmosphere is given. The cut-off in the UV-region at about 0.3 μm is caused by electronic transitions in stratospheric ozon molecules. The possible depletion of the ozon layer at about 25 km height, caused by nitrogen oxides from super-sonic transports or freons from spray-cans and refrigation units has recently been a matter of much concern. At still shorter wavelengths the abundant atmospheric molecules also absorb strongly. Thus the cut-off for normal laboratory spectroscopy at 0.18 μm, calling for vacuum spectroscopy, is caused by electronic transitions in oxygen. The IR vibrational-rotational absorption bands shown in Fig. 6 are mainly caused by water- and

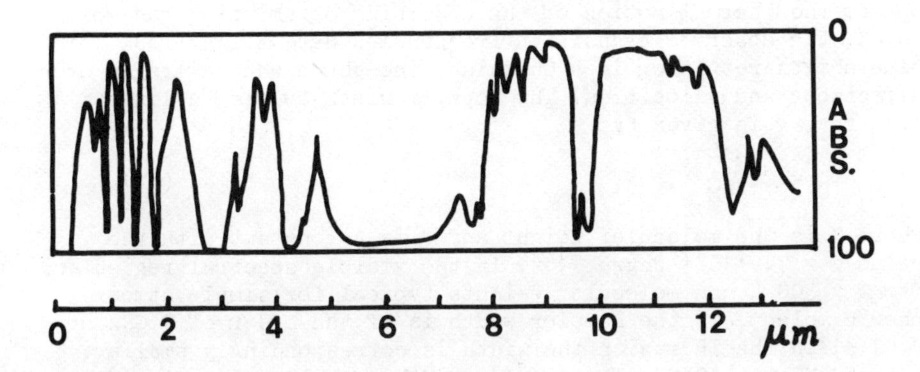

Fig. 6. Absorption of the earth's atmosphere, measured at zenit.

carbon dioxide molecules. However, the band at 9.7 μm is due to ozon. Other minor features are caused by N_2O, CH_4 and other atmospheric constituents. Clearly, the atmospheric absorption at the sea - level will be somewhat different reflecting the actual mixing ratio of the gases.

Widths of spectral lines. The spectrum shown in Fig. 6 is obtained with spectroscopic equipment of very low resolution. However, several modern techniques for atmospheric spectroscopy have an extremely high spectral resolution, and the recorded line-shapes will be determined by intrinsic processes rather than by limitations of apparatus. As the intrinsic width and shape of the lines are of great importance we will briefly consider the origins of line-widths.

a). Natural radiation width

Due to the finite lifetime τ of the excited state, the energy of the state has, according to the Heisenberg uncertainty relation, a finite width. This width is reflected in a natural radiation width $\Delta\nu_N$ of the spectral line

$$\Delta\nu_N = \frac{1}{2\pi\tau} \quad . \tag{20}$$

For a typical lifetime of 10 ns for an electronically excited state we obtain $\Delta\nu_N$ = 15 MHz. The line-shape can be shown to be Lorentzian.

b) Doppler width

Due to the thermal motion of the absorbing or emitting gas molecules, the observed transition frequencies have slight red- or blue-shifts resulting in a Gaussian line-shape when averaged over directions and velocities. The Doppler width $\Delta\nu_D$ for a line of frequency ν is given by

$$\Delta\nu_D = C\nu\sqrt{\frac{T}{M}} , \tag{21}$$

where M is the molecular weight and C is a constant with the value $C = 7.2\cdot10^{-7}$ degree$^{-1/2}$. In the visible spectral region and for T ~ 300 K and molecular weights typical for simple atmospheric molecules, the Doppler width is of the order of 1 GHz or 0.01 Å. In the IR-region the width is correspondingly smaller, e.g. 50 MHz at 10 μm. The Doppler width normally strongly dominates over the natural radiation width.

c) Collisional width

The dominating contribution to line broadening at atmospheric preassures is, at least in the IR region, due to collisions. The collisional width $\Delta\nu_{coll}(P_o,T_o)$ at normal pressures and temperatures is 0.5-5 GHz. The line-shape can be calculated using different approaches and different functions have been derived. The simplest assumptions lead to a Lorentzian line-shape, which in many cases is found to be a good approximation. If the width $\Delta\nu_{coll}(P_o,T_o)$ is known, the width corresponding to other pressures and temperatures can be easily obtained from the approximate formula

$$\Delta\nu_{coll}(P,T) = \Delta\nu_{coll}(P_o,T_o)\,\frac{P}{P_o}\sqrt{\frac{T_o}{T}} \tag{22}$$

The strong pressure dependence of IR absorption lines can be utilized for calculating the height distribution profiles of atmospheric constituents. This technique has been used for calculating the distribution of ozon.

3.2. Rayleigh and Raman scattering

So far we have discussed transitions involving absorption and emission of photons of an energy corresponding to molecular energy-level separations. Light can also be scattered when non-resonant light is irradiated on to molecules, although this occurs with a much lower probability than for resonance scattering. The electric field, E, of the light induces an electric dipole moment, P, in the electronic shell according to

$$P = \alpha E . \tag{23}$$

α is called the polarizability constant. For the general case, α has to be replaced by a polarizability tensor. According to the electro-magnetic theory, the total radiated energy I from the dipole is given by

$$I = \frac{2}{3c^2} \overline{\left(\frac{d^2P}{dt^2}\right)^2} \quad , \qquad (24)$$

where the dash signifies the average value. With $E = E_o \sin 2\pi\nu t$ we then have

$$I = \frac{16\pi^4 c^2}{3\lambda^4} \alpha^2 E_o^2 \qquad (25)$$

The light, reradiated from the molecules at the same frequency as the primary frequency, is called Rayleigh-scattered light. As can be seen from (25) the strength of this elastic scattering is strongly wavelength-dependent ($I \propto 1/\lambda^4$).

If the molecule is vibrating at the frequency ν_{vibr} at the same time as it is subject to the light field, the polarizability constant can be written

$$\alpha = \alpha_o + \alpha_1 \sin 2\pi\nu_{vibr} t \quad ; \quad \alpha_1 \ll \alpha_o \qquad (26)$$

Substituting this expression into (23), we obtain

$$P = \alpha_o E_o \sin 2\pi\nu t + \frac{1}{2}\alpha_1 E_o \left(\cos 2\pi(\nu-\nu_{vibr})t - \cos 2\pi(\nu+\nu_{vibr})t\right) \qquad (27)$$

Rayleigh Stokes Anti-Stokes

Here we see that weak side-bands, shifted by the vibration frequency from the light frequency, ν, are generated. This inelastic scattering is called Raman scattering, consisting of two components, the Stokes and the Anti-Stokes component. For rotating molecules an expression similar to (27) is obtained. However, the Raman shifts are $2\nu_{rot}$ since a certain polarization situation is identical two times in every revolution. According to the phenomenological description presented here, the Stokes and Anti-Stokes components have equal intensities in contrast to experimental observations. On the other hand, a quantum-mechanical approach correctly describes the process in all details. In this theory, the molecules are excited to virtual states of extremely short lifetime. The population in the lower levels determine the strength of the components, giving dominance to the Stokes' component as illustrated in Fig. 7. The quantum-mechanical selection rules are the same as the classical ones, e.g.

$\Delta v = \pm 1$
$\Delta J = 0, \pm 2$

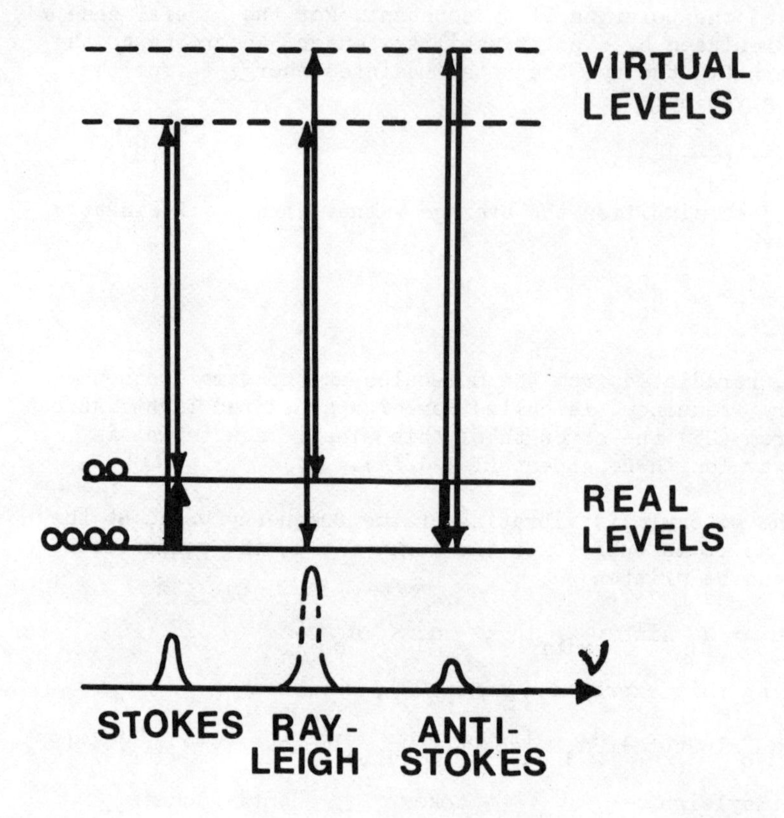

Fig. 7. Illustration of Raman scattering and the origin of the Stokes—Anti-Stokes intensity asymmetry .

Thus in vibrational-rotational Raman scattering the spectrum is given by:

$$\nu = \nu_c \begin{cases} + \frac{4B}{h}(J+3/2) & ; \ J \rightarrow J+2 \ ; \ J = 0, 1,.. \ (S) \\ \pm 0 & ; \ J \rightarrow J \ ; \ J = 0, 1,.. \ (Q) \\ - \frac{4B}{h}(J-1/2) & ; \ J \rightarrow J-2 \ ; \ J = 2, 3,.. \ (O) \end{cases} \quad (28)$$

Again we have assumed the same value for B in the two vibrational states (generally the v = 0 and the v = 1 states). Three branches are generated, the S, Q, and O branches. The same considerations regarding population distribution between the rotational levels apply as discussed in the context of IR resonance transitions, leading to a O-S branch asymmetry in the Raman scattering, similar to that illustrated in Fig. 5. However, the Q-branch, consisting of a single frequency, strongly dominates over the O and S-wings and is the one normally observed in measurements at moderate resolution. The Stokes — Anti-Stokes asymmetry for the Q-branches, or the asymmetry in the Stokes O-S wings, is very useful for temperature determinations.

3.3. Mie scattering

Light scattering from particles, which are large in comparison to the wavelength employed, cannot be treated in the same way as Rayleigh scattering from molecules, in which case the size/wavelength ratio is much less than unity. In light scattering from particles, interference effects are important. As early as 1908, Mie treated the simplest case of spherical particles, such as water droplets, and many situations with other kinds of particles are approximated by Mie scattering , or extensions of the theory [3]. The Mie scattering cross section is a function of two parameters:

$$\sigma_{Mie} = f(x,m) \text{ , with} \qquad x = \frac{2\pi r}{\lambda} \; ; \; m = \frac{m_1}{m_2} \text{ ,} \tag{29}$$

where λ is the wavelength of the elastically scattered radiation and r is the radius of the spherical particles of complex index of refraction $m_1 = n - ik$. m_2 is the corresponding quantity for the ambient medium. For air $m_2 = 1$. σ_{Mie} shows a strong oscillatory behaviour as a function of the x-parameter, due to interference caused by surface waves. For particle-size distributions typical for the atmosphere, the oscillations are smoothed out and the intensity of the scattering only slowly varies with the wavelength. The inverse problem, to calculate the particle-size distribution from the observed wavelength dependence of the scattering, caused e.g. by a laser beam, has been a subject of intense efforts. The atmospheric Mie scattering shows an over-all increase in intensity for shorter wavelengths, although the dependence is not so strong as for the Rayleigh scattering (Eq. 25). The wavelength dependence of the Rayleigh- and Mie-scattering explains the blue sky on sunny days, and red sunsets, especially in dusty atmospheres. Mie scattering generally dominates strongly over Rayleigh scattering in the lower atmosphere, and thus determines the visibility range. This means that the effective Mie-scattering cross section can be determined from the observed visibility range.

3.4. Comparison between radiative interactions

In Table I the cross sections, σ, for the different radiation processes are compared. Clearly, strong variations in intensity exist for an individual process, but the numbers in the table give an indication of the relative strengths. The strongest processes are

Table 1. Strength of processes

Process	σ cm^2
Res. absorption	10^{-15}
Fluorescence	10^{-15}
Fluor. (quenched)	10^{-20}
Rayleigh scatt.	10^{-26}
Raman scatt.	10^{-29}
Mie scatt.	10^{-26}-10^{-8}

resonance absorption and associated fluorescence.
However, strong fluorescence only occurs if the pressure is low as in the stratosphere. At atmospheric pressures the fluorescence is strongly quenched (by a factor ~10^5) due to radiation-less transitions caused by collisions. Rayleigh scattering is generally more than 10^{10} times weaker than resonance absorption and unquenched fluorescence. Raman scattering is still about 10^3 times weaker. The intensity of Mie-scattering varies in a very large range depending on the visibility.

4. METHODS FOR ATMOSPHERIC SPECTROSCOPY

The atmosphere consists of two major constituents: nitrogen and oxygen. Dry air contains 78.1 per cent (by volume) of N_2 and 20.9 per cent of O_2. The inert gases Ar, Ne and He are present in 9300, 18 and 5 ppm (parts per million), respectively. The content of CO_2 is presently 318 ppm and this number is increasing by 0.7 ppm annually, which may result in the so called green-house effect. In addition to these gases, normal air also contains widely varying concentrations of water vapour. The gases N_2O, H_2, CH_4, NO_2, O_3, SO_2, CO, NH_3 and others are naturally present in concentrations ranging from several ppm to fractions of ppb (parts per billion). When such gases occur in increased concentrations due to human activities, they are considered to be pollutants. The main aim of the various atmospheric-spectroscopy methods discussed in this section is to determine the concentration and distribution of the gases in the atmosphere. In addition, temperature and wind speed can be measured with spectroscopic techniques.

Atmospheric measurements can be performed using *point monitoring* or *remote sensing*. In point measurements, the atmosphere in the direct surrounding of the measurement equipment is investigated, whereas in remote-sensing measurements a suitable instrumentation is used to gain information about the conditions at a considerable distance from the equipment. From the remote-sensing concept it follows that a large atmospheric volume can be probed from a single measuring site. A further distinction can be made between *passive* and *active* measurement techniques. In passive remote-sensing measurements, the atmosphere is probed by analyzing the spectral distribution of the back-ground radition from e.g. the sun or the sky. In active techniques radiation is transmitted from the remote sensing unit into the atmosphere, and the back-scattered radiation, carrying information on the constituents, is received and analyzed by the unit. In this section we will briefly discuss several atmospheric spectroscopy methods. For a more detailed information we refer to the references [4-6].

4.1. Passive techniques

The general principle of passive atmospheric spectroscopy measurements is shown in Fig. 8. The absorption of the back-ground sky- or sun radiation by atmospheric constituents at certain frequencies cies is measured by a suitable spectrometer. The measured frequencies are used for gas identification, and the strength of the absorption lines determines the concentration of the absorbing gas, according to the Beer-Lambert law:

$$\frac{I_t(\nu)}{I_o(\nu)} = \exp\left[-\sigma(\nu)\int_o^\infty N(r)dr\right] \qquad (30)$$

From the transmitted intensity fraction, $I_t(\nu)/I_o(\nu)$, the integral number of absorbing molecules can be determined when the absorption cross section $\sigma(\nu)$ at the studied line is known. Generally, the presence of an atmospheric inversion layer limits the distance over which the integration needs to be performed, and thus the mean concentration of the molecules up to the inversion layer can be

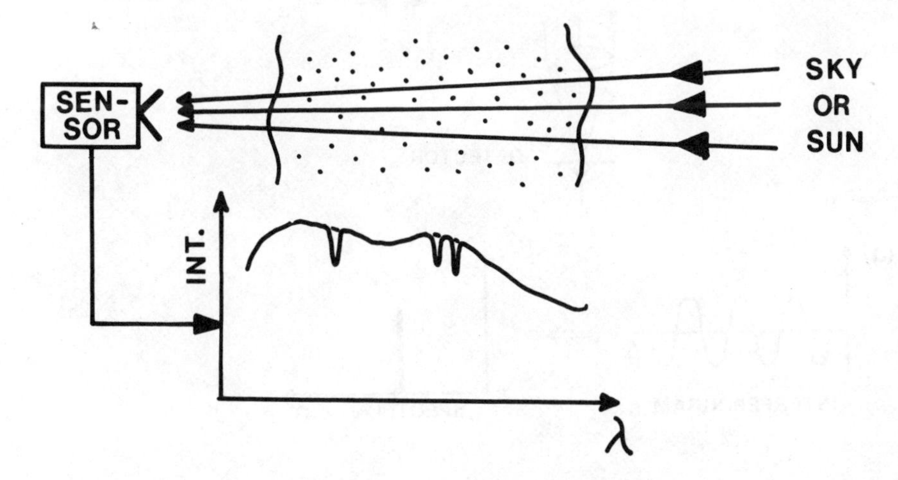

Fig. 8. Schematic diagram of passive remote sensing of atmospheric gases. The absorbing molecules cause absorption dips in the background spectral distribution.

determined. As discussed in Section 3.1, the absorption line shape can be used to determine the height distribution of a gas, which is present at widely differing heights. From satellites limb measurements are important. Besides absorption, IR- or microwave emission from hot gases can be used in passive measurements. We will consider here a few types of spectrometers, useful for absorption and/or emission measurements of atmospheric constituents.

Dispersive spectrometers. Grating or prism spectrometers of suitable resolution can be used for a wavelength analysis of the radiation. In photo-electric recording instruments, the first- or second-derivative spectra may be recorded to obtain an increased sensitivity. Array detectors are also becoming increasingly important.

Fourier spectrometers. The basic principle of a Fourier spectrometer is shown in Fig. 9. The incoming radiation is divided by a beam-splitter into two perpendicular beams which after reflection are brought together again at the detector. One of the

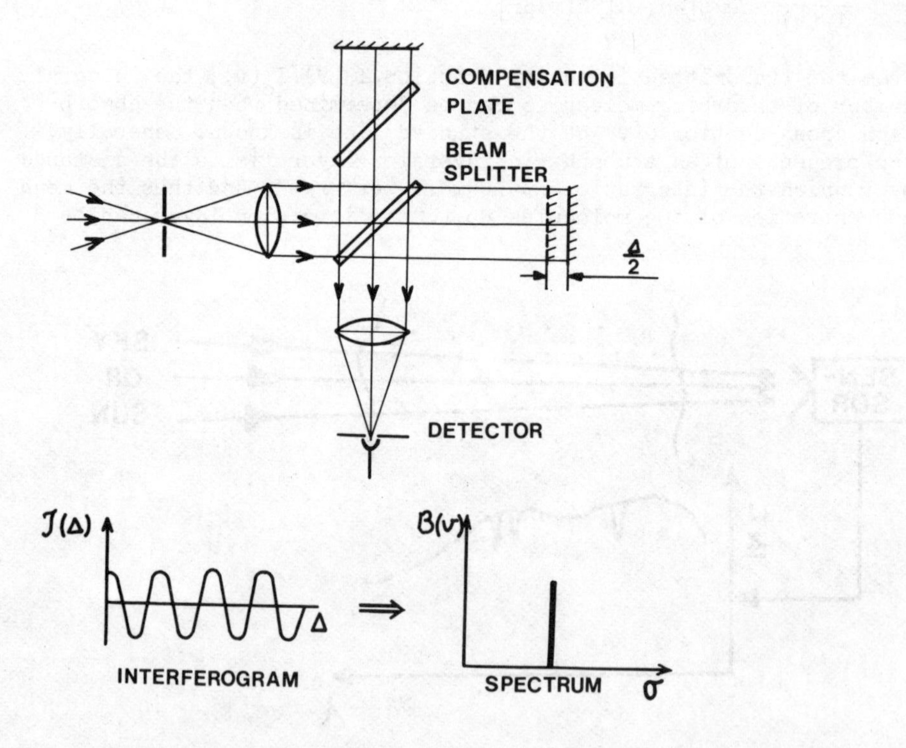

Fig. 9. Schematic diagram of a Fourier spectrometer, based on a Michelson interferometer. An interferogram and the corresponding spectrum for the case of a single line is shown.

reflecting mirrors can be uniformly moved, introducing an optical path difference Δ. If the incoming light consists of only one line, regular interference fringes are produced at the detector as the mirror is moved. The recorded intensity $I(\Delta)$ is given by

$$I(\Delta) = I_o \cos^2 \frac{\phi}{2} \tag{31}$$

where $\phi = 2\pi\Delta/\lambda = 2\pi\Delta\sigma$. For a spectral distribution $B(\sigma)$ we then obtain

$$I(\Delta) = \int_0^\infty B(\sigma)\cos^2\pi\Delta\sigma d\sigma = \frac{1}{2}\int_0^\infty B(\sigma)(1+\cos2\pi\Delta\sigma)d\sigma \tag{32}$$

The part depending on the path difference is called the interferogram $\mathcal{I}(\Delta)$:

$$\mathcal{I}(\Delta) = \frac{1}{2}\int_0^\infty B(\sigma)\cos2\pi\Delta\sigma d\sigma \tag{33}$$

From the recorded interferogram the spectrum $B(\sigma)$ can be mathematically calculated as the Fourier-cosine-transform:

$$B(\sigma) \propto \int_0^\infty \mathcal{I}(\Delta)\cos2\pi\sigma\Delta d\Delta \tag{34}$$

In practice the mirror cannot be moved to infinity and the integration must be truncated. This introduces unwanted oscillations in the Fourier transform, which can however be suppressed by an apodization procedure. The Fourier spectrometer simultaneously records all lines in the spectrum and thus is much more efficient than a normal dispersive instrument with entrance and exit slits (the Fellget advantage). It is especially competitive in the infra-red spectral region, where no photographic plates are available for multi-line registration.

Gas-filter-correlation spectrometers. The principle of a gas-filter-correlation instrument is shown in Fig. 10. The incident radiation is switched between an empty cell and a cell, containing gas of the type to be detected in the atmosphere. Using lock-in techniques, the total intensities measured through the two cells are compared at the common detector. The gas cell contains a sufficient concentration of gas to completely absorb the characteristic wavelengths. For a situation where the atmosphere contains no gas of the kind kept in the cell, a gray wedge is used to attenuate the reference-cell beam to cause a zero indication on the lock-in detector (shaded areas are equal). If now an atmosphere containing the specific gas is probed this will result in no change in transmission through the gas cell, whereas an additional, characteristic absorption is observed in the reference channel. Thus, an unbalance, corresponding to the integrated atmospheric gas burden, is recorded. The spectra of other atmospheric gases will not correlate with the gas-cell spectrum and will thus cause the same influence in both channels. In practice, the accuracy

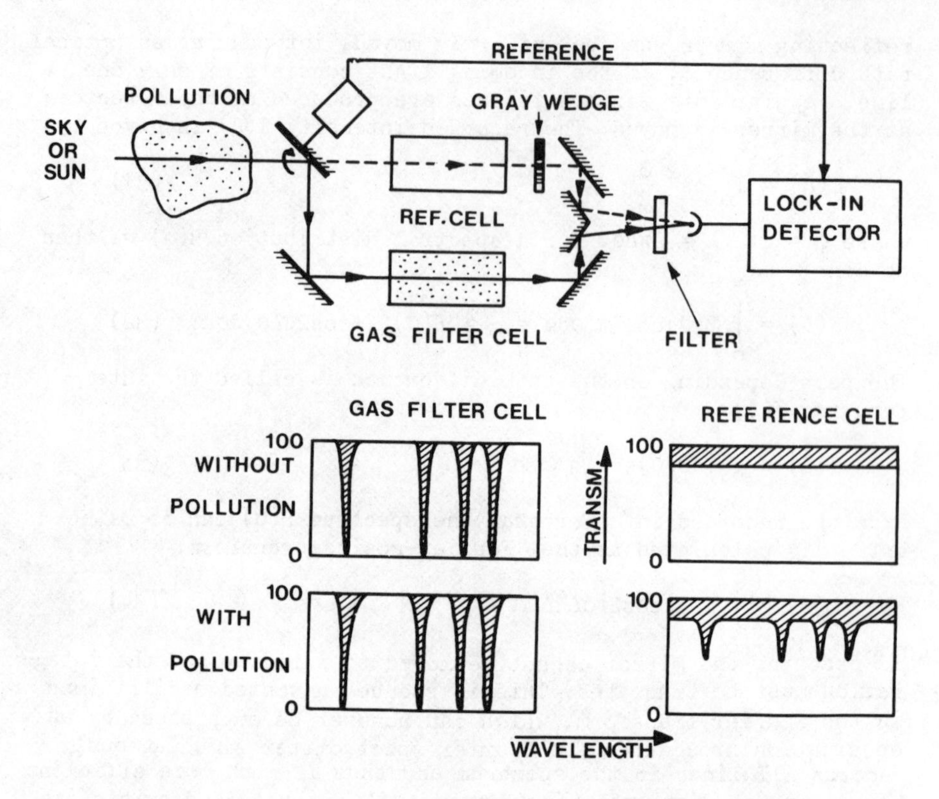

Fig. 10. Principle of a gas-filter correlation spectrometer. The spectral absorption in the two channels for the case of a clean and a polluted atmosphere is shown.

of the instrument is improved by working in spectral bands free from interference. Thus the gas-filter technique is often combined with dispersive techniques. Correlation spectroscopy can also be performed using pure dispersive spectrometers equipped with correlation masks to sense the dispersed spectrum.

Heterodyne spectrometers. The heterodyne principle is very useful for signal detection both in the microwave- and IR regions. An optical heterodyne receiver is shown in Fig. 11. Here the incoming radiation is chopped and mixed by the radiation from a laser, used as a local oscillator. A narrow-band electronic filter of fixed frequency $\nu_{IF} = \nu_S - \nu_L$ defines the intermediate frequency of the receiver, chosen in the radio-frequency region. By sweeping the laser frequency ν_L, the detected signal frequency ν_S, forming the difference frequency ν_{IF} with the laser, will also be swept and the spectrum will be recorded on the lock-in amplifier, operating at an audio frequency. The heterodyne principle offers an extremely high sensitivity due to the mixing of the weak

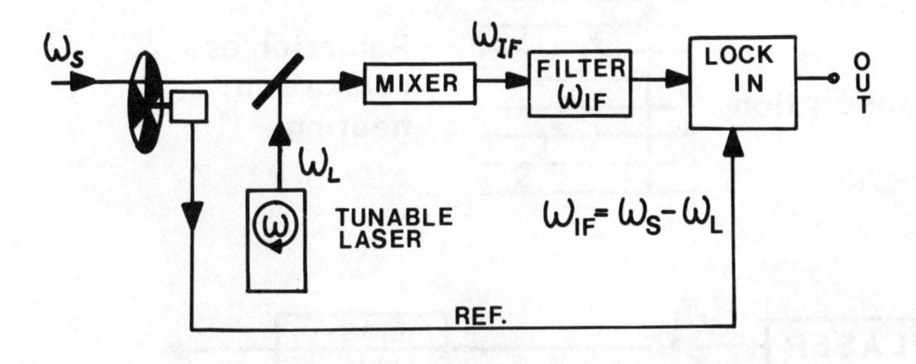

Fig. 11. Schematic diagram of an optical heterodyne receiver.

spectroscopic signal with a strong local-oscillator signal. The power obtained at the mixer is

$$(A_S \cos\omega_S t + A_L \cos\omega_L t)^2 = C + A_S A_L \cos(\omega_S - \omega_L)t \qquad (35)$$

The intensity $A_S A_L$, proportional to A_S, is detected by the system, whereas the high frequencies contained in C are suppressed. As a local oscillator for the heterodyne receiver a CW CO_2- or a diode laser is generally used.

4.2. Active techniques

Lasers are very useful in active remote-sensing techniques. Before considering such techniques in more detail we will discuss a few laser techniques for point monitoring. A laser beam of suitable frequency will, when crossing a probe volume connected with the ambient air, give rise to specific molecular fluorescence, useful for measuring the gas concentration. Collisional quenching can be reduced by working at low pressures. Whereas the number of absorbing molecules is then also reduced, the narrower line-widths give an increased selectivity. The fluorescence can be increased by placing the sample volume inside the laser cavity. In this case the absorption can also be monitored by observing the selective mode-quenching, which is suppressing those modes of a multi-mode laser, capable of otherwise exciting fluorescence in an external monitor cell filled with the gas to be studied. A further point-monitor technique is the opto-acoustic approach, illustrated in Fig. 12. Air is let into the probe cell, which is crossed by a chopped laser beam, tuned to an allowed transition. If the pressure

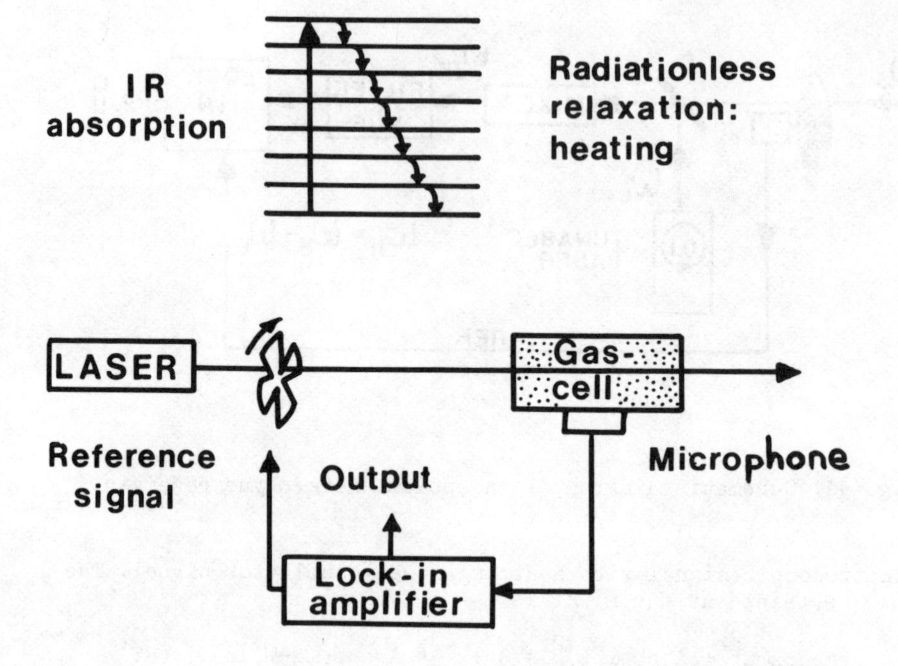

Fig. 12. Illustration of the opto-acoustic measurement method.

is not very low, radiationless transitions to lower levels (relaxation) will occur due to collisions. This means a heating of the gas and a corresponding pressure increase in the closed cell. A microphone, coupled to the cell, will pick up the periodic pressure changes, which are recorded with a lock-in amplifier, operating at the chopping frequency. The technique is extremely sensitive and has recently been used in balloon-borne measurements of stratospheric NO, relevant to the O_3 destruction cycle.

We will now describe two active remote sensing techniques, based on lasers.

Long-path absorption measurements. As laser beams have a divergence that is often only diffraction-limited, they are very useful for probing long absorption paths, as illustrated in Fig. 13. After reflection from a retro-reflector, the beam is received by an optical telescope and detected. The average concentration over the measuring path, which can be several kilometers long, is measured. In measurements of air pollutants it is necessary to operate at those wavelengths, where absorption from the dominant gases does not cause the atmosphere to be opaque. The CO_2 laser, operating at around 10 μm, falls in a suitable

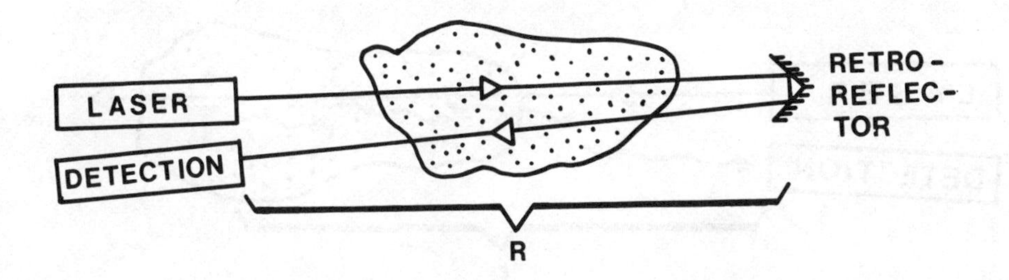

Fig. 13. Principle of long-path-absorption measurements.

optical window, as can be seen from Fig. 6. Such a laser can be grating-tuned to a large number of discrete P- and R-lines, partly or fully coinciding with absorption lines of e.g. ethylen, vinylchloride and freons. The absorption at a frequency ν in the case of n different molecular species of mean concentrations N_i along the measuring path, is given by a Beer-Lambert expression, similar to (30),

$$\ln \frac{I_o(\nu)}{I_t(\nu)} = 2R\left(\sum_{i=1}^{n} \sigma_i(\nu)N_i + K_{ext}\right) \qquad (36)$$

Here $\sigma_i(\nu)$ are the absorption cross-sections at the measuring frequency and K_{ext} is a back-ground extinction coefficient due to e.g. the Mie scattering. In order to determine the n concentration values, N_i, it is necessary to perform measurements for at least n+1 frequencies, where the cross-sections σ_i differ. The frequencies must be chosen within a sufficiently small interval so that K_{ext} can be assumed to be independent of frequency. As discussed in Sec. 3.1, the atmospheric line profiles in the IR region are quite sensitive to pressure and temperature, and sometimes it would be very desirable to be able to tune the laser continuously over the frequently overlapping absorption profiles. This can be done with diode lasers, for which the measuring path is however limited due to low output powers. The development of high-pressure CO_2-lasers with continous tunability gives considerable promise for powerful long-path-absorption applications.

Long-path-absorption measurements in the IR region can be performed employing heterodyne detection (Sec. 4.1). If the same laser is used for the transmitted beam as well as for the local oscillator, a frequency shift corresponding to the intermediate frequency must be introduced, employing e.g. an acousto-optic deflector.

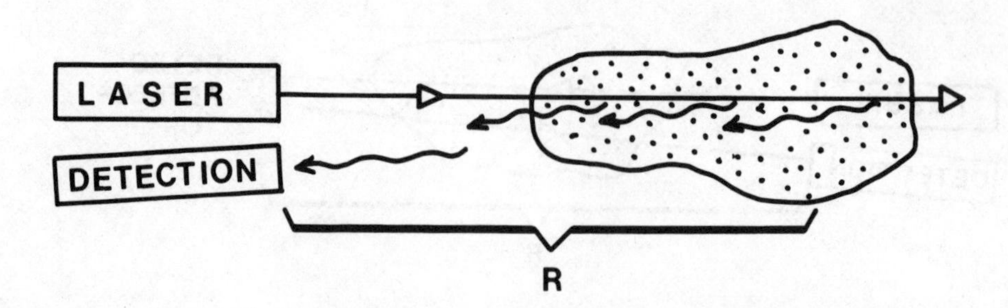

Fig. 14. Principle of Lidar measurements.

Lidar measurements. The Lidar (Light detection and ranging) principle is shown in Fig. 14. A short intense pulse of light is transmitted into the atmosphere and light, back-scattered in e.g. Mie-scattering from particles, is collected with a telescope at the site of the laser transmitter. Light, scattered at a distance R from the equipment, is received at a time

$$t = \frac{2R}{c} \tag{37}$$

after the transmission of the pulse, which travels with the velocity c = 300 m/μs. The laser pulse length Δt_p limits the range resolution ΔR to

$$\Delta R = \frac{c\Delta t_p}{2} \tag{38}$$

The power $P_\nu(R,\Delta R)$, received from the interval ΔR at the range R, is given by the general Lidar equation:

$$P(R,\Delta R) = CW\sigma_b N_b(R)\,\frac{\Delta R}{R^2}\,\exp\left[-2\int_o^R \sigma(\nu)N(r)dr\right] \tag{39}$$

W is the transmitted pulse power and $N_b(R)$ is the density of scatterers with the back-scattering cross section σ_b. The exponential describes the attenuation of the beam and the back-scattered light due to the absorbtion of molecules of density N(r) and absorption cross-section $\sigma(\nu)$. C is a system constant.

In (39) the product $\sigma_b N_b(R)$ determines the strength of the back-scattering, which can be caused by different processes. Thus it can be due to fluorescence from molecules, excited by the laser light. The process is particularly suitable for stratospheric atoms/molecules, as quenching does not then occur. Thus extremely diluted stratospheric layers of alkali atoms have been detected from the ground using fluorescence Lidar systems employing tunable dye lasers. Light can also be returned to the detection telescope because of Raman scattering. As the Raman cross-sections are extremely low, only abundant gases like N_2, O_2 and H_2O yield appreciable signals. Atmospheric temperatures can be measured by Raman Lidar techniques observing intensity asymmetries in the Raman scattering from N_2 or O_2, whereas an utilization of the Raman shift of H_2O yields information about the air moisture.

As we have noticed in Sec. 3.4, Mie scattering generally provides strong signals and thus e.g. particles from smoke-stacks can easily be detected with a Mie Lidar system. In Fig. 15 a Lidar registration curve, obtained in measurements at an iron alloy plant is shown. The scattering in the uniform distribution of atmospheric particles gives rise to a falling curve, showing the $1/R^2$ dependence of Eq. (39). In addition, a sharp peak due to the increased particle concentration in the smoke is shown as well as structures due to emissions through roof windows. The complexity of the Mie scattering calls for a calibration of the elastic plume echo versus the particle contents.

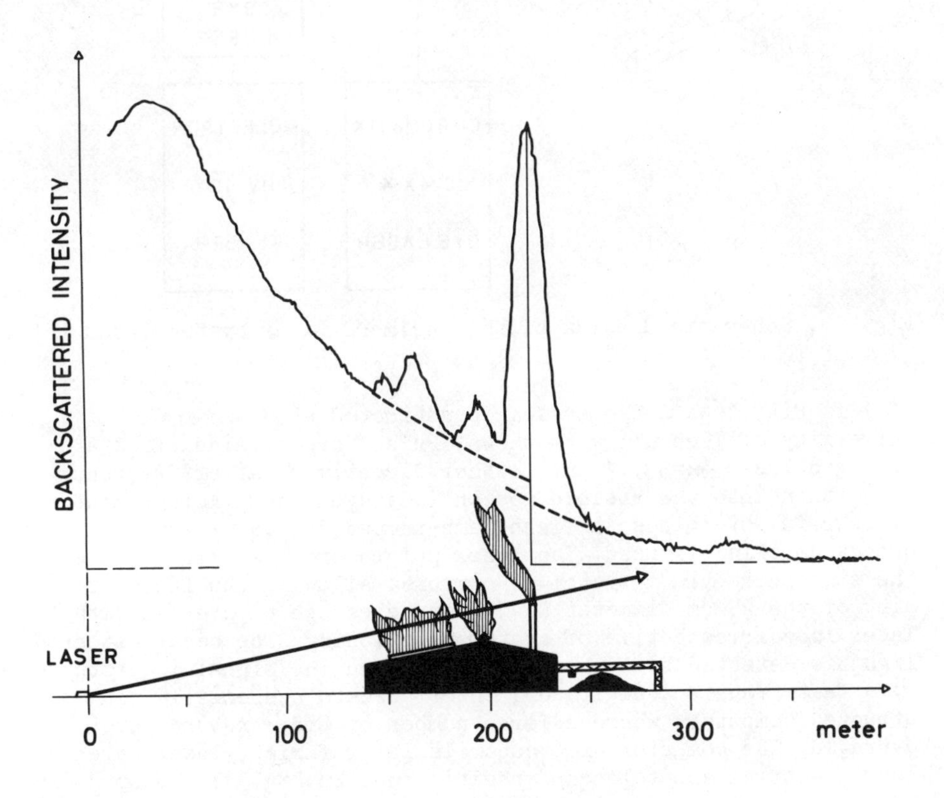

Fig. 15. Lidar measurement of particle concentrations (From [7]).

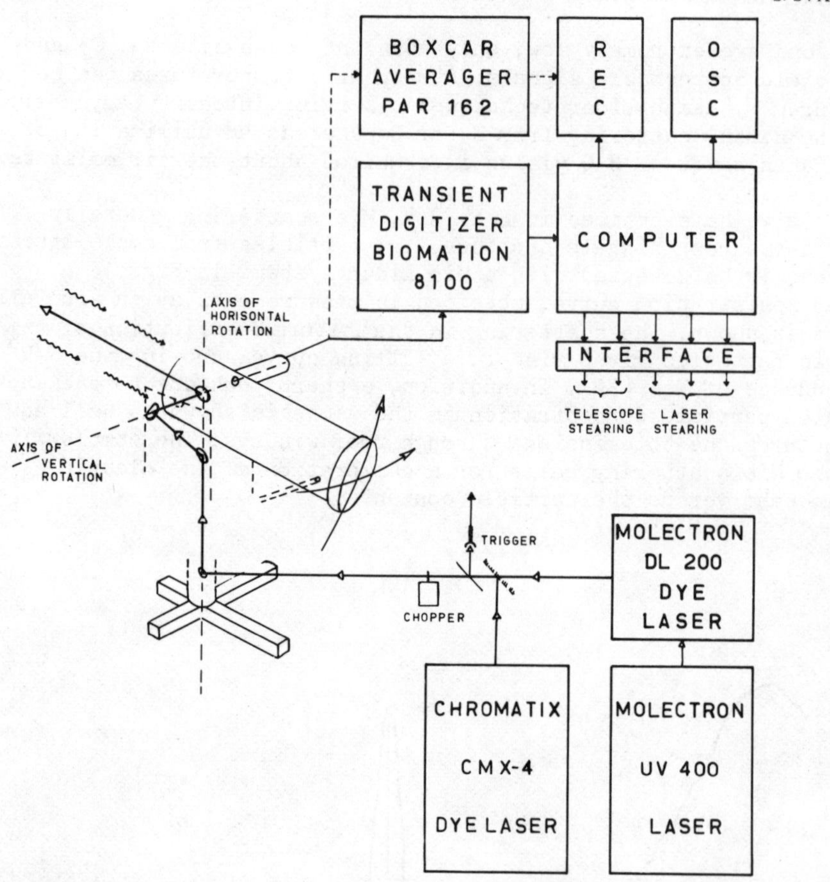

Fig. 16. Schematic diagram of the Chalmers Lidar system (From [8]).

In Fig. 16 a Lidar system, constructed at Chalmers University of Technology is shown. Pulses are obtained from a nitrogen laser (λ=337.1 nm), either directly or after frequency conversion into the visible region in a dye laser. For generation of powerful UV-pulses, a flash-lamp-pumped dye laser with frequency doubling is used. The laser pulses are transmitted into the atmosphere via a system of mirrors, allowing the field of view of the 25 cm diameter Newtonian telescope to overlap with the laser lobe irrespective of system positioning. The back-scattered light is detected by a photomultiplier and the signal is captured by a fast-transient digitizer, interfaced to a specially constructed computer, where a large number of Lidar returns are averaged. The computer also controls laser firing, laser wavelength setting and telescope positioning. In Fig. 17 the spreading of a plume has been monitored with the system shown in Fig. 16. The wind was blowing towards the Lidar system, which was operating

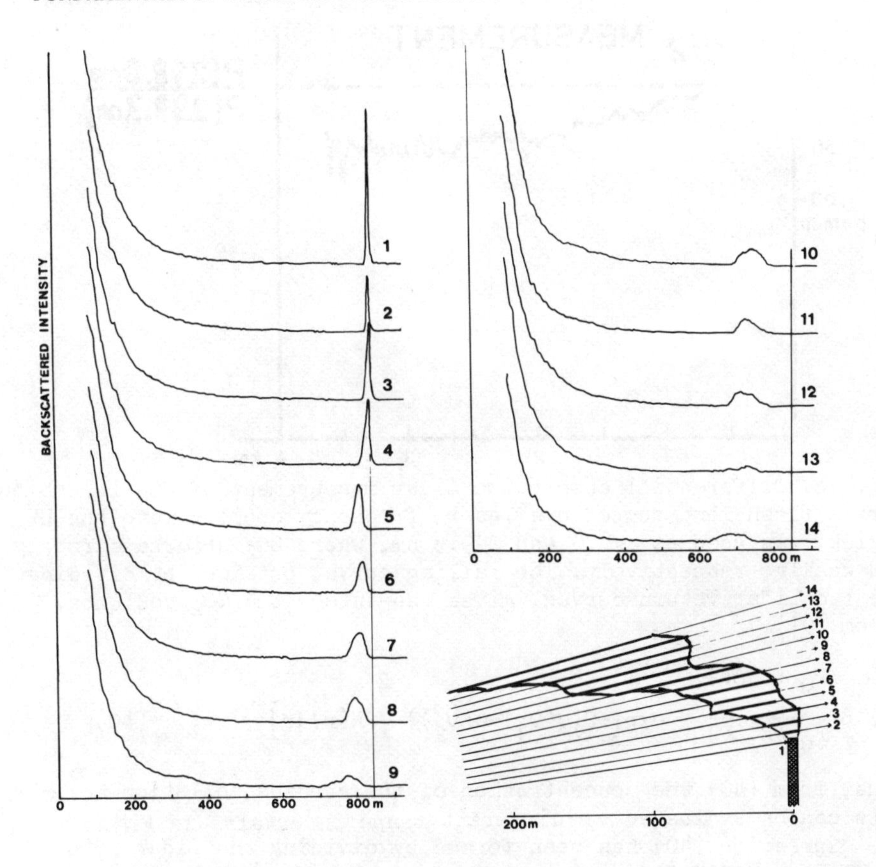

Fig. 17. Scanning of a smoke plume with a Lidar system using a pulsed nitrogen laser (From [8]).

with the nitrogen laser.

In order to measure the concentration of gaseous pollutants with Lidar techniques, the resonance absorption can be utilized in a similar way as in long-path absorption. However, in Lidar measurements no fixed retro-reflector is needed. Instead, the "distributed mirror" provided by the Mie scattering is used, permitting range-resolved measurements to be performed. The differential absorption of the atmosphere at close-lying frequencies as measured in the Mie scattering, is used to identify and quantify the molecules. This method is generally referred to as Dial (Differential absorption Lidar). If we evaluate the expression (39) for two frequencies of strongly differing absorption cross sections and form the ratio, a very simple expression is obtained.

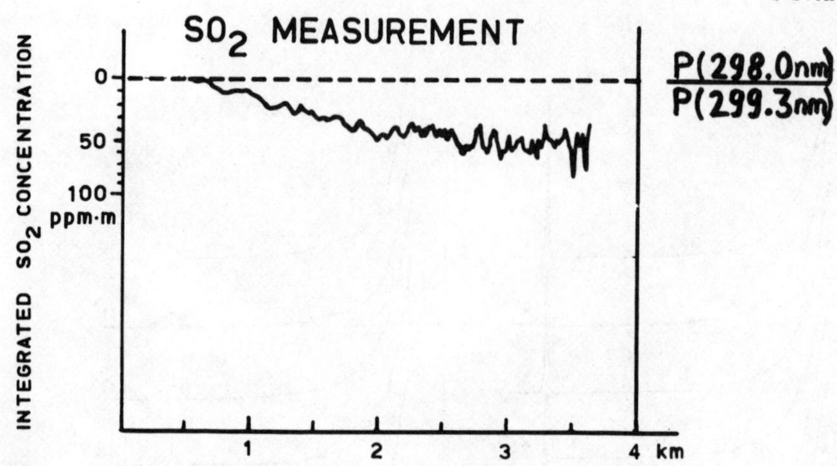

Fig. 18. Differential absorption Lidar measurement of SO_2 in ambient air. A flash-lamp pumped dye laser, frequency doubled into the UV region, was used at 298.0 and 299.3 nm, where SO_2 absorbs strongly and weakly, respectively. The falling curve, obtained by dividing the two Lidar return curves, gives the integrated SO_2 contents. (From [9]).

$$\frac{P_{\nu_1}(R,\Delta R)}{P_{\nu_2}(R,\Delta R)} = \exp\left[-2(\sigma(\nu_1)-\sigma(\nu_2)) \int_0^R N(r)dr\right] \qquad (40)$$

Thus, from (40) the concentration of the studied pollution molecule can be evaluated in different range intervals. In Fig. 18 the expression (40) has been formed by dividing the Lidar returns at 298.0 and 299.3 nm, strongly and weakly absorbed by SO_2, respectively. From the figure, e.g. a mean SO_2 concentration of 30 ppb over the interval 1-2 km, and 10 ppb over the interval 2-3 km can be inferred.

A Dial measurement of NO_2 in a smoke plume at 2 km distance is illustrated in Fig. 19. The NO_2 absorption cross section at 4582 Å is stronger than at 4568 Å resulting in a stronger attenuation through the smoke plume in the former case. The elastic plume echoes are in both cases very strong and have been cut off. From the two curves an integrated NO_2 content of 100 ppm x m through the plume is inferred.

Mie scattering is most prominent in the visible and UV region. With tunable dye lasers, NO_2, SO_2 and O_3 can be measured in this range. Pulsed IR lasers can be used for Dial measurements of other gases. However, the low Mie scattering cross section for such wavelengths calls for the use of high-power lasers.

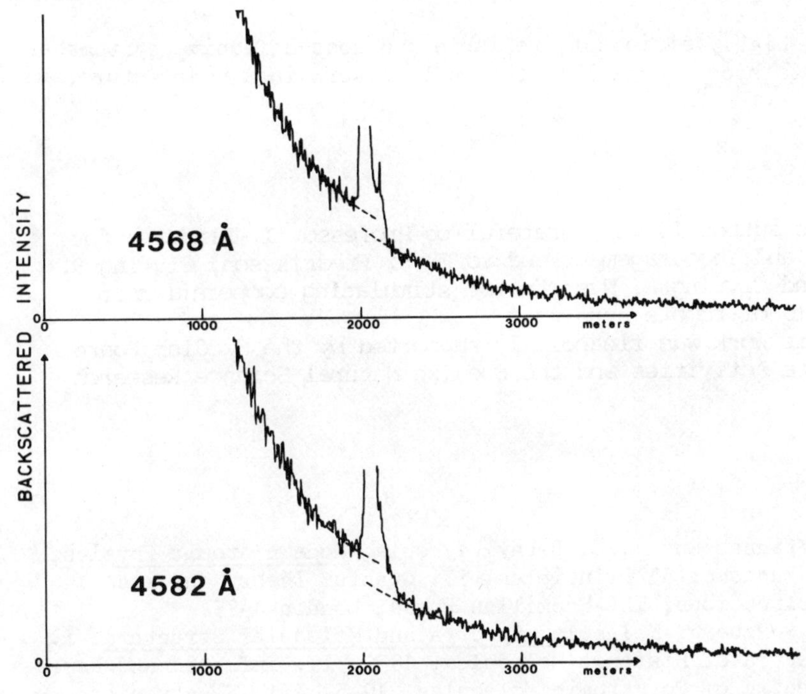

Fig. 19. Dial measurement of the atmosphere surrounding the plume from an oil refinery. The signal is more attenuated by the plume for the wavelength 4582 Å than for 4568 Å because of the stronger absorption cross section for NO_2 at that wavelength. From the curves an integrated plume content of 100 ppm · m NO_2 is inferred.

5. OUT-LOOKS FOR THE FUTURE

An increasing awareness of the fragility of the biosphere has led to strong demands for a limitation of pollution. Possible global effects of atmospheric pollution, such as the "greenhouse effect", which is due to CO_2, and an increasing level of UV radiation due to nitrogen oxide, depleting the stratospheric ozon layer, have been widely discussed. Local effects of air pollutants can be health hazards in urban and industrial areas. Many countries have passed anti-pollution laws and international cooperation is increasing. Obviously, probing the state of the atmosphere will become increasingly important. Satellites are being equipped with sophisticated instruments for probing not only meteorological conditions but also pollution levels on a world-wide basis. A growing number of mobile remote-sensing laboratories using passive- and active instruments will be operated by environmental protection agencies. It is likely that large urban and industrial areas will be continously surveyed from measuring towers, equipped with remote sensors. With the rapid development in laser- and electronic technology, highly efficient, reliable and cost-effective systems are likely to become available for routine work in a near future.

The author is very grateful to Professor I. Lindgren for support and encouragement and to FK K. Fredriksson, Civ ing B. Galle and Civ ing K. Nyström for stimulating cooperation in the field of remote sensing.

This work was financially supported by the Swedish Board for Space Activities and the Swedish Natural Science Research Council.

REFERENCES

1. B. Cagnac and J.-C. Pebay-Peyroula, Modern Atomic Physics, I. Fundamental Principles, II. Quantum Theory and its Applications, The Macmillan Press, London 1975.
2. G. Herzberg, Molecular Spectra and Molecular Structure, I. Spectra of Diatomic Molecules, 1939, II. Infrared and Raman Spectra of Polyatomic Molecules, 1945, III. Electronic Spectra and Electronic Structure of Polyatomic Molecules, 1966, Van Nostrand Reinhold.
3. H.C. van de Hulst, Light Scattering by Small Particles, J. Wiley and Sons, New York, 1957.
4. V.E. Derr, Ed., Remote Sensing of the Troposphere, Wave Propagation Laboratory, NOAA, 1972.
5. E.D. Hinkley, Ed., Laser Monitoring of the Atmosphere, Topics in Applied Physics, Vol. 14, Springer, Berlin-Heidelberg 1976.
6. J. Lintz, Jr. and D.S. Simonett, Eds., Remote Sensing of Environment, Addison-Wesley, Reading 1976.
7. K. Fredriksson, I. Lindgren, S. Svanberg and G. Weibull, "Measurement of the Emission from Industrial Smoke-stacks using Laser Radar Techniques", Göteborg Institute of Physics Reports GIPR-121, 1976.
8. K. Fredriksson, B. Galle, A. Linder, K. Nyström and S. Svanberg, "Laser Radar Measurements of Air Pollutants at an Oil-burning Power Station", Göteborg Institute of Physics Reports GIPR-150, 1977.
9. K. Fredriksson, B. Galle, K. Nyström and S. Svanberg, "Measurements of Air Pollutants in the Trollhättan Area using Lidar Techniques", Göteborg Institute of Physics Reports GIPR-171, 1978.
10. K. Fredriksson, I. Lindgren, K. Nyström and S. Svanberg, "Field Test of a Lidar System for the Detection of Atmospheric Pollutants", Göteborg Institute of Physics Reports GIPR-134, 1976.

SIGNAL-TO-NOISE RATIO OF HETERODYNE LIDAR SYSTEMS IN THE PRESENCE OF ATMOSPHERIC TURBULENCE

H. T. Yura

The Ivan A. Getting Laboratories, The Aerospace Corporation, P. O. Box 92957, Los Angeles, California, 90009, USA

ABSTRACT. A general expression for the signal-to-noise ratio of a heterodyne lidar system in the presence of atmospheric turbulence is derived which is valid both in the near- and far-field of the laser and remoted scattering source. We consider the situation where a laser transmitter directs an optical beam at some remote scattering region of interest. The backscattered light is collected by a receiving aperture and mixed with a suitable coherent local oscillator reference field. Both coaxial and bistatic lidar systems are considered. In both cases we are able to obtain algebraic expressions for the signal-to-noise ratio which are valid for an arbitrary propagation path through the atmosphere. Numerical results are presented for both a 3.7 and 10.6 micrometer lidar system. Additionally, we obtain the conditions under which atmospheric turbulence will limit severely the performance of heterodyne lidar systems.

I. INTRODUCTION

Recently, there has been interest in heterodyne lidar systems at infrared wavelengths operating in the atmosphere (Ref. 1). Examples of such systems include cw lidars operating at 3.7 μm with an output power of the order of one watt and 10.6 μm pulsed lidars with a duty cycle of about 10 pulses per second with pulse widths and powers of the order of 10-100 nsec and 0.1-1 joule, respectively. The performance of heterodyne systems are based on mixing the incident signal field with a coherent local oscillator reference field. In remote sensing situations, in which atmospheric path lengths greater than or of the order of a hundred meters are encountered, the received signal will be degraded by atmospheric

T. Lund (ed.), Surveillance of Environmental Pollution and Resources by Electromagnetic Waves, 67-93.

turbulence. In particular, the coherence of the received signal field will be degraded from the corresponding case in the absence of turbulence. As a result, effective mixing with the local oscillator field will not occur, resulting in a corresponding reduction in the signal-to-noise ratio (S/N).

The analysis presented here develops an expression for the S/N in the presence of atmospheric turbulence. We consider the case where a laser transmitter directs a beam at some remote scattering region of interest. The backscattered light is collected by a receiving aperture that may or may not be co-located with the transmitter. For the case of a coaxial transmitter/receiver system, the direct and backscattered light travel over essentially the same atmospheric path. The characteristic fluctuation time of atmospheric turbulence is of the order or greater than a few msec. Hence for one-way propagation distances less than a few hundred kilometers the atmosphere is essentially "frozen in" and the direct and backscattered light will travel over correlated paths. This is not the case for a bistatic system where the separation between the transmitting and receiving apertures is greater than their respective diameters. Both cases are considered here.

For simplicity, we consider the case where the intervening space between the heterodyne lidar system and the scattering region of interest is characterized by clear-air-turbulence only. The effect of absorption and large angle scattering (e.g., due to aerosols and molecular constituents) on the S/N can be taken into account by a multiplicative factor which gives the round-trip attenuation loss due to these other degrading effects. We consider a scalar monochromatic laser source of frequency ω and omit the time dependent factor $\exp(-i\omega t)$ in the following. In addition, we assume that both the transmitted laser field and local oscillator field have a gaussian shape. This assumption is not necessary but is employed here as it will enable us to obtain a simple analytic expression for the S/N which contains all of the essential physics of the problem.

Before we calculate the S/N ratio in the presence of turbulence we summarize briefly some essential physical characteristics of the index of refraction fluctuations in the atmosphere. Next we present a brief summary of the basic optical statistics resulting from atmospheric turbulence, giving some examples which are relevant to the calculation of the heterodyne lidar S/N.

II. ATMOSPHERIC INDEX OF REFRACTION FLUCTUATIONS

Random fluctuations in the index of refraction of the atmosphere are primarily due to temperature fluctuations. There are corresponding fluctuations in the absorptive part of the index

(Ref. 2) but this is small in comparison with the fluctuations of the real part of the index and is neglected here. These fluctuations are in general functions of the spatial position $\underline{r}$ and time t so that the index of refraction n can be written as

$$n(\underline{r},t) \simeq 1 + n_1(\underline{r},t)$$

where n_1 is the fluctuation in the index of refraction. For clear-air-turbulence, we have $|n_1| \lesssim 10^{-6}$, and $\langle n_1 \rangle = 0$, where angular brackets denote the ensemble average. It is also possible to assume that the temporal dependence of n_1 is primarily due to a net transport of the random inhomogenieties of the medium as a whole past the line of sight (e.g., due to atmospheric winds) so that $n_1(\underline{r},t) = n_1(\underline{r} - \underline{v}t)$, where $\underline{v}(\underline{r})$ is the local wind velocity. This assumption is known as Taylor's frozen flow hypothesis and appears to hold in most practical cases of interest. As is shown below, we are interested in spatial statistics and so we suppress the explicit temporal behavior of n_1.

For atmospheric turbulence we employ the much used Kolmogorov model of index of refraction fluctuations according to which within a particular range of separations between $\underline{r}_1$ and $\underline{r}_2$ that (Ref. 3)

$$\left\langle [n_1(\underline{r}_1) - n_1(\underline{r}_2)]^2 \right\rangle = C_n^2 |\underline{r}_1 - \underline{r}_2|^{2/3}, \quad \ell_o \ll |\underline{r}_1 - \underline{r}_2| \ll L_o, \qquad (1)$$

where ℓ_o and L_o are called the inner and outer scales of turbulence, respectively, C_n is called the index structure constant, and angular brackets denote the ensemble average. In the atmosphere, ℓ_o is of the order of a few millimeters to a centimeter and, for propagation in the lower atmosphere L_o is of the order of the height above ground. Typical values of C_n^2 ranges from 10^{-13} to $10^{-14}\ m^{-2/3}$ near the ground during midday, decreasing in value with increasing height above ground (e.g., 10^{-16} to $10^{-17}\ m^{-2/3}$ for a height of a few kilometers above ground).

For turbulent eddy sizes of characteristic length ℓ within the inertial subrange (i.e., $\ell_o \ll \ell \ll L_o$), the Kolmogorov "2/3" law implies that to a numerical multiplicative factor of the order unity the mean square index fluctuation associated with the scale size increases as the two-thirds power of that scale size. That is,

$$n_1^2(\ell) \sim C_n^2 \ell^{2/3}, \quad \ell_o \ll \ell \ll L_o.$$

For separations larger than L_o, the mean square index fluctuation does not increase with increasing separation, rather it levels off to a value of the order $C_n^2 L_o^{2/3}$. Conversely, for separations that are small compared with ℓ_o, friction effects due to viscosity result in a very rapid decrease in the index fluctuation and for most applications (as is the case at hand) can be taken as being equal to zero. The inner scale ℓ_0 is frequently much smaller than any length of interest in a propagation problem resulting in negligible contributions due to the smallest turbulent eddies.

III. OPTICAL WAVE STATISTICS

The study of the statistics of optical wave propagation through atmospheric turbulence has received increased interest in recent years. This interest is due primarily to the advent of laser systems operating in the atmosphere. For example, the design and development of optical imaging and radar systems operating in the atmosphere should be based on the results of the analysis of an optical wave field propagating in turbulent media. Other areas of interest where the interaction of an optical beam with turbulent media is important include astronomy, laser communication and remote sensing. During the last twenty years a great deal of progress has been made in this problem area. In particular, theoretical results which compared well with experiments have been obtained for the coherence properties of an optical wave as a function of propagation distance z, optical wave number k (which is equal to 2π divided by the optical wavelength λ) and the environmental parameter C_n^2. Here we summarize those results that relate to the S/N in a heterodyne lidar system only. The interested reader is referred to Refs. 4 and 5 for excellent general reviews on this subject that list several hundred references.

Consider an optical wave field propagating through the atmosphere at distance z. We denote this field by $U(\underline{r})$, where $\underline{r} = (\underline{p}, z)$ and $\underline{p}$ is a two-dimensional vector in a plane transverse to the optic-axis at propagation distance z. One of the most important optical wave statistical quantities is the mutual coherence function (MCF) which is defined as the cross-correlation function, i.e., the second statistical moment, of the complex field U in a plane perpendicular to the mean direction of propagation.

We have

$$M(\underline{\rho}) \equiv \left\langle U(\underline{r}_1) U^*(\underline{r}_2) \right\rangle \tag{2}$$

where $\underline{\rho} = \underline{r}_1 - \underline{r}_2$. The dependence of M on the difference of spatial position vectors results owing to the assumed stationary behavior of the medium. In addition, owing to the isotropic nature of the Kolmogorov model for separations between the inner and outer scale of turbulence, the MCF is independent of direction, i.e., it is a function of $\rho = |\underline{\rho}|$. The MCF is important for a number of reasons. It describes the loss of coherence of an initially coherent wave propagating in the medium. In addition, the MCF of a spherical wave is the central quantity which determines both optical beam spread and the S/N in heterodyne lidar systems. Below we present results for the MCF of a spherical wave only, the general case being treated in Ref. 6.

For spherical waves of unit strength and the Kolmogorov model of refractive index fluctuations, it can be shown that (see Ref. 3)

$$M(\rho,z) = \exp\left[-(\rho/\rho_o)^{5/3}\right] \tag{3}$$

where ρ_o, the "lateral coherence length" is given by

$$\rho_o \simeq \left[1.45\ k^2 \int_0^z C_n^{\,2}(s)\ (s/z)^{5/3} ds\right]^{-3/5} \tag{4}$$

In Eq. (4) the integration is along a straight line path with the origin of integration being at the location of the spherical wave source, and $C_n^{\,2}(s)$ is the structure constant profile that applies to the propagation path of interest. The quantity ρ_o is the lateral separation where the MCF is reduced by the factor 1/e from its initial value and can be taken as a measure of the lateral coherence length of the field. For separations much less (greater) than ρ_o the field can be considered to be mutually coherent (incoherent). Physically, ρ_o is the source separation in a Young's interferometer where the mean fringe visibility is down by 1/e in comparison to its value for small separations. Note that ρ_o is a monotonic decreasing function of propagation distance, wave number and strength of turbulence.

Examination of Eq. 4 reveals that the numerical value of ρ_o is geometry dependent. For example, consider the cases where we have a turbulent slab located near the spherical wave source and near the plane of observation, respectively. Examination of Eq. (4) reveals that the value of ρ_o for the case where the turbulent slab is located near the source is larger than the corresponding value where the slab is located near the observation plane. In other words, the coherence length of a spherical wave for the case where the turbulence is located near the source is larger than the corresponding value where the turbulence is located near the receiver, as expected intuitively from geometrical expansion considerations. Of practical interest we note that spherical wave degradation for propagating up and down through the atmosphere are quite different implying corresponding differences in the ultimate achievable resolution. This effect is sometimes referred to as the "shower curtain" effect in the literature (you can see the girl behind the shower curtain but she can't see you).

For uniform turbulence conditions (e.g., horizontal paths above level ground) Eq. (4) reduces to

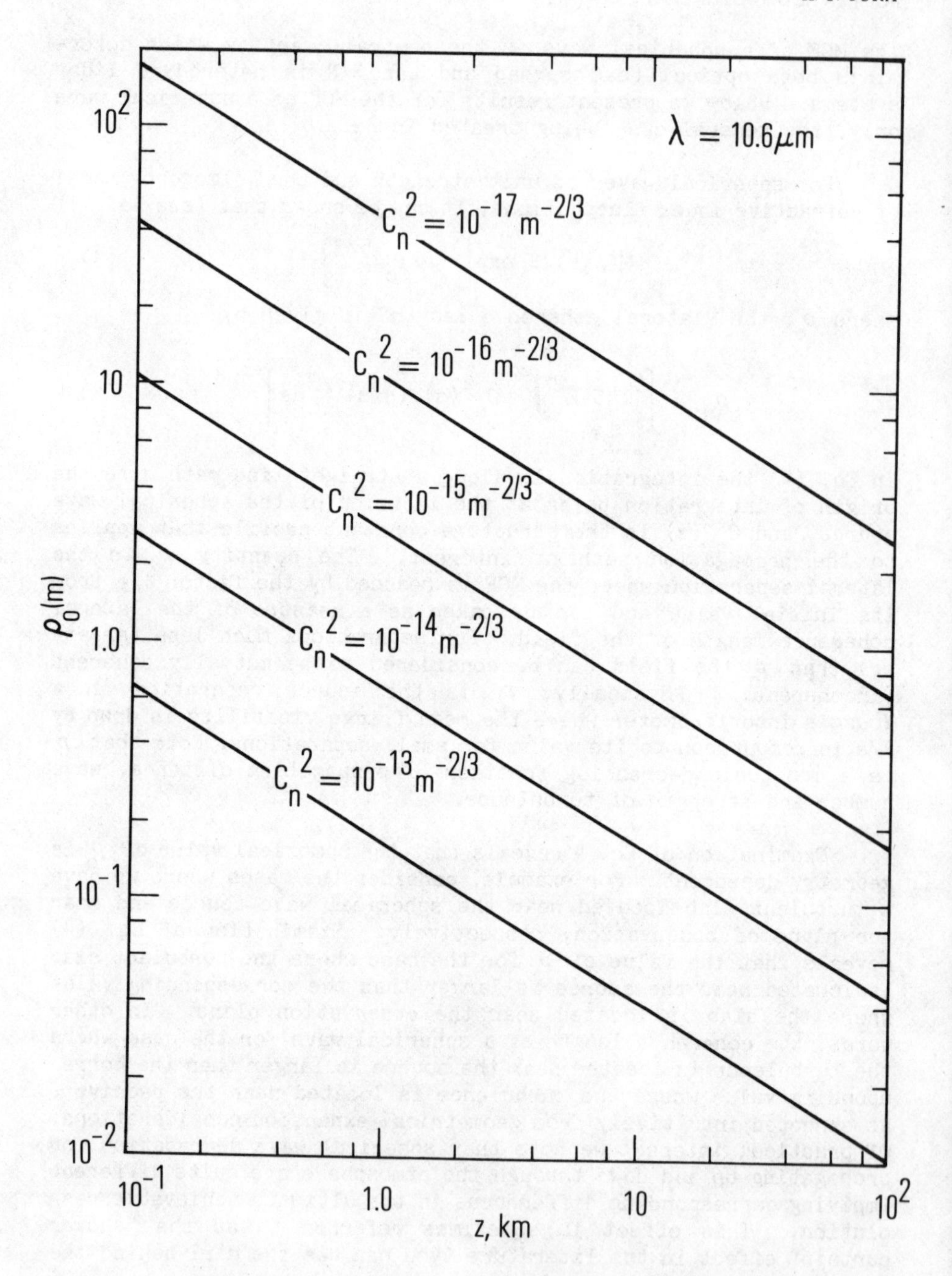

Fig. 1. The spherical wave lateral coherence length at 10.6 μm as a function of propagation distance for various values of C_n^2.

$$\rho_o \simeq \left[0.545\ k^2 C_n^{\ 2} z\right]^{-3/5} \tag{5}$$

and is plotted in Fig. 1 as a function of z for λ = 10.6 μm and various values of C_n^2. For propagation near the ground over distances of the order of a few kilometers we obtain a value of ρ_o of the order a meter (at 10.6 μm). The corresponding value of ρ_o at other wavelengths is obtained from Eq. 5 as

$$\rho_o(\lambda) = \rho_o(10.6\ \mu m)\left[\frac{\lambda(\mu m)}{10.6}\right]^{6/5} . \tag{6}$$

For example, $\rho_o(3.7\ \mu m) \simeq 0.28\ \rho_o(10.6\ \mu m)$.

The expressions for ρ_o given by Eqs. (4) and (5) pertain to a one-way path. In lidar applications we are concerned with a round-trip path and we present the corresponding expressions for ρ_o for round-trip situations. As will be shown below, the round-trip coherence length ρ_R appropriate to lidar heterodyne systems results from the case where the point source is located at the scattering region of interest and evaluated at the lidar location.

First consider the case of uncorrelated paths as would be obtained, for example, in a bistatic lidar system. In this case the effective path length is 2z and from Eq. (4) we obtain that

$$\rho_R = \rho_o(2z) = 2^{-3/5}\rho_o(z) \simeq 0.66\ \rho_o(z), \tag{7}$$

where $\rho_o(z)$ is the one-way coherence length. For the case of correlated paths, where each turbulent eddy is traversed twice by the direct and backscattered light we note the following. It can be shown (Ref. 3) that the logarithm of the MCF is essentially a sum of the mean-square phase fluctuation contributions, of the independent turbulent eddies. Inasmuch as each eddy is traversed twice for the case of correlated paths, this sum over eddies contains an additional factor of 2^2 = 4 as compared with the one way case. That is, $\ell n[M(\text{Rd. trip})] = 4\ell n[M(\text{one-way})]$ from which it follows for the Kolmogorov spectrum that (see Eq. 4 or 5)

$$\rho_R = 4^{-3/5}\rho_o(z) \simeq 0.44\,\rho_o(z), \tag{8}$$

where $\rho_o(z)$ is the one-way coherence length. As expected, the round trip coherence length for the correlated path case is less than the corresponding coherence length for the uncorrelated path case.

IV. BEAM WAVE PROPAGATION

Since all laser systems that operate in the atmosphere employ beams of finite cross-section, it is necessary, in general, to have a quantitative understanding of beam wave propagation. For homogeneous media (e.g., vacuum), the Huygens-Fresnel formulation is an excellent approximation for the optical field in space when the propagation distance is much larger than the physical size of the transmitting aperture (as it is for all cases of practical interest). The Huygens-Fresnel principle represents the optical wave function as a coherent sum of elementary spherical wave contributions over the transmitting aperture, each of which are weighted by the (complex) transmitting aperture wave function. In this manner, the resulting optical wave function is related directly to the source variables. The Huygens-Fresnel principle is general, in that it applies for an arbitrary source distribution and is valid both in the near and far field of the transmitting aperture. The Huygens-Fresnel formulation is particularly suitable for atmospheric propagation because it can be extended directly to an inhomogeneous or turbulent medium (Ref. 7).

Denote the (scalar) optical field at propagation distance z and transverse coordinate $\underline{p}$ by U ($\underline{p}$,z). From Ref. 7 we have that

$$U(\underline{p},z) = \left(\frac{k}{2\pi i}\right) \int \frac{\exp\left[iks(\underline{r},\underline{p})\right]}{s(\underline{r},\underline{p})} U_s(\underline{r},\underline{p}) U_A(\underline{r}) d^2r, \qquad (9)$$

where the integration extends over the transmitting aperture, $s(\underline{r},\underline{p})$ is the distance between the field coordinate ($\underline{p}$,z) and the aperture coordinate ($\underline{r}$,0), U_A ($\underline{r}$) is the (assumed known) transmitting aperture wave function and U_s ($\underline{r}$,$\underline{p}$) contains all of the effects of the turbulent medium on spherical wave propagation. The quantity (exp iks)/s is the spherical wave function in the absence of turbulence and U_s represents the additional spherical wave function perturbation produced by the turbulent medium. In practice the paraxial approximation is used where in the exponential term of Eq. (9) we set

$$s(\underline{r},\underline{p}) = z + \frac{1}{2z} (\underline{r} - \underline{p})^2$$

and set s = z in the denominator. Note that the case of propagation in the absence of turbulence is obtained by setting U_s = 1 in Eq. 9.

In many cases we are concerned with the mean irradiance profile $I(\underline{p},z) = \langle |U(\underline{p},z)|^2 \rangle$, where, in general, the angular brackets indicate that we must average over the medium variables and source variables (e.g., partially coherent source). We assume that the source and medium are uncorrelated from which it follows that the indicated averaging reduces to a product of averages, one for the medium and one for the source. The final result is that

$$I(\underline{p},z) = \left(\frac{k}{2\pi z}\right)^2 \int d^2\rho \; M_A(\underline{\rho}) M_s(\rho,z) \exp\left(- i \frac{k}{z} \underline{p} \cdot \underline{\rho}\right) \quad (10)$$

where

$$P = \int |U_A(\underline{r})|^2 d^2r = \text{transmitted power} \quad (11)$$

$$M_A(\underline{\rho}) = \frac{1}{P} \int \left\langle U_A(\underline{R} + \tfrac{1}{2}\underline{\rho}) U_A^*(\underline{R} - \tfrac{1}{2}\underline{\rho}) \right\rangle_{Source} e^{-ik\underline{\rho}\cdot\underline{R}/z} d^2R \quad (12)$$

$$M_s(\rho) = \left\langle U_s(\underline{r}_1, p) U_s^*(\underline{r}_2, \underline{p}) \right\rangle_{Medium}, \quad (13)$$

where

$$\underline{\rho} = \underline{r}_1 - \underline{r}_2$$

Examination of Eq. (12) reveals that M_s is just the spherical wave MCF where the point source is located at the point of observation (i.e., $\underline{p}$) and evaluated over the transmitting aperture. For propagation in the absence of turbulence $M_s = 1$ and Eq. (10) reduces to the well known Huygens-Fresnel principle in homogeneous media. Note that for a deterministic (e.g., a laser) source the indicated average over source variables in Eq. (12) becomes just the product of the indicated deterministic wave functions.

As is indicated in Eq. 10, the spherical wave mutual coherence function plays a central role in propagation through turbulence. Basically, it describes the reduction in lateral coherence between different elements of the transmitting aperture, effectively transforming it into a partially coherent radiator, with the degree of coherence decreasing with increasing propagation distance through the medium. In general, the resulting mean irradiance distribution in space is characteristic of a coherent aperture of dimension $\rho_o(z)$, where ρ_o is the lateral coherence length of the spherical wave whose origin is at the point of observation. If ρ_o is smaller than the physical aperture diameter D, the resulting mean radiation pattern is characteristic of that of an aperture of size ρ_o rather than D. Thus, for example, in the far field one obtains an angular spread of the order of λ/ρ_o for $\rho_o < D$ rather than of the order of λ/D.

We now give some illustrative examples of irradiance profiles which will be used later in obtaining the S/N in heterodyne lidar systems. Consider the case where the source wave function U_A is a (deterministic) gaussian function:

$$U_A(r) = U_o \exp\left(-\frac{r^2}{2}\left[\frac{1}{a^2} + \frac{ik}{f}\right]\right) , \tag{14}$$

where U_o is a constant, $k = 2\pi/\lambda$, a is the 1/e intensity radius and f is the (positive) focal length. Negative values of f correspond to a divergent beam whose angle of divergence is of the order a/f. As discussed in Secs. II and III it is customary to assume the Kolmogorov model for the turbulent index of refraction fluctuations and hence the spherical wave MCF is given by Eq. 3:

$$M_s(\rho) = \exp\left[-(\rho/\rho_o)^{5/3}\right], \tag{15}$$

where ρ_o is given by Eq. 4.

The resulting mean irradiance profile is obtained by substituting Eqs. 14 and 15 into Eq. 10 and performing the indicated integrations. With M_s given by Eq. (15) closed form expressions can be obtained in terms of confluent hypergeometric functions. The resulting expression is awkward in that it is difficult to extract a physical interpretation. In order to obtain an analytic approximation in terms of well known functions, we approximate the 5/3 in the exponent of Eq. 15 by 2. The resulting integrals then become quadratic in the exponent permitting the integrations to be performed by completing the square. The result is that

$$I(p,z) = \frac{P}{\pi a_t^2(z)} \exp\left[-\frac{p^2}{a_t^2(z)}\right] , \tag{16}$$

where

$$P = \pi a^2 |U_o|^2 = \text{transmitted power}$$

and

$$a_t^2(z) = a^2\left(1 - \frac{z}{f}\right)^2 + \left(\frac{z}{ka}\right)^2 + \left(\frac{2z}{k\rho_o}\right)^2 . \tag{17}$$

The result given in Eq. 16 is an excellent approximation, as can be seen from Fig. 2 where we have plotted the integral that results by retaining the 5/3 in the exponent in Eq. 15 and compared it to the corresponding integral with the 5/3 replaced by 2.

A physical interpretation of the irradiance pattern can now be readily obtained from an examination of Eq. 16. The mean irradiance profile is to an excellent approximation gaussian with a 1/e "spot-size" given by $a_t(z)$. Furthermore, examination of Eq.

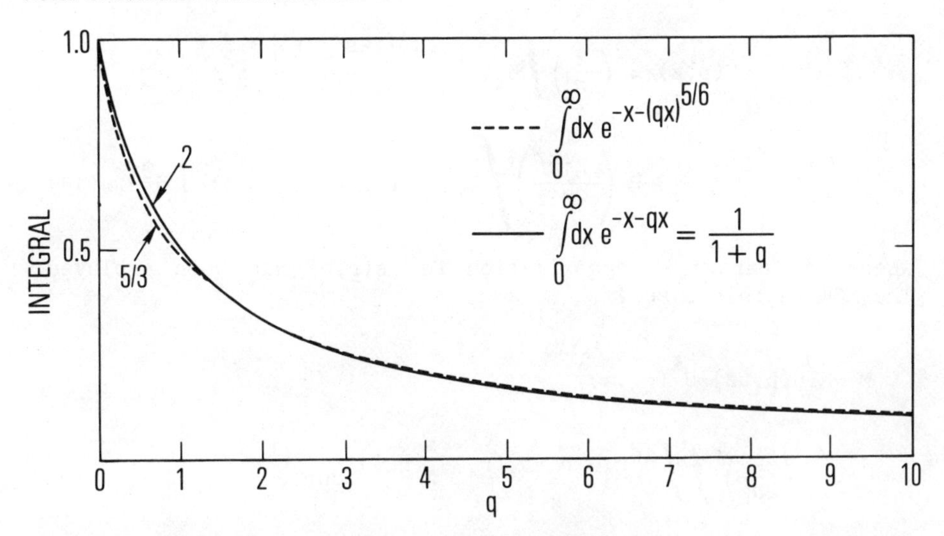

Fig. 2. Comparison of the integral that results by retaining the 5/3 in the exponent of Eq. 15 to the corresponding integral obtained by replacing the 5/3 by 2. The quantity q is a dimensionless parameter that is the same for both cases.

17 reveals that $a_t(z)$ consists of three independent terms. The first term on the right-hand side of Eq. 17 is the effect of the turbulent medium and for $\rho_o \ll a$ it is seen that for $z \gg ka^2$ the resulting far field spot size is of the order of $\lambda z/\rho_o$, in agreement with the qualitative discussion given previously. We note that the corresponding irradiance profile in the absence of turbulence is obtained from Eq. 17 by setting $\rho_o = \infty$. That is, the mean irradiance is gaussian with a 1/e spot-size $a_o(z)$ given by

$$a_o^2(z) = a^2\left(1 - \frac{z}{f}\right)^2 + \left(\frac{z}{ka}\right)^2 . \qquad (18)$$

As another example of the utility of the use of the Huygens-Fresnel principle, and one that is needed in the derivation of the S/N of a heterodyne lidar system, is the calculation of the mutual coherence function that results from an incoherent source distribution of finite extent in the absence of turbulence. The field from an arbitrary source in the absence of turbulence is obtained from Eq. 9 by setting $U_s = 1$:

$$U(p,z) = \left(\frac{k}{2\pi i}\right)\int U_A(\underline{r})\ \frac{\exp[iks(\underline{r},\underline{p})]}{s(\underline{r},\underline{p})}\ d^2r$$

$$\cong \left(\frac{ke^{ikz}}{2\pi iz}\right)\int U_A(\underline{r})\ \exp\left[\frac{ik}{2z}(\underline{r}-\underline{p})^2\right] d^2r, \quad (19)$$

where the paraxial approximation for $s(\underline{r},\underline{p})$ has been employed. The MCF in this case is

$$M = \left\langle U(\underline{p}_1,z)\ U^*(\underline{p}_2,z)\right\rangle_{Source}$$

$$= \left(\frac{k}{2\pi z}\right)^2 \iint d^2r_1 d^2r_2 \left\langle U_A(\underline{r}_1)U_A^*(\underline{r}_2)\right\rangle_{Source}$$

$$\times \exp\left[\frac{ik}{2z}\left\{(\underline{r}_1-\underline{p}_1)^2 - (\underline{r}_2-\underline{p}_2)^2\right\}\right]. \quad (20)$$

For a statistically stationary source we have that

$$\left\langle U_A(\underline{r}_1)U_A^*(\underline{r}_2)\right\rangle_{Source} = B_s(\underline{r}_1-\underline{r}_2), \quad (21)$$

where B_s, the source correlation function, is a function of $\underline{r}_1 - \underline{r}_2$. For a given source correlation function one can obtain from Eqs. 20 and 21 the resulting MCF. Here, however, we consider an incoherent source where

$$B_s(\underline{r}_1-\underline{r}_2) = \delta(\underline{r}_1-\underline{r}_2)I_s(\underline{r}_1), \quad (22)$$

$\delta(\underline{r})$ is the Dirac delta-function, and $I_s(\underline{r})$ is proportional to the source brightness distribution. Upon substituting Eq. 22 into Eq. 20 we obtain that

$$M = \left(\frac{k}{2\pi z}\right)^2 \exp\left[\frac{ik}{2z}\ \left\{p_1^{\ 2} - p_2^{\ 2}\right\}\right]$$

$$\times\int I_s(\underline{r})\exp\left[-i\frac{k}{z}\underline{p}\cdot\underline{r}\right]d^2r, \quad (23)$$

where $\underline{p} = \underline{p}_1 - \underline{p}_2$. In order to proceed further we choose a gaussian function for I_s:

$$I_s \sim \exp\left[-\frac{r^2}{r_s^2}\right], \qquad (24)$$

where r_s is the 1/e spot radius of the source distribution. Upon substituting Eq. 24 into Eq. 23, performing the integration we obtain, after normalizing the MCF to unity for p = 0, that

$$M(p,z) = \frac{\left\langle U(\underline{p}_1 z)U^*(\underline{p}_2,z)\right\rangle}{\sqrt{|U(\underline{p}_1,z)|^2}\sqrt{|U(\underline{p}_2,z)|^2}}$$

$$= \exp\left[-\frac{\underline{p}^2}{p_s^2}\right], \qquad (25)$$

where

$$p_s = \frac{2z}{kr_s}. \qquad (26)$$

We obtain a gaussian shaped MCF with a coherence length p_s given by Eq. 26. We see that the radiation field due to an incoherent source becomes partially coherent, with a coherence length that increases with increasing propagation distance. One can understand the cause for this from the following physical consideration. Very close to the source (i.e., $z \rightarrow 0$) the field is essentially incoherent, as is exhibited by Eq. 26 (i.e., $p_s \rightarrow 0$). On the other hand, consider the source to be at very large (i.e., infinite) distance. In this case the source can be considered to be a point and the resulting radiation field is completely coherent, again in accord with Eq. 26 (i.e., $p_s \rightarrow \infty$). For a numerical example we consider the sun at $\lambda = 10.6\,\mu m$ and obtain that $p_s \simeq 0.2$ mm. Equation 23 is essentially a statement of the well known van Cittert-Zernike Theorem (Ref. 8).

V. SIGNAL-TO-NOISE RATIO IN HETERODYNE LIDAR SYSTEMS

Consider the situation that is illustrated in Fig. 3. A laser beam is transmitted toward a remote scattering region of interest (e.g.; pollutants which are emitted from a smoke stack, topographical regions such as forests or vegetation, etc.). The backscattered light is collected by a receiving aperture and mixed with a suitable local oscillator reference field, as indicated by the dashed line in Fig. 3. As discussed in the Introduction, the receiver may be coaxial with the transmitter. Depending on the

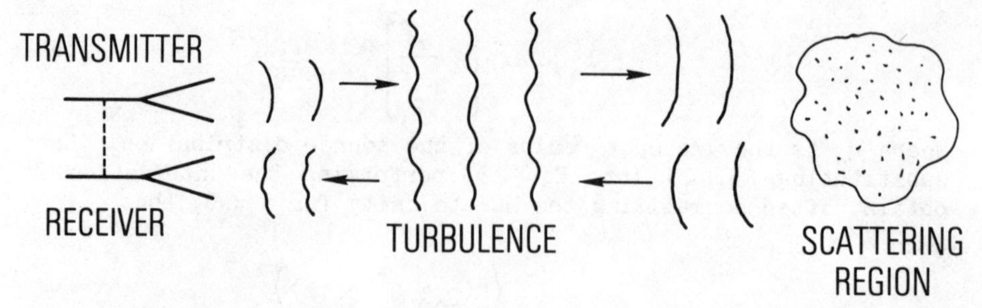

Fig. 3. Schematic illustration of a heterodyne lidar remote sensing configuration.

laser, continuous wave or pulsed transmitting systems may be employed. In any case one averages over times long compared to atmospheric fluctuation times to obtain a favorable S/N (many pulses for a pulsed system). Here we assume that the intervening space between the lidar system and the scattering region is characterized by clear-air-turbulence only.

In heterodyne detection it can be shown that (Refs. 1,9)

$$S/N \sim \langle |i|^2 \rangle, \tag{27}$$

where i is the information carrying part of the current in the photodetector load. Implicit in Eq. 27 is an average over a time that is long compared to the reciprocal (doppler) bandwidth of the signal field. The information carrying part of the current is proportional to the product of the signal field and the local oscillator field, integrated over the receiving aperture. That is,

$$i = \text{constant} \int U(\underline{r}) U_R^*(\underline{r}) d^2 r, \tag{28}$$

where $U(\underline{r})$ is the signal field, U_R is the reference local oscillator field, and the integration is carried out over the receiving aperture.

Upon substituting Eq. 28 into Eq. 27, and performing the average over the statistics of the medium yields

$$S/N = \text{constant} \iint U(\underline{r}_1) U^*(\underline{r}_2) U_R^*(\underline{r}_1) U_R(\underline{r}_2) d^2 r_1 d^2 r_2$$

$$= \text{constant} \iint M(\underline{r}_1 - \underline{r}_2) U_R^*(\underline{r}_1) U_R(\underline{r}_2 d) d^2 r_1 d^2 r_2, \tag{29}$$

where $M(\underline{r})$ is the mutual coherence function of the signal field at the receiver aperture plane. Examination of Eq. 29 indicates that the MCF plays a central role in the determination of the S/N. Changing the integration variables to sum and difference coordinates permits the integration over the sum coordinate to be performed directly with the result that

$$S/N = \text{constant} \int M(\underline{\rho}) W_R(\underline{\rho}) d^2\rho \ , \tag{30}$$

where

$$W_R(\underline{\rho}) = \int U_R(\underline{r} + \tfrac{1}{2}\underline{\rho}) U_R^*(\underline{r} - \tfrac{1}{2}\underline{\rho}) d^2 r \tag{31}$$

is the receiver weighting function. Although it is possible to carry through the analysis for an arbitrary U_R, it is convenient to take U_R to have a gaussian shape. This choice for U_R, together with the gaussian shaped wave function of Eq. 14 for the laser output field, permits us to obtain analytic results in terms of algebraic functions. Specifically we assume that

$$U_R \sim \exp\left[-\frac{r^2}{2b^2}\right] \ , \tag{32}$$

where b is the 1/e intensity radius of the local oscillator reference field. Substituting Eq. 32 into Eq. 31, performing the integration yields that

$$W_R(\rho) \ = \pi b^2 \exp\left[-\rho^2/4b^2\right] \tag{33}$$

Inasmuch as we are primarily interested in turbulence effects we first compute the reduction factor $\psi (\leq 1)$ by which the S/N is reduced in the presence of turbulence as compared to the corresponding case in the absence of turbulence. Denote by the subscript t and o the case in the presence and absence of turbulence, respectively. The reduction factor ψ is defined as

$$\psi \equiv \frac{(S/N)_t}{(S/N)_o} \ . \tag{34}$$

Substituting Eqs. 30 and 33 into Eq. 34 yields that

$$\psi = \frac{\int M_t(\rho)\exp[-\rho^2/4b^2]\,d^2\rho}{\int M_o(\rho)\exp[-\rho^2/4b^2]\,d^2\rho} . \tag{35}$$

where M_t and M_o are the MCF at the receiving aperture that result in the presence and absence of turbulence, respectively.

Consider first the case of the absence of turbulence. We assume an output laser field as given by Eq. 14. This beam illuminates the scattering region which is assumed to be located at a propagation distance z. The light which is (incoherently) backscattered at every point in the scattering region is, among other things, proportional to the incident irradiance at that point. Since the scattering particles are at random positions with respect to each other, the backscattering source distribution can be taken as being mutually incoherent and having a brightness distribution proportional to the incident irradiance distribution. In the absence of turbulence, and for a gaussian laser source distribution we have shown that

$$I_o(p,z) = \frac{P}{\pi a_o^2(z)} \exp\left[-\frac{p^2}{a_o^2(z)}\right]$$

where $a_o(z)$ is given by Eq. 18. That is, as far as the backscattered light is concerned, we have essentially a gaussian-shaped incoherent source distribution of 1/e spot-radius $a_o(z)$. Hence we may apply directly the results given by Eqs. 25 and 26 with $r_s = a_o(z)$ to obtain the MCF as

$$M_o(\rho) = \exp\left[-\frac{\rho^2}{\rho_{so}^2}\right] , \tag{36}$$

where

$$\rho_{so} = \frac{2z}{ka_o(z)} \tag{37}$$

and $a_o(z)$ is given by Eq. 18.

For the case in the presence of turbulence we may obtain M_t by the following considerations. With turbulence present the mean irradiance pattern is not given by Eq. 16, where the spot-size $a_t(z)$, given by Eq. 17, is greater than the corresponding spot-size $a_o(z)$ in the absence of turbulence. The turbulence has spread the beam and a larger spot-size results at the scattering region of interest. We note that the appropriate coherence length ρ_o that appears in the expression for a_t is for the case where the point source is located at the scattering region of interest (see the discussion following Eq. 13). In addition, as the backscattered light travels back toward the receiver it now propagates through the turbulent atmosphere and a loss of coherence of the wave is incurred. Therefore, the resulting MCF in the presence of turbulence is given by

$$M_t(\rho,z) = M_{t1}(\rho,z)\, M_{t2}(\rho,z)$$

where M_{t1} is the MCF that results from an incoherent source distribution of spot size $a_t(z)$, and M_{t2} is the additional MCF that results as the backscattered light propagates through the turbulent atmosphere to the receiver. From our previous discussion we have that

$$M_{t1} = \exp\left[-\frac{\rho^2}{\rho^2_{st}}\right], \tag{38}$$

where

$$\rho_{st} = \frac{2z}{ka_t(z)} \tag{39}$$

$a_t(z)$ is given by 17, and

$$M_{t2} = \exp\left[-\left(\frac{\rho}{\rho_o}\right)^{5/3}\right], \tag{40}$$

where ρ_o, given by Eq. 4, is the lateral coherence length of a point source located in the scattering region. Note both the ρ_o's that appear in Eqs. 39 and 40 have their origins in the scattering region.

Upon approximating the 5/3 in the exponent of Eq. 40 by 2, substituting Eq. 36 to 40 into Eq. 35, and performing the resulting gaussian integrals we obtain that

$$\psi = \frac{a^2[1-(z/f)]^2 + (z/ka)^2 + (z/kb)^2}{a^2[1-(z/f)]^2 + (z/ka)^2 + (z/kb)^2 + (2z/k\rho_R)^2}, \tag{41}$$

where ρ_R is the round-trip coherence length of a point source located at the scattering region and is given explicitly for both uncorrelated and correlated paths by Eqs. 7 and 8, respectively. Equation 41 is general in that it applies for receiver and transmitter aperture radii of different sizes, it is valid both in the near and far field of the transmitting aperture, for laser beams of arbitrary focal length, and for an arbitrary propagation path through the turbulent atmosphere.

For illustrative purposes we consider the special cases of both focused ($f = z$) and collimated ($f = \infty$) coaxial systems ($a = b$) in the presence of uniform turbulence conditions where the one-way coherence length $\rho_o(z)$ is given by Eq. 5. The results presented below are expressed in terms of the more commonly used $1/e^2$ intensity radii a_2 ($= \sqrt{2}a$).

A. Focused Coaxial System ($z = f$)

We obtain that

$$\psi = \frac{1}{1 + (a_2/\rho_R)^2} , \tag{42}$$

where ρ_R is given by Eq. 8, and is plotted in Fig. 4 as a function of normalized radius a_2/ρ_R. We note that when the round trip coherence length is less than the aperture radius, $\psi \sim (\rho_R/a_2)^2 \ll 1$, and a significant reduction in the S/N will be obtained. In Figs. 5 through 8 we have plotted ψ as a function of a_2 for 10.6 and 3.7 μm for coaxial focused systems for various values of C_n^2. Examination of these figures reveal that a greater reduction in S/N is expected both at the shorter wavelength and greater propagation distance.

B. Collimated Coaxial System ($f = \infty$)

We obtain that

$$\psi = \frac{1 + \frac{1}{2}\delta^2}{1 + \frac{1}{2}\delta^2\left[1 + (a_2/\rho_R)^2\right]} , \tag{43}$$

where $\delta = 4z/ka_2^2$, and is plotted in Fig. 9 as a function of a_2/ρ_R for various values of δ. The cases $\delta \ll 1$ and $\delta \gg 1$ correspond to the near and far field, respectively. Examination of Eq. 43 reveals (and which is readily exhibited in Fig. 7) that in the far field the reduction factor for the collimated case reduces to that of the focused case.

Having derived the turbulence induced reduction factor ψ we now return to Eq. 30 and calculate the signal-to-noise ratio $(S/N)_o$ in the absence of turbulence. From Eqs. 30 and 33 we have

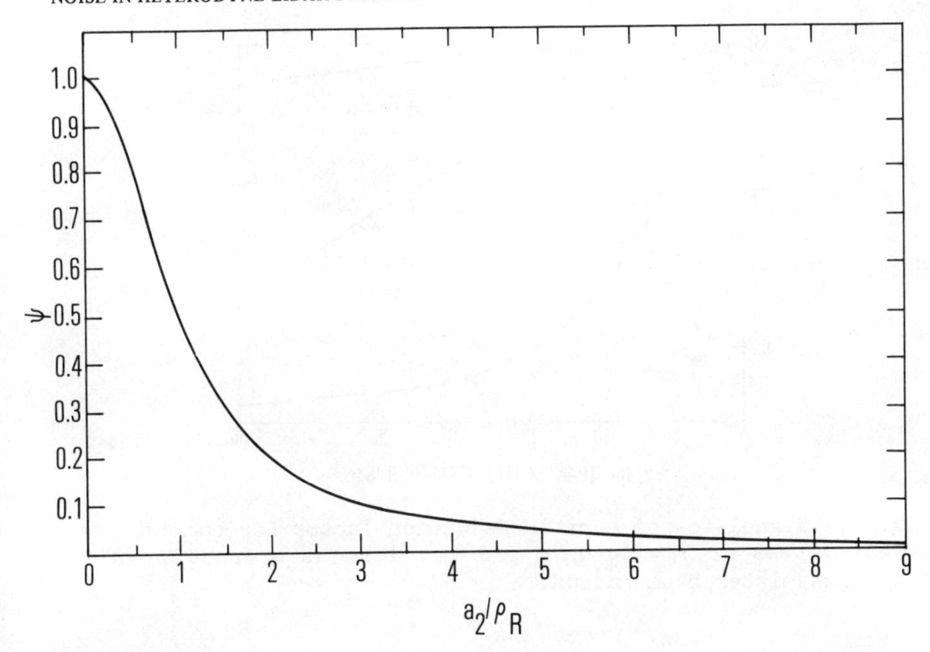

Fig. 4. Signal-to-noise ratio reduction factor for coaxial focused systems as a function of normalized receiver radius.

$$(S/N)_o = \text{constant} \int M_o(\rho)\exp[-\rho^2/4b^2]\,d^2\rho\,, \tag{44}$$

where M_o is given by Eq. 36.

To proceed further we note that if the signal field were completely coherent with respect to the local oscillator reference field over the receiving aperture one would obtain the optimum, or best, signal-to-noise ratio that could be achieved. The optimum signal-to-noise ratio, obtained when the signal field mutual coherence function is constant, has been derived by many authors (see Ref. 1,9,10) and for a shot noise-limited cw system is given by

$$(S/N)_{opt} = \frac{\eta P_s}{h\nu B} \tag{45}$$

where η is the photodetector quantum efficiency, h is Planck's constant, ν is the optical carrier frequency (Hz), B is the detection bandwidth, P_s is the received signal power and is given by

$$P_s = P_o\beta(\pi)\,d\Delta\Omega \tag{46}$$

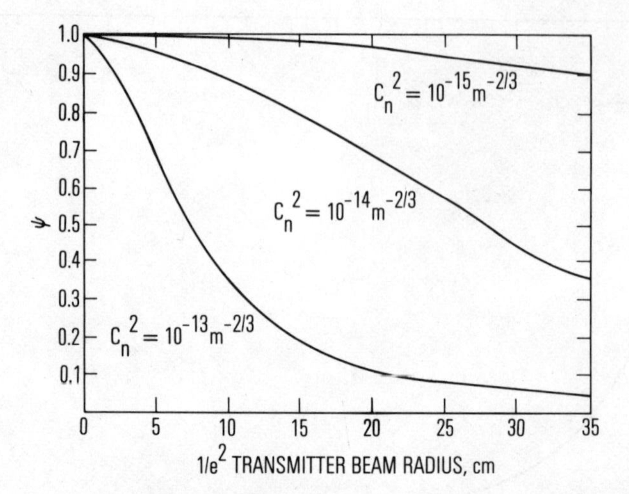

Fig. 5. Signal-to-noise ratio reduction factor for coaxial focused systems at $\lambda = 10.6\ \mu m$ and z = 1 km as a function of the $1/e^2$ transmitter beam radius.

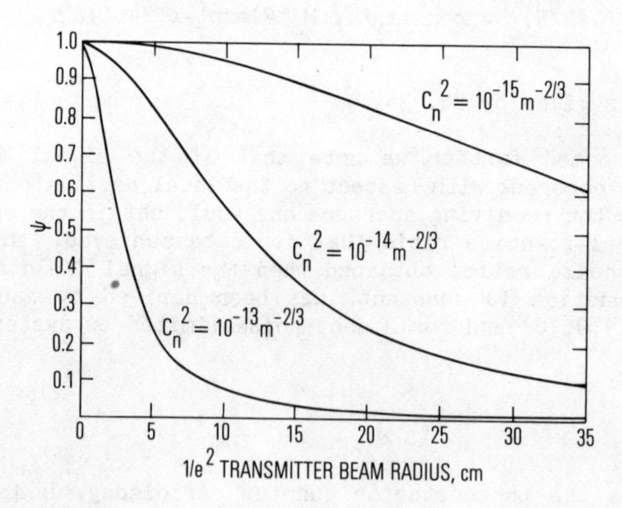

Fig. 6. Signal-to-noise ratio reduction factor for coaxial focused systems at $\lambda = 10.6\ \mu m$ and z = 5 km as a function of the $1/e^2$ transmitter beam radius.

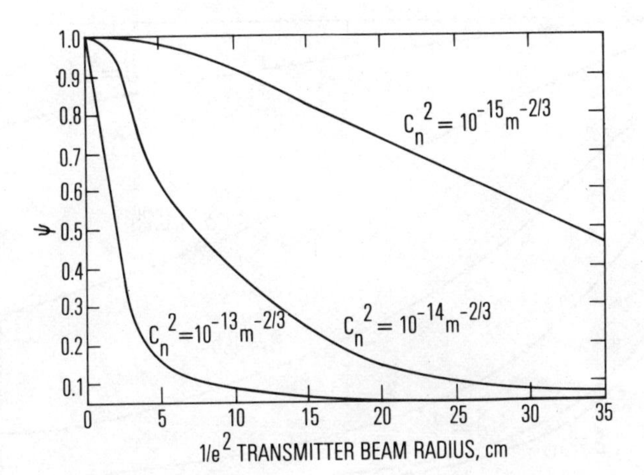

Fig. 7. Signal-to-noise ratio reduction factor for coaxial focused systems at $\lambda = 3.7\ \mu m$ and $z = 1$ km as a function of the $1/e^2$ transmitter beam radius.

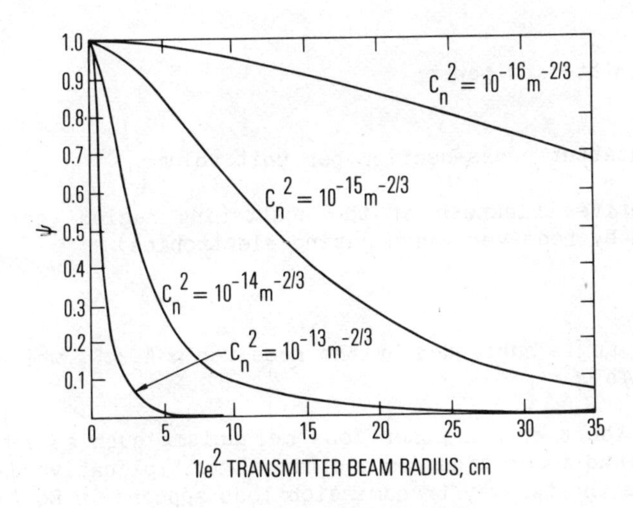

Fig. 8. Signal-to-noise ratio reduction factor for coaxial focused systems at $\lambda = 3.7\ \mu m$ and $z = 5$ km as a function of the $1/e^2$ transmitter beam radius.

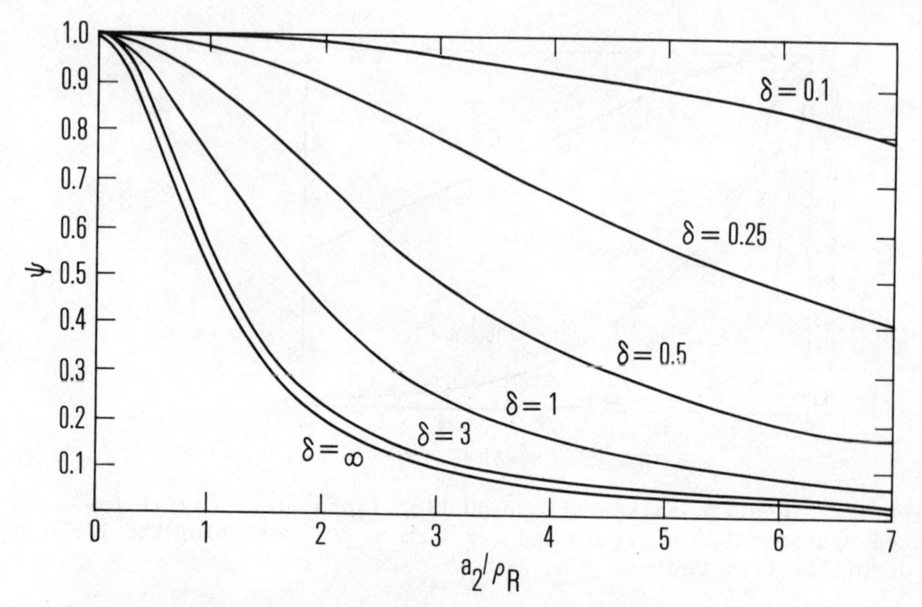

Fig. 9. Signal-to-noise ratio reduction factor for coaxial collimated systems as a function of normalized transmitter radius for various values of $\delta (= 4z/ka_2^2)$.

where

P_o = transmitted cw power

$\beta(\pi)$ = backscatter cross-section per unit volume

d = effective thickness of the scattering region (determined by receiver range gating electronics)

and

$\Delta\Omega$ = solid angle subtended by the receiver $\simeq A_R/z^2$, where $A_R = \pi b^2$.

We note that if there exists other loss mechanisms such as large angle scattering and absorption, an additional multiplicative factor that reflects the two-way transmission loss appears in Eq. 46. We also note that for a pulsed lidar system, which averages over N_p pulses, Eq. 46 is modified by an additional multiplicative factor of $\sqrt{N_p}$ and P_o is now interpreted as the transmitted power per pulse.

Returning to Eq. 44 we find that

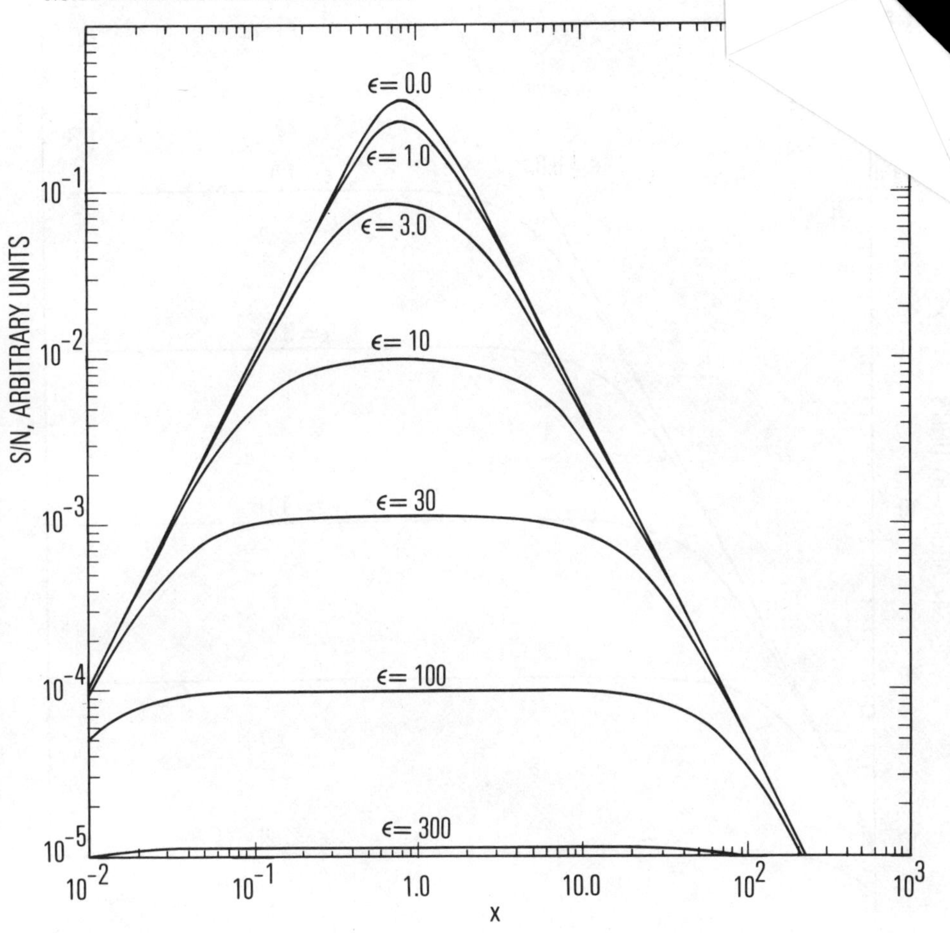

Fig. 10. Signal-to-noise ratio for coaxial collimated systems as a function of normalized beam radius x (= $a_2/\sqrt{4z/k}$) for various values of the turbulence parameter ϵ (= $\sqrt{4z/k}/\rho_R$). The figures shown above are plots of the quantity in the square brackets on the right-hand side of Eq. 50 for various values of ϵ.

$$(S/N)_o = (S/N)_{opt} \frac{\int M_o(\rho)\exp\left[-\rho^2/4b^2\right] d^2\rho}{\int \exp\left[-\rho^2/4b^2\right] d^2\rho} . \tag{47}$$

Substituting Eq. 36 into Eq. 47, performing the integration yields

$$(S/N)_o = \left(\frac{\eta P_o \beta(\pi) d\Delta\Omega}{h\nu B}\right)\left[\frac{(z/kb)^2}{a^2\left[1-(z/f)\right]^2+(z/ka)^2+(z/kb)^2}\right], \tag{48}$$

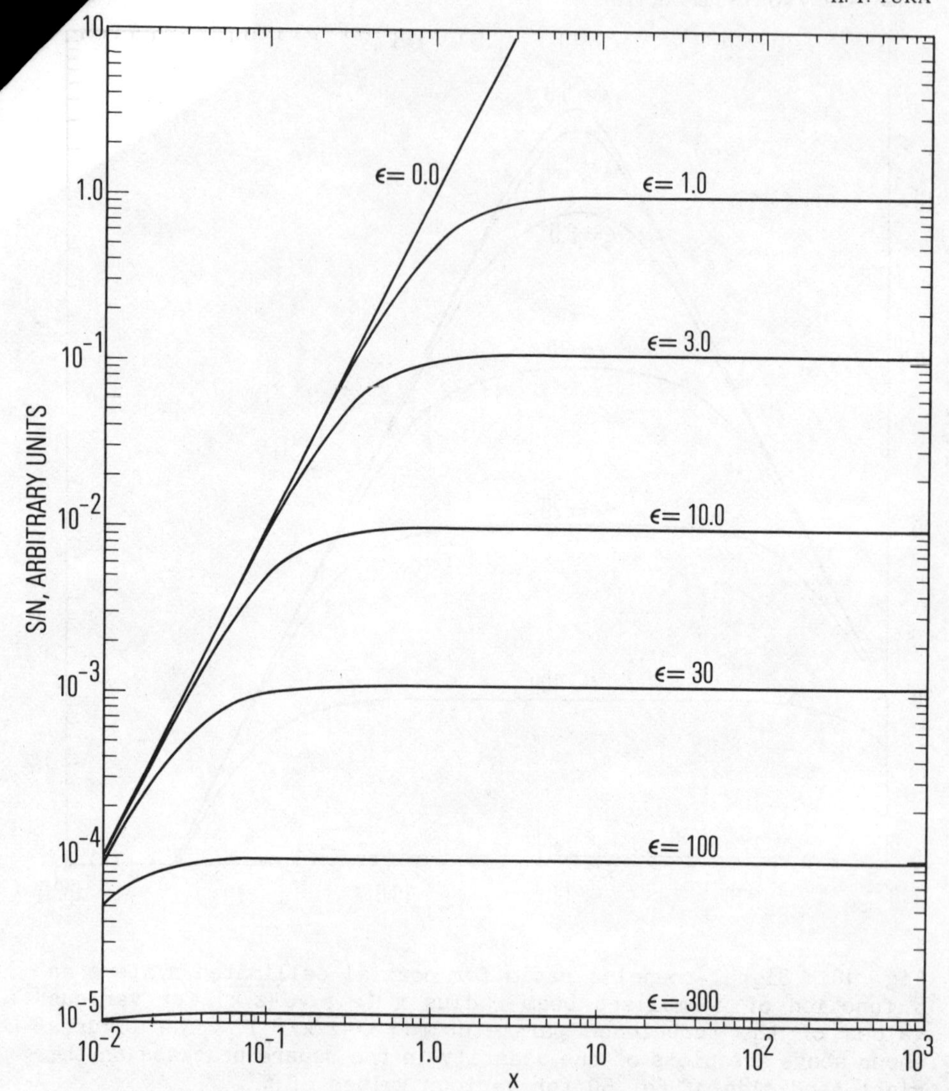

Fig. 11. Signal-to-noise ratio for coaxial focused systems as a function of normalized beam radius x $(= a_2/\sqrt{4z/k})$ for various values of the turbulence parameter ϵ $(= \sqrt{4z/k}/\rho_R)$. The figures shown above are plots of the quantity in the square brackets on the right-hand side of Eq. 52 for various values of ϵ.

The first factor on the right-hand side of Eq. 48 is the optimum signal-to-noise ratio, as discussed above, while the second factor is a reduction factor, relative to the optimum S/N, that results from the finite size scattering region (i.e., it reflects the fact that the signal field is partially coherent over the receiving aperture and hence optimum mixing between the signal and local oscillator fields does not occur).

We now combine Eq. 34, 41, and 48 to obtain the final general result for the signal-to-noise ratio of a heterodyne lidar system in the presence of turbulence:

$$(S/N)_t = (S/N)_o \psi$$

$$= \left(\frac{\eta P_o \beta(\pi) d \Delta\Omega}{h\nu B}\right)\left[\frac{(z/kb)^2}{a^2[1-(z/f)]^2+(z/ka)^2+(z/kb)^2+(2z/k\rho_R)^2}\right], \quad (49)$$

where all the terms appearing in Eq. 49 have been defined previously.

In general, atmospheric turbulence will eventually limit severely the performance of heterodyne lidar systems. Indeed, examination of Eq. 49 reveals that for strong turbulence conditions such that $\rho_R \ll$ a and b, the $(S/N)_t \sim \rho_R^2$, where we have used that fact that the solid angle subtended by the receiver optics is proportional to the area of the receiver aperture. This is in contrast to the case of the absence of turbulence where the $(S/N)_o$ increases with increasing area of the receiver aperture optics, reaches a maximum value when the receiver aperture area is of the order the square of the Fresnel length (i.e., ~ z/k), then decreases with increasing values of receiver aperture area. This latter behavior is expected because in the near field of the source the received signal field becomes progressively more incoherent as the receiver aperture area increases relative to the square of the Fresnel far field, the signal-to-noise ratio in the presence of turbulence tends asymptotically to a constant with increasing receiver aperture area.

These results can be demonstrated clearly by considering the case of a coaxial (a = b) collimated system. Upon substituting $\Delta\Omega = \pi b^2$ and $f = \infty$ into Eq. 49 we obtain that the signal-to-noise ratio can be written as

$$(S/N)_t = C\left[\frac{x^2}{1 + 2x^4 + \epsilon^2 x^2}\right] , \quad (50)$$

where

$$\left.\begin{aligned} C &= \frac{2\pi z\eta P_o \beta(\pi) d\Delta\Omega}{kh\nu B} \\ x &= \frac{a_2}{\sqrt{4z/k}} \\ \epsilon &= \frac{\sqrt{4z/k}}{\rho_R} , \end{aligned}\right\} \quad (51)$$

and a_2 is the $1/e^2$ transmitter/receiver aperture radius. The quantity x is essentially the initial beam radius divided by the Fresnel length. The effects of turbulence on the signal-to-noise ratio is contained in the parameter ϵ. The case of the absence of turbulence corresponds to $\epsilon = 0$ while strong turbulence conditions correspond to $\epsilon x = a_2/\rho_R \gg 1$. In Fig. 10 we present a plot of the relative signal-to-noise ratio as a function of the normalized receiver aperture radius x for various values of the turbulence parameter ϵ.

Examination of Eq. 50 and Fig. 10 reveals that the $(S/N)_t$ reaches a maximum value equal to $(2\sqrt{2} + \epsilon^2)^{-1}$ at $x = (2)^{-\frac{1}{4}}$ i.e., $a_2 = (2\sqrt{2}\ z/k)^{\frac{1}{2}}$. For $\epsilon \gg 1$, the maximum achievable value of the signal-to-noise ratio is inversely proportional to ϵ^2 (i.e., $\rho_R^2 \to 0$). Furthermore, for strong turbulence conditions, the signal-to-noise ratio remains essentially constant over a wide range of values of receiver radius a_2, in strong contrast to the case of the absence of turbulence (see Fig. 10). For a numerical example, consider the case of propagation at $\lambda = 3.7\ \mu m$, a one-way path length z of 5 km, $C_n^2 = 10^{-14} m^{-2/3}$, and $a_2 = 15$ cm. From Eqs. 51 and 8 we obtain that $x \simeq 1.38$ and $\epsilon \simeq 4.9$. Hence from Eq. 50 we obtain a signal-to-noise ratio, relative to that of the constant C, of 0.035, as compared to the corresponding value in the absence of turbulence of 0.23.

In addition, we present results for the signal-to-noise ratio for a coaxial (a = b) focused system. Upon substituting $\Delta\Omega = \pi b^2$ and z = f into Eq. 49 we obtain that the signal-to-noise ratio can be written as

$$(S/N) = C\left[\frac{x^2}{1 + \epsilon^2 x^2}\right] , \quad (52)$$

where C, x, and ϵ are given by Eqs. 51. In Fig. 11 we present a plot of the relative signal-to-noise as a function normalized the receiver aperture radius x for various values of the turbulence parameter ϵ. In the absence of turbulence ($\epsilon = 0$) the signal-to-noise ratio increases as the square of the receiver aperture radius (i.ew., receiver aperture area). For finite ϵ, the signal-to-noise ratio levels off for $\epsilon x > 1$ (i.e., $\rho_R^2 < a^2$), and tends asymptotically to a constant (proportional to ρ_R^2) which tends to zero for strong turbulence conditions. For a numerical example we employ the same values of the parameters as assumed for the collimated case given above. We obtain a signal-to-noise ratio, relative to the constant C, of 0.041, as compared to the corresponding value in the absence of turbulence of 1.89.

In conclusion, we have derived a general expression for the signal-to-noise ratio of a heterodyne lidar system in the presence of atmospheric turbulence which is valid both in the near and far field of the laaser and remote scattering source. Although we have employed a gaussian shaped-laser source and receiver wave function, which enabled us to obtain the S/N in terms of algebraic functions, the corresponding qualitative behavior and numerical results for an arbitrary shaped laser source and receiver wave function are not expected to differ significantly as all of the essential physics of the problem are contained herein.

REFERENCES

1. Hinckley, E. D., "Laser Monitoring of the Atmosphere", Springer Verlag, Berlin (1976).
2. Lee, R. W. and J. C. Harp, Proc. IEEE, 57, 375 (1969).
3. V. I. Tatarskii, "The Effects of the Turbulent Atmosphere on Wave Propagation", National Technical Information Service, U.S. Department of Commerce, Springfield, Va., (1971).
4. Fante, R. L., Proc. IEEE, 63, 1669 (1975).
5. Prokhorov, A.M., F. V. Bunkin, KL. S. Gochelasvily, and V. I. Shishov, Proc. IEEE, 63, 790 (1975).
6. Yura, H. T., Appl. Opt. 11, 1399 (1972).
7. Lutormirski, R. F. and H. T. Yura, Appl. Opt. 10 1652 (1971).
8. Born, M. and E. Wolf, "Principles of Optics", Pergamon Press, New York (1965).
9. Seigman, A. E., Proc. IEEE 54, 1350 (1966).
10. Fried, D. L, Proc. IEEE 55, 57 (1967).

REMOTE MEASUREMENTS OF ATMOSPHERIC PROPERTIES FROM SATELLITES

G.E. Peckham

Department of Physics, Heriot-Watt University,
Edinburgh.

ABSTRACT. Techniques for the remote sensing of atmospheric temperature from satellites are reviewed. Methods for retrieving atmospheric temperature profiles from measurements of the emitted thermal radiation are described. The technique of limb sounding and its use both for temperature measurement and for stratospheric composition measurements is included.

1. INTRODUCTION

During the last two decades, a great deal of research has been devoted to ways in which the properties of the atmosphere may be inferred from measurements made by instruments mounted on satellites. Some measurements can be made satisfactorily only from such a platform (e.g. large scale cloud features, radiation balance), but even in the case where more detailed measurements can be made by instruments placed in situ, the satellite measurement offers the great advantage of covering a large part of the earth's surface with a single instrument. The simplicity of a polar orbiting satellite carrying an instrument to remotely measure atmospheric temperature is very attractive when compared with the world wide network of radiosonde stations and associated communications network. However, as I shall show, there are limitations inherent in the techniques available for making remote measurements, so that such measurements are unlikely to replace the use of in situ instruments, but they do have an important role in filling in the gaps in geographical coverage. The present network of radiosonde stations is concentrated on the continents of the northern hemisphere and the density of observations is inadequate in the arctic, the tropics, over oceans and in most of the southern hemisphere. This fact is becoming more apparent as

T. Lund (ed.), Surveillance of Environmental Pollution and Resources by Electromagnetic Waves, 95–110.

attempts are made to increase the period of forecasts using numerical models which cover a whole hemisphere or even the entire globe.

The most widespread use of satellites in meteorology has been in observing large scale cloud features. Since 1960, cloud photographs from the TIROS and NIMBUS satellites have been used extensively in forecasting, and now almost continuous observations are made from the American, European and Japanese geostationary meteorological satellites. The equipment is straightforward in principle. Cloud images are formed from reflected sunlight using either a vidicon camera tube or an elemental detector with a spinning mirror to scan the image. In the case of the geostationary satellite, the spin of the satellite itself is used for the line scan, the frame scan being accomplished by rocking a mirror (or the entire telescope). Images may also be formed in the infrared from thermal emission in the 11 μm atmospheric window. Low emission from the cold cloud top contrasts with greater emission from the warm surface. These images are available both day and night and also allow an estimate of cloud top altitude from the deduced cloud top temperature.

A number of satellites, including the European geostationary Meteosat, have carried instruments sensitive to infrared radiation at about 6.7 μm, where there is a strong absorption due to water vapour. Thermal emission from the cool water vapour is weaker than that from the warm surface so that images show a global distribution for water vapour.

In these lectures, I shall concentrate on the use of satellite instruments to measure (a) atmospheric temperature, and (b) the concentration of minor constituents in the stratosphere.

2. THE MEASUREMENT OF ATMOSPHERIC TEMPERATURE

The potential of satellite infrared radiometry is demonstrated by the emission spectrum (Figure 1) recorded by the Fourier Transform spectrometer (Iris) on Nimbus 4. In spectral regions where the atmosphere is transparent, the emission corresponds to that from a black body at the temperature of the earth's surface (290°K). However, where the atmosphere absorbs, the emitted power corresponds to a black body at some lower temperature characteristic of the region of the atmosphere from which emission occurs. Water vapour emission at 1400 cm^{-1} is from the troposphere at 260°, somewhat colder than the surface. CO_2 emission corresponds to temperatures characteristic of the lower stratosphere (220°K), while O_3 at higher altitudes emits more radiation, indicating the higher temperatures in the upper stratosphere. Within the CO_2 absorption band, there is an increase in emission corresponding to the very strongly absorbing Q branch at the centre of the band.

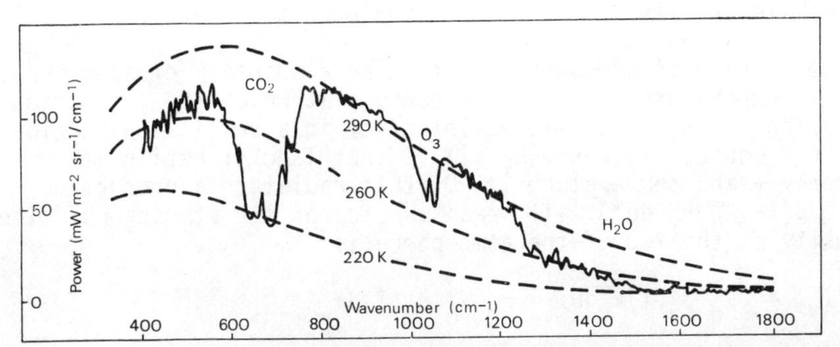

Figure 1. Power spectral density plot from Nimbus 3 Iris data (full line) compared with black body spectrum (broken line) (from Nimbus 4 users handbook, NASA and General Electric).

Radiation at wavelengths strongly absorbed by the atmosphere must originate at high altitude as that from lower altitude is reabsorbed before leaving the atmosphere, so this increase in emission within the Q branch again indicates increasing temperature with higher altitude in the stratosphere.

The emitted radiation depends on the concentration of the absorber and on its temperature. Measurements of this radiation may be used to investigate either factor. For temperature sounding, the absorber should be uniformly mixed so that the most suitable bands are the 4.3 µm CO_2 band, the 15 µm CO_2 band and the 5 mm O_2 band. Of these, the 15 µm CO_2 band seems the most promising at first sight as this wavelength is close to the maximum emission for a black body at atmospheric temperatures. However, the 5 mm oxygen band has the advantage of using long wavelength radiation for which clouds show far less absorption and the very low thermal emission in the tail of the black body spectrum is compensated by the excellence of the coherent detection techniques available for microwaves. The 4.3 µm CO_2 band has also been used by some instruments.

If the properties of the absorber and the temperature structure of the atmosphere are known, the calculation of the emitted radiation spectrum is relatively straightforward. The inverse problem, however, is ill-posed in that there is a multiplicity of atmospheric states which give identical values to the radiation intensities measured by the small number of spectral channels of the instrument. There is an estimation problem to choose the "best" solution from all the possible solutions consistent with the measurements, taking account of instrumental noise.

Although an accurate calculation must be based on line by line integration of the absorption spectrum over the band of wavelengths to which the instrument is sensitive, it is instructive to obtain approximate results by the use of band models.

Consider a slab of atmosphere containing a quantity of absorber du at a temperature T. At a frequency v at which the absorption coefficient is k_v, the emitted intensity in a vertical direction will be k_v du B_v (T) where B_v (T) is the Planck function at frequency v and temperature T. Of this radiation a proportion $\tau_v = \exp(- \int_u^\infty k_v \, du')$ will reach the top of the atmosphere. The intensity at the top of the atmosphere is

$$I_v = \int_0^\infty B_v(T) k_v du \tau_v = \int_0^1 B_v(T) d\tau_v$$

$$= \int_0^\infty B_v(T) \frac{d\tau}{dy} dy \qquad \text{.... (1)}$$

in terms of an altitude variable y = -lnp, where p is the pressure. The intensity I_v is the weighted average of black body emission from the atmosphere, the weighting function describing the height from which radiation is emitted being $K(y) = (d\tau_v/dy)$. K(y) must be known accurately to interpret the observed radiances in terms of a vertical temperature profile. This function may be found by numerical integration from tabulated line positions and strengths combined with laboratory measurements of the transmission of carbon dioxide.

However, the form of K(y) can be illustrated using simple models for the CO_2 absorption spectrum.

(1) *Monochromatic case*. For a single collision broadened spectral line of strength S centred at frequency v_0 the weighting function is

$$K(y) = 2\left(\frac{p}{p_0}\right)^2 \exp - \left(\frac{p}{p_0}\right)^2$$

where $p_0^2 = 2\pi(v - v_0)^2 (S\gamma_0 a)^{-1}$, γ_0 being the line halfwidth at 1 atm pressure and a the amount of absorber in a vertical column of atmosphere. This function has a peak at atmospheric pressure p_0 depending on the particular value of the absorption coefficient. The larger the absorption coefficient, the higher the peak. Its width at half height, however, is a constant 1.22 in the height variable y. This is equivalent to a width of 8.3 km assuming a scale height of 6.8 km.

(2) *Wide band case*. To accept sufficient radiation to make the measurement in the presence of detector noise, it is necessary in practice to include a spectral interval of 3 - 10 cm^{-1}. This interval will include a number of absorption lines. The Elsasser model in which lines are assumed equally spaced may be used to approximate the CO_2 absorption spectrum in this case. In the strong line limit this model gives for the weighting function

$$K(y) = (2/\pi)^{\frac{1}{2}} (p/p_0) \exp - p^2/(2p_0^2)$$

where $p_0^2 = \delta^2/\pi S a \gamma_0$, δ being the line spacing. The function peaks at $p = p_0$, but now the half width is 1.80 in γ (12.2 km).

This increase in width over the monochromatic case is understood easily as spectral regions near line centres will have high absorption coefficients and hence weighting functions peaking high in the atmosphere whereas in the wings of the lines the weighting functions will be lower. Averaging over a frequency band then produces a broad function.

Many of the problems of vertical temperature sounding are immediately apparent from this simple calculation. The measurement has a vertical resolution of 12 km and even given infinite spectral resolution, the height resolution is decreased to only 8 km. This is comparable with the depth of the troposphere so that it is almost impossible to obtain information about fine details of the tropospheric temperature profile. So far we have assumed a clear atmosphere, whereas much of the time the field of view of the instrument will include clouds which are usually thick enough to be opaque at 15 μm. This is probably the most serious problem of all in that reliable estimates of the cloud amount and cloud top temperatures are necessary to correct the measurement to a clear column radiance.

3. THE RETRIEVAL OF ATMOSPHERIC TEMPERATURE PROFILES

As we have previously noted, the retrieval problem is ill-posed in that there are many temperature profiles consistent with the observed radiances. It is necessary to impose constraints on the problem in order to obtain a unique solution. The constraints may take a number of forms. For instance, the profile may be expressed as a linear combination of a small number of functions with unknown coefficients. Various functions have been proposed including the weighting functions themselves, and linear interpolation. However, it has proved difficult to describe atmospheric profiles adequately in this way and most methods in use are based on statistical estimation methods.

At this stage it is useful to rewrite equation (1) in a discrete form.

$$y_i = \sum_j K_{ij} x_j \qquad \text{.... (2)}$$

Here y_i are the measured radiances, K_{ij} are a discrete version of the weighting functions, and x_j are the unknown black body radiances corresponding to the unknown temperatures at the discrete levels specified by index j. Equation (2) may be written in a matrix notation

$$\underline{y} = \underline{K}\, x \qquad \text{.... (3)}$$

Suppose that we have a non unique solution of (3)

$$\underline{x}_1 = \underline{D}\, \underline{y} \qquad \text{.... (4)}$$

so that D satisfies $\underline{K}\,\underline{D} = \underline{I}$ (the unit matrix). If $\underline{S}_y$ is the covariance matrix of the measured quantities $\underline{y}$, then the covariance of $\underline{x}_1$ is $\underline{S}_1$, given by

$$\underline{S}_1^{-1} = \underline{K}^T\, \underline{S}_y^{-1}\, \underline{K} \qquad \text{.... (5)}$$

Note that since the solution $\underline{x}_1$ is non unique, $\underline{S}_1^{-1}$ is singular.

The constraints which it is necessary to impose to obtain a unique solution are most conveniently included by supposing that they define on a priori estimate $\underline{x}_o$ of $\underline{x}$, with an associated covariance matrix $\underline{S}_o$. $\underline{x}_o$ is to be combined with the estimate $\underline{x}_1$ obtained from the measurements to give $\hat{\underline{x}}$ the final estimate of $\underline{x}$. $\underline{x}_o$ and $\underline{x}_1$ are weighted by the inverse of their covariances.

$$\hat{\underline{x}} = (\underline{S}_o^{-1} + \underline{S}_1^{-1})^{-1}\, (\underline{S}_o^{-1}\, \underline{x}_o + \underline{S}_1^{-1}\, \underline{x}_1)$$

or

$$\hat{\underline{x}} = (\underline{S}_o^{-1} + \underline{K}^T\, \underline{S}_y^{-1}\, \underline{K})^{-1}\, (\underline{S}_o^{-1}\, \underline{x}_o + \underline{K}^T\, \underline{S}_y^{-1}\, \underline{y}) \qquad \text{.... (6)}$$

The covariance of this estimate is given by

$$\hat{\underline{S}}^{-1} = \underline{S}_o^{-1} + \underline{S}_1^{-1} \qquad \text{.... (7)}$$

or

$$\hat{\underline{S}}^{-1} = \underline{S}_o^{-1} + \underline{K}^T\, \underline{S}_y^{-1}\, \underline{K}$$

Note that the expression (6) for the best estimate is now independent of the particular choice of $\underline{D}$ and can be calculated from a knowledge of the weighting functions, the a priori estimate $\underline{x}_o$, the measured values of $\underline{y}$ and the appropriate covariance matrices.

By making use of the matrix identity

$$\underline{K}^T(\underline{1} + \underline{S}_y^{-1}\, \underline{K}\, \underline{S}_o\, \underline{K}^T) = \underline{K}^T + \underline{K}^T\, \underline{S}_y^{-1}\, \underline{K}\, \underline{S}_o\, \underline{K}^T =$$

$$(\underline{1} + \underline{K}^T\, \underline{S}_y^{-1}\, \underline{K}\, \underline{S}_o)\, \underline{K}^T$$

we can put equation (6) into the form

$$\hat{\underline{x}} = \underline{x}_0 + \underline{S}_0 \underline{K}^T (\underline{K} \underline{S}_0 \underline{K}^T + \underline{S}_y)^{-1} (\underline{y} - \underline{K} \underline{x}_0) \quad .... (8)$$

where there is only one relatively small matrix to be inverted. Also the expression (7) for the covariance of $\hat{\underline{x}}$ can be reduced to a computationally simpler form, although it is algebraically more complex:

$$\hat{\underline{S}} = \underline{S}_0 - \underline{S}_0 \underline{K}^T (\underline{K} \underline{S}_0 \underline{K}^T + \underline{S}_y)^{-1} \underline{K} \underline{S}_0 \quad .... (9)$$

In practice, the estimate $\underline{x}_0$ may be a climatological mean profile on a forecast profile. Considerable difficulty has been experienced in obtaining a satisfactory covariance matrix $\underline{S}_0$ and many workers have used a highly simplified form

$$\underline{S}_0 = \sigma_0 \underline{I}$$

For historical reasons this assumption together with equation (9) is known as the minimum information method. The covariance matrix for the measurements $\underline{S}_y$ is usually diagonal.

If sufficient profiles are available from radiosonde ascents it is possible to determine a linear predictor $\underline{D}$

$$\hat{\underline{x}} = \underline{D}\, \underline{y} \quad .... (10)$$

which results in a minimum variance in the estimate $\hat{\underline{x}}$ without a knowledge of the weighting functions. We choose $\underline{D}$ to minimise

$$E\,\{(\underline{x} - \hat{\underline{x}})^T (\underline{x} - \hat{\underline{x}})\}$$

The solution is

$$\underline{D} = E\,\{\underline{x}\, \underline{y}^T\} [E\,\{\underline{y}\, \underline{y}^T\}]^{-1} \quad .... (11)$$

Both the expectation values are covariance matrices which may be estimated from an independent knowledge of a range of profiles, $\underline{x}$ and the corresponding radiance measurements $\underline{y}$. The estimate defined by equation (11) may be shown to be equivalent to that of equation (8) by substituting equation (3).

4. NONLINEAR PROBLEMS

The general remote sounding retrieval problem is nonlinear. It is only by making simplifying assumptions that we can construct a linear problem. The main sources of nonlinearity in the equation are:

(a) Temperature dependence of the atmospheric transmission;

(b) Wavenumber dependence of the Planck function across a spectral band:

(c) Wavenumber dependence of the Planck function between spectral bands;
(d) Clouds;
(e) Nonlinear constraints.

The first two of these usually lead to relatively small nonlinearities, whilst the rest may cause large nonlinearities.

Formally we may write the nonlinear problem as a generalisation of the linear problem. We wish to know the value of an unknown vector $\underline{x}$. We can measure a vector $\underline{y}$, which is related to $\underline{x}$ in a known way:

$$\underline{y}_1 = \underline{F}_1(\underline{x}) \qquad .... (12)$$

with an error covariance $\underline{S}_1$. We may have a priori information $\underline{y}_0$ about $\underline{x}$, which is similarly a known function of $\underline{x}$

$$\underline{y}_0 = \underline{F}_0(\underline{x}) \qquad .... (13)$$

with error covariance $\underline{S}_0$. The process of solution is in principle the same as for linear problems, although it may not be possible to find an explicit algebraic form for the best estimate $\hat{\underline{x}}$. We must first ensure that there are enough measurements and constraints to determine $\hat{\underline{x}}$ uniquely (i.e. with finite covariance), then set up a set of equations which $\hat{\underline{x}}$ must satisfy, such as the maximum likelihood condition that $\hat{\underline{x}}$ minimises

$$\sum_i (\underline{y}_i - \underline{F}_i(\underline{x}))^T \underline{S}_i^{-1} (\underline{y}_i - \underline{F}_i(\underline{x})) \qquad .... (14)$$

The solution of this kind of equation is usually the most difficult part of a nonlinear estimate. When the solution has been found, we must characterise its error bounds by finding its error covariance matrix.

The degree of nonlinearity in a problem can be classified in terms of the approximations that may be used to solve it. Problems which are slightly nonlinear are best solved by some form of Newtonian iteration. That is, linearise the problems and use a linear solution method, iterating as required. A second class of moderately nonlinear problems are those which need other methods to find the solution efficiently, but are sufficiently linear in the neighbourhood of the solution for linearisation to be used in the error analysis. A third class is of grossly nonlinear problems, which are the most difficult kind to solve or to understand their solutions. Solutions may always be found in principle simply by minimising expression (14) by a brute force numerical method, but this should be regarded as a last resort, as in any particular case it is usually possible to find some ad hoc method which exploits the algebraic form. A review of a number of methods is given by Rodgers, 1976.

Clouds present a severe problem in the troposphere. A number of methods have been proposed to estimate clear column radiances from measurements made in the presence of cloud. The most successful are based on the fact that clouds have a considerable amount of structure in the horizontal so that we may assume that in two adjacent fields of view, the temperature profiles and cloud top temperatures are the same, but the cloud amounts differ. Of course, no method can give reliable temperatures below continuous cloud cover. A potential solution is to use the 5 mm O_2 band for temperature sounding where clouds are much more transparent.

5. THE INSTRUMENTS

The characteristics of some successful temperature sounding instruments are listed in Table 1. The instruments differ from each other mainly in the techniques used to select appropriate spectral intervals. All incorporate means for in-flight calibration. This is achieved by directing radiation from a black body emitting a known intensity into the instrument. Space is used as a zero reference.

Absolute calibration must be better than 1^oK imposing severe conditions on stray radiation and reference black body design. The random noise level is a function of entrance aperture, field of view, spectral bandwidth, detector performance and signal integration times.

A single sounding from the Selective Chopper Radiometer on Nimbus 4 is shown in Figure 2, compared with a radiosonde ascent. The weighting functions for this instrument, the top two of which were achieved by selective chopping, are also shown. Such profiles are obtained every 16 s. Recorded on magnetic tape, the satellite derived temperature patterns form a more complete data base for global scale atmospheric studies than has ever been available before. This is particularly so for the high stratosphere and the southern hemisphere where few radiosonde data are available.

Table I. Characteristics of some successful temperature sounding instruments

Experiment	Spectral selection	Number of channels	Spectral resolution	Field of view
IR interferometer spectrometer	Michelson interferometer †	Continuous, 5 - 25 μm ‡	5 cm^{-1}	8°
		Continuous, 6.25 - 50 μm §	3 cm^{-1}	5°
Satellite IR spectrometer	Grating spectrometer with array of detectors	7 in CO_2 band, ‡ 1 in 11 μm window	5 cm^{-1}	12°
		14 including CO_2 § and water vapour	5 cm^{-1}	12° (scanned)
Selective chopper radiometer	Selective chopping with interference filters	6 in CO_2 band, § 1 in 11 μm window	0.01 - 10 cm^{-1}	10°
		16 including water ‖ vapour and window channels	0.01 - 10 cm^{-1}	1.5°
Vertical temperature profile radiometer	Interference filters	6 in CO_2 band 2 window 1 water vapour	3.5 - 10 cm^{-1}	2.2° (scanned)
IR temperature profile radiometer	Interference filters	4 in CO_2 band ‖ 2 window 1 water vapour		1.5° (scanned)
Nimbus microwave spectrometer	Microwave techniques ‖	3 in 5 mm O_2 band 2 at 1 cm (water and liquid water)	220 MHz	10°
Pressure Modulation radiometer	Selective chopping by pressure modulation X	2	.001	20° x 4° (Doppler scanning)

†Interferogram is Fourier transformed to give spectrum
‡Nimbus 3, §Nimbus 4, ‖Nimbus 5, Tiros, XNimbus 6

Two of these instruments make use of a technique for improving spectral performance known as selective chopping. The radiation is modulated by passing it alternately through two cells, one containing CO_2 and the other empty. Modulation will occur only at wavelengths corresponding to strong absorption in the CO_2 gas and the corresponding weighting functions will peak at high altitude (up to 45 km). Figure 2 shows the weighting functions for one of these instruments, and a comparison between a retrieved profile and a rocket sonde.

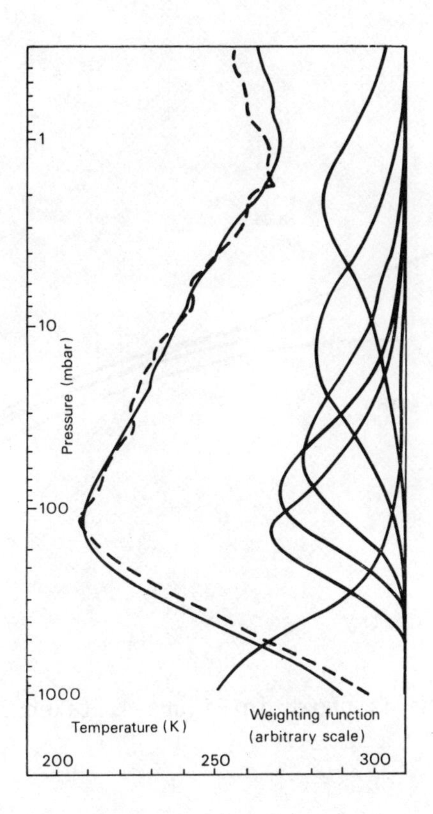

Figure 2. Comparison between a profile computed from Nimbus 4 SCR data (full line) and radio and rocket sonde (broken line) over Wallops Island, 27 August, 1970. The weighting functions for the six channels of this instrument are also shown.

6. LIMB SOUNDING

The height resolution of the vertical temperature sounders so far described is determined by the optical properties of the atmosphere and is of order 12 km. An entirely different approach is to view the limb of the earth with a narrow field of view so that the radiation originates along a tangent path through the atmosphere, the lowest point of which is defined geometrically by the direction of the instrument's field of view (see Figure 3). Only one spectral channel is now required.

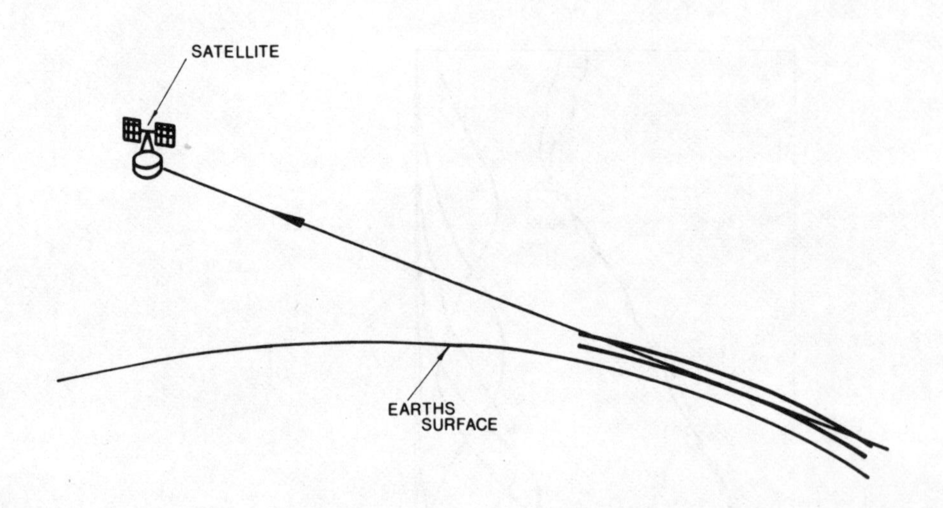

Figure 3. Limb sounding geometry

A set of weighting functions is shown in Figure 4, taken from Rodgers, 1976.

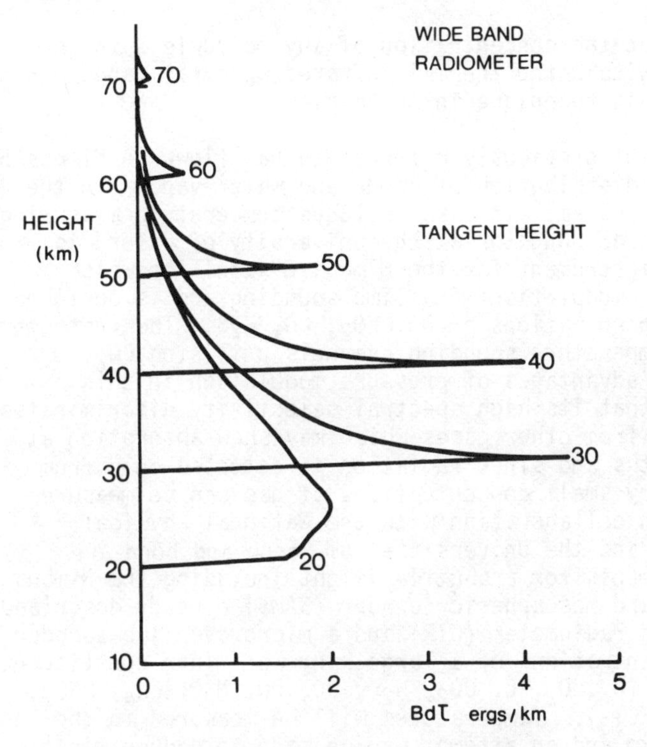

Figure 4. Limb sounding weighting function

The technique is appropriate only for the stratosphere as the atmosphere rapidly becomes opaque due to the presence of clouds at lower altitudes for such a long tangential path.

The spacecraft attitude is not usually known well enough to define altitude sufficiently accurately, so that is becomes necessary to design the experiment to find its own reference level. This has been done in the case of the LRIR instrument (Sissala, 1975) by measuring limb radiance in two spectral intervals, a narrow band and a broad band. To a first approximation, the ratio of the two signals is independent of the temperature profile, and is equal to the ratio of the absorbtivities of the emitting gas (carbon dioxide), which is a known function of atmospheric pressure. Thus the direction of a known pressure level may be found.

7. STRATOSPHERIC COMPOSITION MEASUREMENT

The limb sounding geometry is particularly advantageous for measuring composition because (a) a long atmospheric path at a given level is available so that weak emission is enhanced; and (b) a zero radiation background is provided by space. Again measurements are restricted to the stratosphere because of the cloud opacity, but the concentration of any molecule with an absorption band within the thermal infrared spectral range can be measured by this technique in principle.

The LRIR experiment previously referred to has flown on Nimbus 6 and measures the distribution of ozone and water vapour in the height range 15 - 60 km. It also includes temperature measuring channels. Professor Houghton at the University of Oxford is constructing an instrument for the Nimbus G satellite which combines pressure modulation with limb sounding and is designed to detect the concentrations of NO, NO_2, CO, H_2O. The instrument also includes temperature sounding channels utilising CO_2 absorption. The advantages of pressure modulation in this application are that its high spectral selectivity discriminates against emission from other gases which may show absorption at similar wavelengths and since radiation is detected only from line centres, very small concentrations of gas can be measured. The same group in collaboration with the National Physical Laboratory, JPL, and the Universities of Berne and Bonn are proposing instruments for a shuttle flight including the Nimbus G Stratospheric and Mesospheric Sounder (SAMS) already described, a cooled infrared radiometer (CIR) and a microwave limb sounder (MLS). The concentrations of a large range of minor constituents will be measured (O_3, O_2, O, CO_2, H_2O, CO, NO, N_2O, CH_4, HNO_3, ClO, CF_2, Cl_2, $CFCl_3$). Temperatures will be measured in the range 15 to 120 km and an attempt may be made to deduce winds from Doppler shifts.

8. THE FUTURE

The difficulties encountered in retrieving reliable tropospheric temperature profiles from the many vertical temperature sounding instruments flown so far have stimulated much discussion about the usefulness of such measurements in operational weather forecasting. The poor vertical resolution and the problems in dealing with clouds mean that such instruments are unlikely to replace the radiosonde. However, there is a role for remote sounding in increasing the geographical coverage provided by more conventional instruments, particularly over oceans and in the southern hemisphere so that future operational weather satellites are almost certain to carry vertical sounding radiometers. The relative merits of infrared and microwave instruments have still to be established. Although the stratosphere is not modelled for short term forecasting, there is much scientific interest in this region of the atmosphere and recently there has been much discussion about the possible effects of man made pollutants on the photochemical processes which occur in the stratosphere. A permanent reduction in ozone content would cause an increase in ultra-violet radiation at the earth's surface which might be directly harmful to human life. Sounding from satellites provides an excellent method for continuously monitoring the state of the stratosphere and is likely to continue. Cloud photographs are likely to remain the main source of meteorological information obtained via satellites and global coverage now established by an international ring of geostationary satellites will continue.

Of possibilities not mentioned in this paper, the most interesting are the development of active techniques. The state of the sea surface can be related to the surface wind - a very important meteorological parameter. Some measurements of sea state have been made by radar techniques. Another possibility is the use of an accurate measurement of atmospheric transmission in the wings of the oxygen line at 5 mm wavelength to deduce atmospheric surface pressure. Such measurements could be made using a transmitter on the satellite reflecting back from the ocean surface and are the subject of a development program at Heriot-Watt University in Edinburgh.

9. REFERENCES

Cahine, M.T., 1976: Remote sounding of cloudy atmospheres. 1. The single cloud layer, J. Atmos. Sci., 31, 233-243.

Gille, J.C. and House, F.B., 1971: On the inversion of limb radiance measurements. 1. Temperature and thickness, J. Atmos. Sci., 28, 1427.

Houghton, J.T., Peskett, G.D., Rodgers, C.D. and Williamson, E.J., 1972: A proposal to employ the pressure modulator radiometer to sound the temperature and composition of the stratosphere and mesosphere, A.A.F.E. Programme Submission by Atmospheric Physics Department, Clarendon Laboratory, Oxford.

Houghton, J.T. and Smith, S.D., 1970: Remote sounding of atmospheric temperature from satellites. 1. Introduction, Proc. Roy. Soc., London, A320, 23-33.

Houghton, J.T. and Taylor, F.W., 1973: Remote sounding from artificial satellites and space probes of the atmospheres of the earth and the planets, Rep. Prog. Phys., 36, 827-919 (lists many further references).

Kaplan, L.D., 1959: Inference of atmospheric structure from remote radiation measurements, J. Opt. Soc. Am., 49, 1004-7.

Rodgers, C.D., 1976: Retrieval of atmospheric temperature and composition from remote measurements of thermal radiation, Rev. Geophysics and Space Physics, 14, 609-624.

Sissala, J.F., Ed., 1975: The Nimbus 6 Users Guide, Goddard Space Flight Centre, Greenbelt, Maryland, U.S.A.

Smith, W.L., Woolf, H.M. and Fleming, H.E., 1972: Retrieval of atmospheric temperature profiles from satellite measurements for dynamical forecasting, J. Appl. Meteor., 11, 113.

Staelin, D.H., 1969: Passive remote sensing at microwave wavelengths, Proc. IEEE, 57, 427-39.

INFRARED LASER AUTOMATED FIELD INSTRUMENTATION FOR MONITORING OF THE ATMOSPHERE+

S. T. Eng

Department of Electrical Measurements
Chalmers University of Technology
Gothenburg, Sweden

ABSTRACT. The performance of computer-automated laser systems to be used in atmospheric gas monitoring is discussed. Factors to be considered are types of lasers to be used, the stability, and the repeatability of these lasers. Special attention is paid to the fundamental mechanism and limitations of CO_2 gas and Pb-salt diode lasers. Also considered are the choice of computers, automation, molecular absorption of laser radiation by atmospheric gases, and on-off absorption line switching techniques suitable for field experimentation. A few spectroscopic systems using CO_2 and Pb-salt lasers are presented to illustrate useful automated field instrumentation.

1. INTRODUCTION

Laser spectroscopic instrumentation is becoming very important in global, regional, and industrial pollution detection and lately also in mineral prospecting [1, 2]. However, without computer automation, laser remote sensing systems will not be performing useful functions in the field. There is clearly a need for rapid response, sophisticated experimental control, and elimination of tedious, inefficient manual operations. In addition, these improvements will enable us to perform hitherto impossible experiments.

The automation requirements put severe limitations on the reliability, stability, and repeatability of the lasers to be used. The laser technique chosen for gas monitoring should be the most efficient one to fulfill the objective of the measurements. The tuning method and the tuning range of the laser are also important in computer control of the laser. Factors to be considered

+This work was supported by the Research Institute of the Swedish National Defence (FOA),Stockholm.

T. Lund (ed.), Surveillance of Environmental Pollution and Resources by Electromagnetic Waves, 111-143.

in the overall system instrumentation are *e.g.* high sensitivity, simplicity, eye safety, and suitability for field tests.

This paper will explore these factors, which affect the performance of computer-automated laser systems in air pollution monitoring. The fundamental mechanism and limitations of the most popular lasers, such as the CO_2-gas and Pb-salt diode lasers, will be examined as well as the choice of computers and spectroscopic techniques suitable for field experimentation and automation. Finally, a computerized CO_2 system for monitoring of ethylene and a Pb-salt laser laboratory and long-path field system for SO_2 detection above sulfide mineral deposits, will be presented.

2. INFRARED TUNABLE LASERS

Tunable lasers in air pollution spectroscopy are capable of measuring atmospheric molecular absorption line parameters such as position, width, intensity, and line shape to a much higher degree of accuracy than conventional lamp-monochromator systems. The improvements arise from the coherence and tunability properties of the laser radiation. At the present there are several types of tunable lasers being used, as outlined in Table I [3, 4].

Pb-salt semiconductor diode lasers are the smallest of the tunable lasers. Lasing takes place by stimulated emission between the conduction and valence bands of an electrically pumped semiconductor device. The output wavelength can be altered between 3-25 μm by changing the material composition, which in turn changes the energy gap. On a single device, fine tuning can be accomplished by changing the pressure, temperature (current), or magnetic field. By heterodyning a laser of this type with a CO_2 laser, the line width was shown to be around 100 kHz [1]. The CW power of < 1 mW is too low for spectrophone absorption measurements.

Fixed frequency, low pressure CO_2 lasers can be tuned discontinuously over several hundred lines by cavity adjustment. The continuous tuning range is then around 50 MHz ($1.7 \cdot 10^{-3}$ cm^{-1}). However, by increasing the pressure, the tuning can be broadened to a GHz or more. Another possibility for tuning is shifting the molecular levels by magnetic fields. These lasers can be designed for CW power output in the mW to kW range. High pressure CO_2 waveguide lasers have been built in our laboratory. However, they have not performed reliably to date for serious consideration in automated tunable laser instrumentation.

Another alternative is parametric oscillation, where a pumping laser beam excites a resonant cavity containing nonlinear material

Table I. Types of infrared tunable lasers.

Device	Operating Principle	Wavelength µm	Highest Resolution cm^{-1}	CW Power mW
Pb-salt	Bandgap tuning	3-25	$3\cdot10^{-6}$	1.0
CO_2-gas	Step-tuning resonator. Tuning within gain profile	9-11	$3\cdot10^{-8}$	10^2-10^5
Parametric oscillator	Nonlinear polarization in crystal	1-11	$3\cdot10^{-2}$	10
Spin-flip Raman	Photon-electron inelastic scattering causes electron spin-flip. Magnetic tuning	5-6 9-14	$3\cdot10^{-6}$ $3\cdot10^{-2}$	10^3
Difference freq.gen.	Use of nonlinear polarization. Var. freq.tuning	2-6	$5\cdot10^{-4}$	$1\cdot10^{-3}$

to produce two beams having frequencies that add up to the original input beam frequency. All three frequencies involved are required to satisfy the energy and momentum conservation for the parametric process. Since the indices of refraction generally depend on frequency, applied field, crystal orientation, and temperature, the tuning can be accomplished by varying these quantities. Using e.g. a crystal of proustite (Ag_3AsS_3) pumped by a Q-switched neodymium laser, mW of CW power with a linewidth of 1 cm^{-1} and tuning range of 2-8 µm could be obtained.

The spin-flip Raman laser is a device that uses a fix-tuned laser (CO, CO_2, or HF gas laser) to pump a semiconductor crystal (n-type indium antimonide or mercury cadmium telluride) in a magnetic field. The pump laser photons lose energy when they collide with an electron in a crystal and flip the spin of the electron. The output frequency of a spin-flip Raman laser is separated in frequency from the pump photon frequency by the frequency corresponding to the change in electron spin energy which can be tuned in a magnetic field. The linewidth can be very small, the tuning range wide, and 1 W CW power obtainable. However, good frequency stability may be difficult to achieve, and the cryogenically cooled laser with a 50-100 kilogauss magnetic field is rather complex.

The final possibility shown in the table uses mixing of a fixed frequency laser with a tunable one in a nonlinear crystal. A ruby laser and a dye laser might be mixed e.g. in a proustite crystal to produce tunable radiation in the 10-13 μm range. The output power expected is rather low (1 μW). Further developments in nonlinear materials with improved optical quality are needed.

In spite of rapid progress in laser developments the last few years, infrared lasers such as the parametric oscillator, spin-flip Raman laser, and difference frequency generation laser systems are rather complex, costly, and do not have the reliability for field test and automation at the present time. We will therefore here focus our attention on the low pressure CO_2 gas laser and the Pb-salt laser, since these have been commercially available for some time and are the only possible choices for field test instrumentation.

3. CO_2 LASERS IN INSTRUMENTATION

In most cases the spectroscopist will build his own CO_2 discrete tunable low cost laser. It will therefore be appropriate to review the major design considerations [5]. The CO_2 laser is representative of molecular lasers in which the energy levels of concern involve the internal vibration and rotation of the molecule. Vibrational levels in the infrared derive from three fundamental modes as shown in Fig. 1. In the symmetric mode of frequency ν_1, the two oxygen atoms move in opposite directions, while the carbon atom is stationary. The asymmetric mode with frequency ν_3 shown has the two O atoms moving together, while the C atom moves in the opposite direction. The bending mode with frequency ν_2 consists of two degenerate vibrations perpendicular to and in the plane of the page. These vibrational frequencies are typically in the near and middle IR range. The vibrational states are defined by $(n_1 . n_2^{\ell} . n_3)$, where n and ℓ are quantum numbers. The molecule as a whole can also be slowly rotated about axes in space at rotational rates characteristic of far infrared frequencies. Thus a series of rotational levels are superimposed on each of the vibrational states. The quantum number of these levels is given by the label J. In addition, there are electronic energy states. However, these will be disregarded since the frequency separation between these states lies in the visible and ultraviolet region. The total energy of the molecule can be written as the sum of the vibrational and rotational energy

$$E_{TOT} = (n_1 + \tfrac{1}{2})h\nu_1 + (n_2 + \tfrac{1}{2})h\nu_2 + (n_3 + \tfrac{1}{2})h\nu_3 + BJ(J + 1), \quad (1)$$

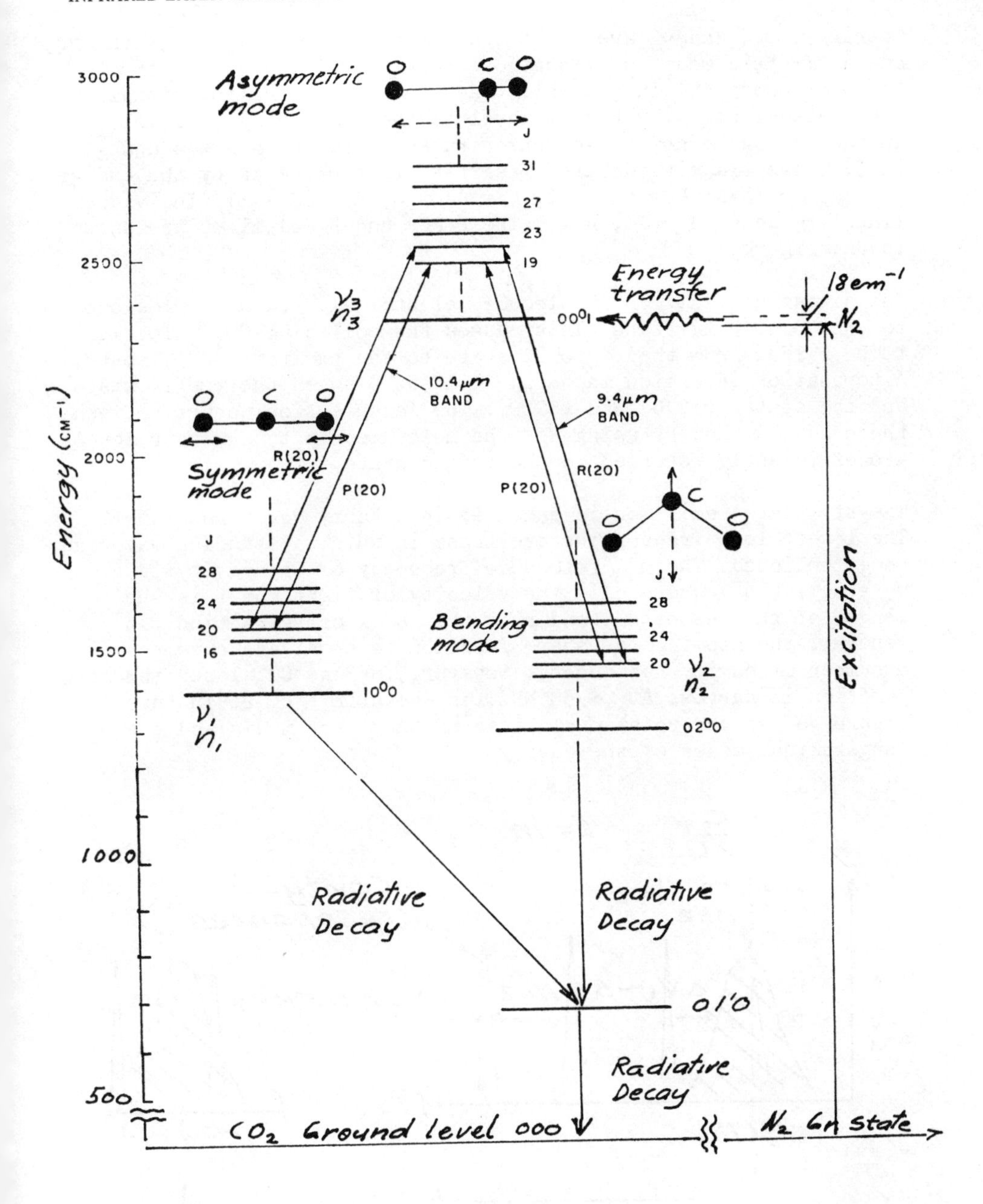

Fig. 1. CO_2-N_2-He laser transition diagram. n is the vibrational and J is the rotational quantum number. The label (00^01) corresponds to ($n_1 \cdot n_2^{\ell} \cdot n_3$).

where ν_1, ν_2, and ν_3 are the frequencies of the symmetric, bending, and asymmetric modes of vibration, respectively, and B is the rotational constant. In general, transitions from the rotational sublevels of one vibrational level to the rotational sublevels of another are governed by selection rules as to the allowed change in J. These transitions are classified into branches by the change in J as follows: $P = \Delta J = -1$, $Q = \Delta J = 0$, $R = \Delta J = +1$. Individual lines are denoted P(J), R(J). The P(20) and R(20) lines are shown in the figure.

The discharge excites N_2 molecules electrically to a level close to the CO_2 (00^01) state. This causes the colliding CO_2 molecules to be raised from their ground state to the excited (00^01) state. A population inversion is being built up between the (00^01) state and the states (10^00) and (02^00). The laser action occurs between these levels, and by using He, the molecules in these lower levels are efficiently returned to the ground state.

The stimulated emission is enhanced in a Fabry-Perot resonator. The lowest loss frequencies are those in which a standing wave can be established. The mth resonance frequency of the cavity is $\nu_c = (cm)/(2nL)$, where c is the velocity of light and L is the length of the resonator, and n is the index of refraction. In general, the atomic line shape function is broad and encompasses a number of cavity resonances. However, for the CO_2 laser the gain function is narrow, $\Delta\nu_a \sim 50$ MHz, as shown in Fig. 2. In this figure we see the rotational lines of the laser media and the longitudinal modes of the cavity.

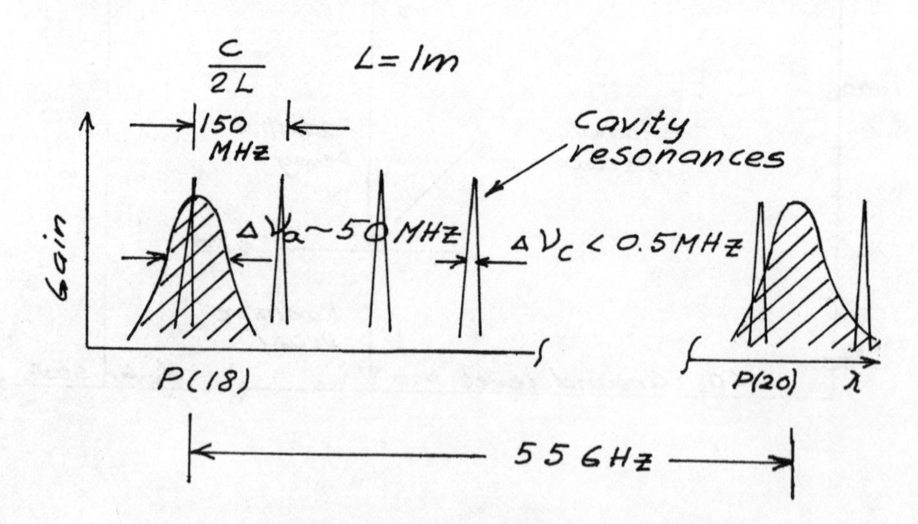

Fig. 2. Gain vs. wavelength for two laser lines of the CO_2 laser. The cavity resonances are also indicated.

The frequency stability may be defined as $s = \nu_L/\Delta\nu_L(\tau)$, where ν_L is the average frequency and $\Delta\nu_L(\tau)$ is some measure of the frequency fluctuations measured during the period τ. Sometimes, only $\Delta\nu_L$ is given in data sheets. If τ is much greater than the limit of the time resolution of the detector, then the equation gives the long-term stability. The short-term stability is given by the same formula, except in this case τ is of the order of the time resolution of the detector. The long-term wavelength and frequency stability of a laser is mainly determined by the stability of the Fabry-Perot resonator. This may be seen from the equation for the oscillation frequency ν_L of a laser as given by [6]

$$\nu_L = \nu_c + (\nu_c - \nu_a)\frac{\Delta\nu_c}{\Delta\nu_a}, \tag{2}$$

where ν_a is the resonance frequency of the atomic transition and $\Delta\nu_c$ and $\Delta\nu_a$ are, respectively, the half-width at half-maximum intensity of the cavity and atomic resonances. Usually, $\Delta\nu_c \ll \Delta\nu_a$. The latter restriction may be violated in CO_2 lasers because of the small $\Delta\nu_a$. The cavity bandwidth is given by

$$\Delta\nu_c = \frac{c(1 - R)}{2\pi nL}, \tag{3}$$

where R is the reflectivity of the mirror. Practical values are around 0.5 MHz or less for the cavity bandwidth. If the cavity resonance coincides with the atomic line center, it is clear that the oscillation takes place at $\nu_L \simeq \nu_c$. For an ideal laser, the spectral width would be limited by phase fluctuations caused by quantum noise. For a homogeneously broadened line, we get according to Schawlow-Townes,

$$\Delta\nu_L = \frac{8\pi h\nu_L\Delta\nu_c^2}{P_L}, \tag{4}$$

where P_L is the power generated by the laser oscillator. For typical systems the last equation predicts $\Delta\nu_L < 1$ Hz. At a given time the deviation of frequency $\Delta\nu_c$ from the resonance frequency ν_c is determined by the fluctuations in the refractive index and cavity length, as given by

$$\frac{\Delta\nu_c}{\nu_c} = \frac{\Delta n}{n} + \frac{\Delta L}{L}. \tag{5}$$

As an example, 1Å displacement of mirrors in an 0.5 m long cavity will cause a shift of ≃ 6 kHz. 0.5-1 MHz/mA and 5-10 MHz/Torr are expected as a result of the change in the refractive index.

In the following we will discuss parameters that are important in spectroscopy such as single wavelength, stability, repeatability, resonance linewidth, and power output.

First, one can simply include transverse-mode control by increasing losses for higher transverse modes by a limiting aperture (iris). In addition, the cavity length can be decreased so that only one longitudinal resonance can lie within the lasing linewidth of the inverted medium. This technique is, however, not always possible to use since the laser dimensions may be too small and the output power low.

Next, steps should be taken to control the optical path length between the mirrors. These steps include use of cavity materials with low expansion coefficients, cavity temperature control, and isolation from mechanical and acoustic disturbances. Also, it is necessary to use a highly stabilized power supply in order to maintain a constant index of refraction in the discharge and a stable population inversion.

Some form of active (electronic) stabilization using a piezoelectric mirror and servo circuitry is required for long-term stability. The plate separation of the Fabry-Perot etalon is modulated at an audio frequency, and the laser beam becomes frequency modulated. Typical FM sidebands are generated inside the laser. Demodulation of the beam by direct detection external to the laser will have beat frequencies between the various sidebands of the FM signal. There will be no beats in the case when the laser transition frequency coincides with a resonance of the etalon because of the rounded Doppler gain curve. If the beat increases to some nonzero value, then one mirror of the laser is moved so that the carrier is once again centered. This is accomplished by controlling the voltage of the piezoelectric transducer by feedback electronics. This stabilization technique is mandatory in stable frequency operation of CO_2 lasers.

A more sophisticated, but commonly used, approach to derive the error signal is to use a molecular transition. The laser frequency in this case can be locked to the resonance frequency of a saturated cell with CO_2 gas. The absorber saturates when the laser increases in intensity. This drop in absorption is similar to a sudden increase in Q. The absorption cell may be placed externally or internally to the laser.

The wavelength control to distinguish between the different molecular transitions from one sublevel in a vibrational band to another sublevel in the adjacent vibrational band is in most cases done by an internal grating. The alignment of the grating selector is easily controlled by a computer. The average separation for the 10.6 μm line is around 55 GHz. However, other techniques such as interferometry and Fabry-Perot etalon may also be very useful and even better in some cases.

The CO_2 laser in its simplest form is unsuitable for spectroscopic applications, since it usually exhibits multiple wavelength operation and frequency as well as amplitude fluctuations. However, if most of the techniques discussed above are applied, a single-frequency stabilized laser suitable for computer control should not be too hard to make with the characteristics shown in Table II.

Table II. Typical CO_2 characteristics with piezoelectric servo control.

λ	9-11 μm
$\Delta\nu_L$-stability, short term	100 kHz
$\Delta\nu_L$-stability, long term	500 kHz
Repeatability	1 MHz
Doppler width	50 MHz
Computer line selection and control	Good
Output power (CW)	5 W
Laser length	1 m

Some of the gases of interest in our planned field experiment, which can be detected by use of CO_2 lasers, are listed in Table III.

Table III. Atmospheric pollutants and the appropriate CO_2 laser lines and absorption coefficients [7].

Pollutant	Laser line	Freq. (cm^{-1})	Wavelength (μm)	Absorp. coeff. ($atm.\ cm)^{-1}$
Ammonia	R-(30)	1084.63	9.2197	77
Ethylene	P+(14)	949.43	10.5326	30
Trichloroethylene	P+(20)	944.18	10.5912	15.18
Benzene	P-(30)	1037.44	9.6391	2.01
Freon 11	R-(30)	1084.63	9.2197	33.5
Freon 12	P+(42)	922.85	10.8360	92
Freon 113	P-(26)	1041.29	9.6035	20.8
Ozone	P-(14)	1052.13	9.5045	12.7
Vinyl chloride	P+(22)	942.35	10.6118	8.69

The + refers to the 00^01-10^00 transitions.
The - refers to the 00^01-02^00 transitions.

The coincidence between the laser line and the molecular absorption line is not exact, so there will be a dependence on pressure as well as temperature of the molecular absorption, which might cause some error in the atmospheric absorption data.

4. Pb-SALT LASERS IN INSTRUMENTATION

Another solution to atmospheric monitoring instrumentation is the use of Pb-salt tunable lasers. They have mainly been used up to now in laboratory scientific applications, particularly in trace gas monitoring and basic molecular infrared spectroscopy. Since these lasers are now available commercially, it is worthwhile to examine the basic principle of operation, the experimental characteristics, and the limitation of these devices for use in field instrumentation.

Binary semiconductors of $Pb_{1-x}Sn_xTe$, $PbS_{1-x}Se_x$, and similar materials have a composition-dependent energy gap which can go to zero [8]. In the first-mentioned material e.g., as the Sn content increases, the energy gap first decreases, and the band states (L_6^- and L_6^+) approach each other and go to zero at a certain composition. The states then reverse and the energy gap increases. This can be explained on the basis of the difference in the relativistic effects in Pb and Sn. The electrons in a crystal are not free (like in metals) but are acted upon by the atoms and the other electrons in the crystal. In the band theory it is thought than an electron moves in a static periodic potential representing the nucleus and in an average potential representing the other electrons. Otherwise it would be too complex to take into consideration all the 10^{23} cm^{-3} particles. Using the conventional quantum mechanical notation for the relativistic Hamiltonian, one gets

$$H = \frac{p^2}{2m} + V + \frac{h}{4m^2c^2}\,\bar{\sigma}\,[(\nabla V)X\,\bar{p}] + \frac{h^2}{8m^2c^2}\,(\nabla^2 V) - \frac{1}{8m^3c^2}\,p^4 . \quad (6)$$

The first two terms represent the nonrelativistic Hamiltonian. The third term is the spin orbit coupling term. The fourth and fifth terms represent the Darwin and the mass-velocity correction, respectively. The relativistic corrections seem to describe the composition-dependent band model shown in Fig. 3.

Lasers made of Pb-salt materials work in the spectral region from about 2.5 µm to beyond 33 µm (see Fig. 4). The emission wavelength of such a laser can thus be tailored to a specific value by the choice of composition, and lasers with a fixed composition may be tuned over appreciable spectral intervals by varying external factors such as magnetic field, temperature, pressure, or the diode current.

A schematic diagram of a p-n junction Pb-salt laser is shown in

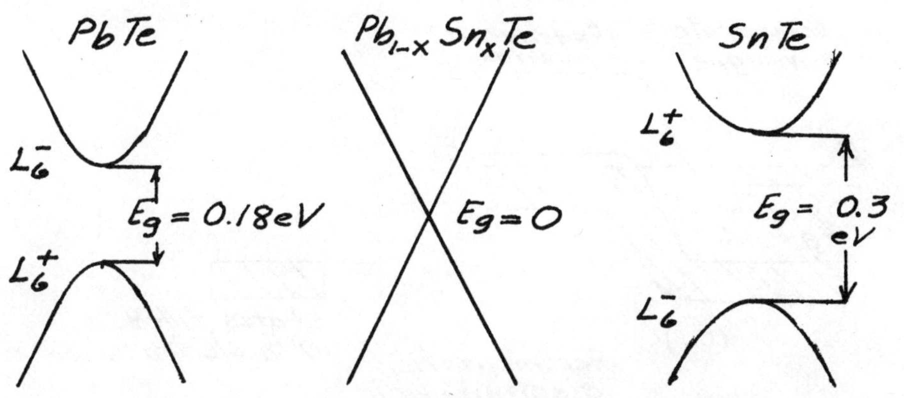

Fig. 3. Composition tuning of Pb-salt infrared lasers.

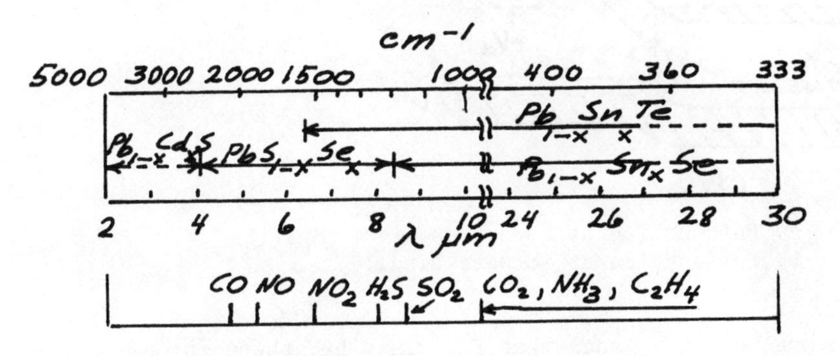

Fig. 4. Pb-salt alloy semiconductor lasers and their emission wavelengths. The absorption lines of a few gases are also indicated.

Fig. 5. A degenerate population of electrons and holes exists on both sides of the junction. In other words, the n- and p-type concentrations are sufficiently high so that the Fermi level is within the conduction and valence bands, respectively. When there is no bias applied to the device, the Fermi level is constant throughout the device. The cross-hatched regions symbolize electrons and the blank space above E_F and E_{FV} denotes the holes in the P-type material. Between the homogeneous n-type and p-type regions is a junction, depleted of carriers by the barrier potential V_B. Now, if a forward bias voltage V_a is applied, quasi-Fermi levels are formed. The barrier is reduced, and electrons can flow over to the p-side. They can then make a transition to an empty state in the valence band, emitting photons with energy approximately equal to E_G. It is also possible to have holes flow to the n-side, where they recombine with electrons.

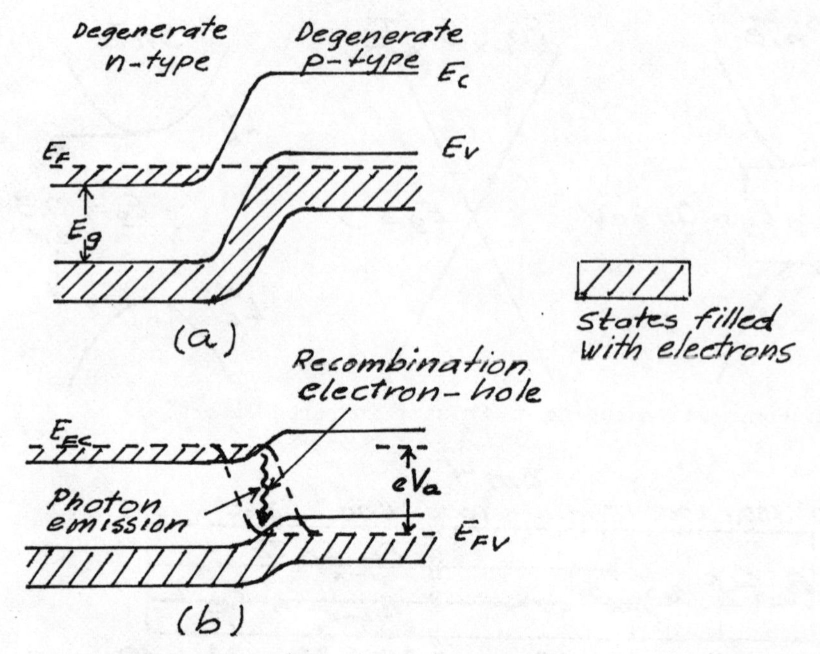

Fig. 5. a. Pn junction at zero bias.
b. Pn junction at forward bias.

For low values of V_a such that $E_{FC}-E_{FV} < h\nu$, there is only spontaneous radiation emitted of energy $h\nu = E_G$. As the applied voltage increases further, a point is reached at which $E_{FC} - E_{FV} > h\nu$. Nonequilibrium electrons and holes can exist in the same region and upon recombination emit stimulated radiation of energy $h\nu$.

At higher forward voltages, a direct diffusive flow of carriers into the region of the opposite type can occur, as illustrated in Fig. 5 b. In the vicinity of the junction an inverted population is indicated.

Let us now examine the laser tuning characteristics [1, 3]. We will consider the simplest case of a single mode laser as shown in Fig. 6. In Fig. 7, the spontaneous emission gain and cavity parameters are sketched. As usual, the subscript 'c' refers to the cavity and 'a' to the gain spectrum. The laser frequency (wavenumber in cm^{-1}) in each continuous portion of the curve in the upper figure is intermediate between ν_a and ν_c in the simplest model and is locked to the cavity frequency in the case where the bandwidth of the cavity is much smaller than the bandwidth of the gain spectrum. Since ν_a and ν_c typically tune at different rates (ν_a may tune 10 times faster than ν_c), there is a relative motion between ν_a and ν_c during tuning. As ν_a passes the midpoint between

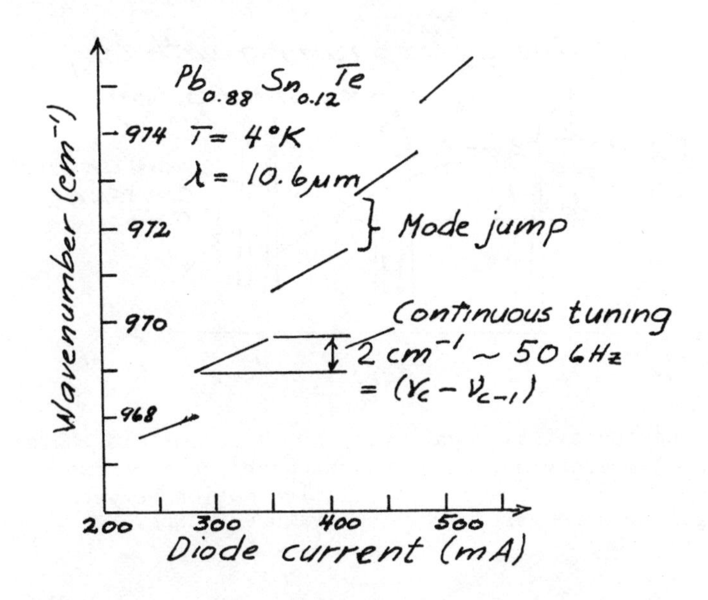

Fig. 6. Current tuning for a single mode laser. This is mainly caused by an increase in energy gap with temperature. The cavity tunes at a much slower rate [3].

two cavity resonances, the laser frequency ν_L discontinuously shifts towards the next mode (Fig. 7). This effect is called mode jumping. The size of the discontinuity depends on the relative sizes of $\Delta\nu_a$ and $\Delta\nu_c$. The axial mode spacing can be expressed in the cavity length L and the optical index of refraction n by [3]

$$\Delta(\nu_c - \nu_{c-1}) = \frac{1}{2L(n + \nu\frac{dn}{d\nu})}, \tag{7}$$

where ν is as before in wavenumber units (cm^{-1}). The current through the diode affects the energy gap and hence the spontaneous gain frequency, the spontaneous emission linewidth, and the index of refraction. Since $\Delta\nu_a \simeq 25\ cm^{-1}$ and far exceeds $\Delta\nu_c \simeq 1\ cm^{-1}$, tuning of the laser frequency is primarily the result of changes in the refractive index. Under these conditions, the continuous tuning range represents the change in laser output frequency as the spontaneous line shifts by an amount equal to $\Delta(\nu_c - \nu_{c-1})$. The largest continuous tuning range is obtained when the ratio of the cavity tuning rate and the spontaneous tuning rate is unity. The difficulties with incomplete wavelength tuning resulting from mode hopping may be eliminated by using magnetic tuning in addition to current tuning. The laser linewidth may in practical cases

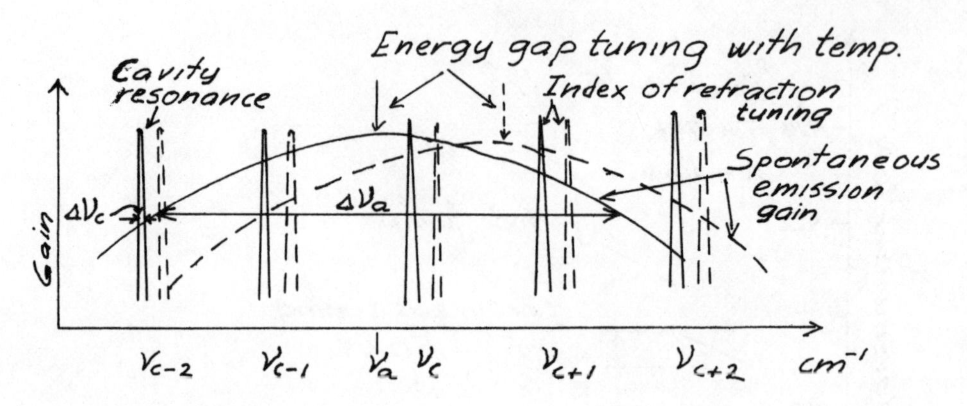

Fig. 7. The spontaneous emission gain with $\Delta\nu_a \simeq 10\ cm^{-1}$ is shown for two different temperatures. Also indicated are cavity resonances with $\Delta\nu_c \simeq 0.5\ cm^{-1}$. There is a relative motion between the cavity and gain resonances during temperature tuning.

be around 2.5 MHz ($10^{-4}\ cm^{-1}$). Further flexibility in tuning may be achieved with the use of an external cavity and incorporating a distributed feedback structure into the diode laser.

The first Pb-salt lasers were of the homojunction type, and the junction extended across the entire device. These devices had bouncing modes involving all four sides of the semiconductor chip. This resulted in poor mode quality and low output power. Then followed a stripe geometry homojunction structure as shown in Fig. 8. The bouncing modes are suppressed by the lossy bulk regions that are located next to the active region.

A second important improvement is the confinement of the injected carriers and the light wave to an active region, which is bounded by epitaxially grown regions of wider energy gap. This results in a substantial decrease in laser threshold current. In this way, single and double heterojunction structures were formed.

In a typical fabrication process the pn-junction is formed by diffusing an n-region into a p-type substrate through an SiO_2 mask. The heterostructure junction can be formed by evaporating a layer of n-type PbTe on a p-type $Pb_{1-x}Sn_xTe$ substrate. Other techniques are annealing processes, where a crystal of one x-value is annealed with a source of another x-value, and a compositional interdiffusion is performed [9], or molecular beam epitaxy [10], which has successfully been used for double heterostructure lasers as well as distributed feedback structures. It is interesting to note that excesses of metal, e.g. PbSn, act as donors and excesses of non-metal, e.g. Te, are acceptors.

I. Homojunction:

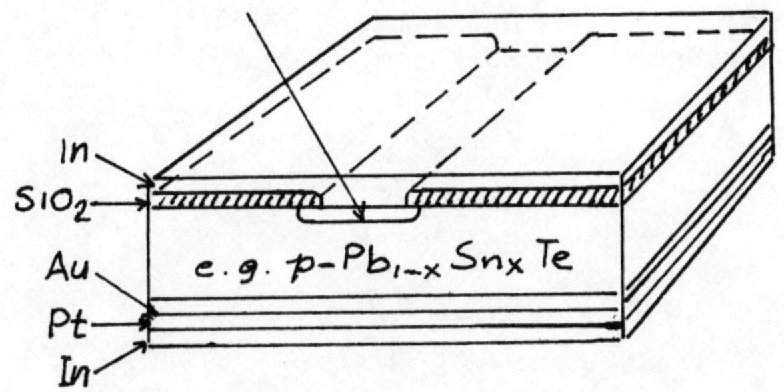

Donor: Excesses of metall (PbSn)

Acceptor: Excesses of nonmetall (Te)

II Single heterostructure: By compositional interdiffusion

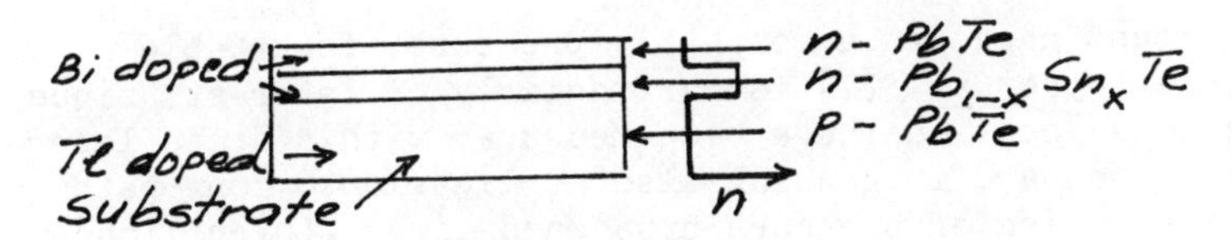

III Double heterostructure: By molecular beam epitaxy

Bi doped
Tl doped
Substrate
n-PbTe
n-$Pb_{1-x}Sn_xTe$
p-PbTe
n

Fig. 8. Stripe geometry lead-salt laser structures. In the single and double heterostructures, the light wave is confined to the active region by energy gap (index of refraction) changes in the materials used.

In Table IV an attempt has been made to list some typical characteristics of Pb-salt laser diodes. The data have been taken from Laser Analytics Inc. data sheets and from the literature.

In Table V are listed the absorption cross sections $\sigma(cm^2)$ for

Table IV. Typical characteristics of Pb-salt laser diodes.

Homojunction (Laser Analytics)	
CW temperature	<30K
Single frequency power	0.1-1 mW
Single device tuning range	20-40 cm^{-1}
Linewidth	10^{-4} cm^{-1}
Material	$Pb_{1-x}Sn_xTe$
	$PbS_{1-x}Se$
	$Pb_{1-x}Sn_xSe$
	$Pb_{1-x}Cd_xS$
Continuous tuning	0.5-2 cm^{-1}
Single Heterostructure (Laser Analytics) (best results reported)	
CW temperature	100K
CW power	3 mW (10K)
Efficiency	2-3%
Material	$Pb_{1-x}Sn_xSe$
Single device tuning range	9-12 µm
Low threshold current	300 mA
Double Heterostructure (Lincoln Lab.) (best results reported)	
CW temperature	114K
Material	$Pb_{1-x}Sn_xTe$
Single device tuning range	8.5-16 µm
Output power	0.1 mW

several important gases at atmospheric pressure. This table taken from Hinkley et al. [1] is particularly of interest, since some of the cross sections have been measured with a diode laser and all the values are in general also applicable to long-path derivative monitoring of pressure-broadened lines (Lorentzian).

5. COMPUTER AUTOMATION

We have already in the introduction pointed out some of the advantages of automatic measurements such as rapid response, sophisticated experimental control, and elimination of tedious and inefficient manual operation. However, another important improvement is the increased sensitivity by real time digital processing, including averaging of several hundred measurements and real time error estimation. The error can be minimized for different pollution measurement situations.

Table V. Absorption cross sections for several important gases. These values are applicable to pressure-broadened lines obtained in long-path monitoring.

Molecule	Formula	$\nu(cm^{-1})$	$\sigma(10^{18} cm^2)$
Acetylene	C_2H_2	719.9	9.2
Ammonia	NH_3	1084.6	3.6
Carbon monoxide	CO	2123.7	2.8
Carbon tetrachloride	CCl_4	793.	4.8
Ethylene	C_2H_4	950.	1.7
Freon 11	CCl_3F	847.	4.4
Freon 12	CCl_2F_2	920.8	11.
Methane	CH_4	3057.7	2.0
Nitric oxide	NO	1900.1	0.6
Nitrogen dioxide	NO_2	22311.	0.2
Ozone	O_3	39425.	12.
		1051.8	0.9
Sulfur dioxide	SO_2	33330.	1.0
		2499.1	0.02
		1126.	0.2
Vinyl chloride	C_2H_3Cl	940.	0.4

Selection of a computer should be determined by the particular measurement task. There are many computer choices and tradeoffs. Some of the important ones are shown in Table VI. In our case a

Table VI. Computer considerations for automated field systems.

	μ-computers (spec. purp.)	Minicomputers (general)	Minicomputers + disc memory
Price	\$ 2 k + \$ 6 k for complete system	\$ 15 k	\$ 30 k
Software	Simple Assembler	Assembler Fortran	Assembler Fortran Real-time
Data storage	1-4 kbits	30 kbits	2.5 Mbits
Speed	3 μs	1 μs	1 μs

multipurpose computer system has been chosen, which can handle many different tasks and a large amount of data. Therefore, the disc memory is an important part of the computer system. By changing the software and some of the interfacing instruments, one can upgrade the system at any time. However, in the case of field instrumentation dedicated to a special task, the microcomputer might be very desirable. The emphasis in the design is then on hardware and a routine and efficient test instrumentation.

Another factor to be considered is how the operator will interact with the instruments. It would be advantageous if the necessary instruments such as analog-to-digital converters, plotters, and display real-time clocks could be using standard interface cards and be easily programmed in a high-level language (Fortran). Such a standard interface kit with an input/output card, cable, and connector for the instruments as well as software is available on request from the major instrument manufacturers, e.g. the HP-IB bus kit.

The principle of data processing is as follows [11, 12]. All requested laser wavelengths and tuning requirements are automatically set, and the lasers are trimmed for maximum output powers. The lasers are continuously checked during the measurement. If any laser power is lower than its nominal value, a new laser adjustment is automatically performed. The absorption data to be used and the inaccuracy in these data for numerous pollutants and background gases which might interfere in the measurement, e.g. H_2O, are stored in the computer disc memory for easy access during the measurement. The system works with one or more measurement strategies. A measurement strategy consists of M wavelength pairs and N expected pollutants in the case of fixed frequency CO_2 laser systems. In the Pb-salt laser system it has to be decided which part of the pulse should be used, and the unnecessary pulsewidth has to be masked out. In other words, the initial measurement conditions may be altered at the end of a few measurement cycles. These steps are indicated in Fig. 9.

The measurement program is written in a combination Fortran/Assembler language, the composition of which is determined by whatever is the most efficient programming solution for the task at hand. When all the measurements are compiled, the concentrations are calculated from 100-500 measurement values, an error estimation is performed, and the result is presented on a display. The flow chart in Fig. 10 shows the principal steps in the measurement program.

In the CO_2 system, when only two wavelengths are used, the measurement time is around 5 s, and with the use of ten wavelengths, it may be around 3 min before a measurement value is available. The measurements can be repeated a requested number of times, and when they are finished, a diagram of concentration as a function of time will be drawn by a special utility program.

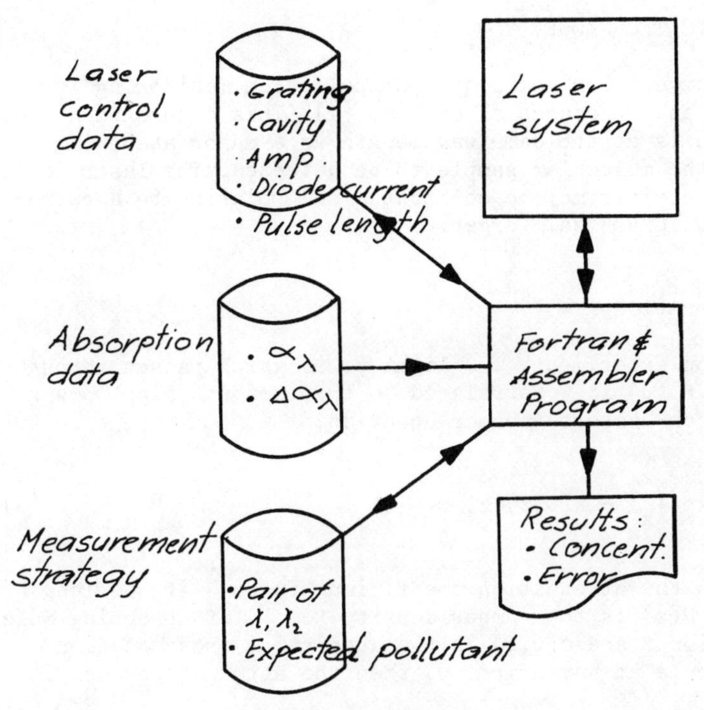

Fig. 9. Principles of data control, storage, and processing.

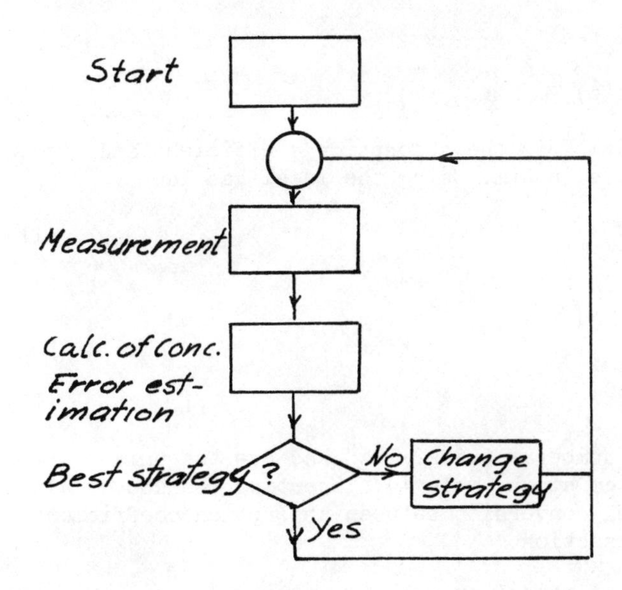

Fig. 10. Flow chart of principal steps in the measurement program.

6. SPECTROSCOPIC TECHNIQUE

The most sensitive laser technique for pollution monitoring is based on long-path resonance absorption [1]. This occurs when the laser radiation is at the same wavelength as a major absorbing transition of the molecular sample to be detected. The laser source and telescope receiver may be co-located and pointing to a retro-reflector or topographical target.

General considerations

In an absorption experiment, the light power which passes through the sample gas $P_r(\nu)$ in W is related to the incident light power $P_t(\nu)$ by the Beer-Lambert-Bouguer equation,

$$P_r = P_t \exp \int_{z_1}^{z_2} - k(\nu,z)dz, \tag{8}$$

where $k(\nu,z)$ is the attenuation coefficient in cm^{-1} if the length z is in cm. If $N(z)$ is the number density (cm^{-3}) of absorbing molecules at location z and $\sigma(\nu,z)$ is the absorption cross section (cm^2) per molecule at wavenumber ν, then the attenuation coefficient becomes

$$k(\nu,z) = N(z)\sigma(\nu,z), \tag{9}$$

which can also be written

$$k(\nu,z) = \alpha(\nu,z)p(z), \tag{10}$$

where $\alpha(\nu,z)$ in $cm^{-1}atm^{-1}$ is the absorption coefficient and $p(z)$ is the partial pressure in atm. Using the ideal gas law

$$p(z) = N(z)\frac{R}{N_A} T, \tag{11}$$

we get

$$\alpha(\nu,z) = \frac{N_A}{RT} \sigma(\nu,z), \tag{12}$$

where N_A = Avogadros number = $0.6022 \cdot 10^{24}$ and R = the universal gas constant = 82.0 ($cm^3 atm\ k^{-1} mole^{-1}$). Inserting the above values, one obtains the useful conversion between absorption coefficient and absorption cross section

$$\alpha(atm^{-1}cm^{-1}) = 2.6 \cdot 10^{19} \sigma(cm^2) \tag{13}$$

for a temperature of zero ^{o}C and a pressure of 1 atm.

Concentrations of gaseous pollutants can be expressed in the number of pollutant molecules per million molecules of air (ppm) or the mass concentration, which is given in mg/m^3. For convenience, we will indicate the formula which will relate these units. If M is the molecular weight in grams of one mole of a gaseous pollutant and one mole of air (25°C, 1 atm) would occupy a volume of $24.5 \cdot 10^{-3}\ m^3$, then

$$1\ mg/m^3 = (\frac{24.5}{M})\ ppm. \tag{14}$$

As an example, for SO_2 with M = 64, one obtains 1 mg/m^3 = 0.38 ppm at 25°C.

Data on spectral line parameters have been compiled on magnetic tape. The theoretical background is given in [13].

Switched on-off absorption system

In the case of the CO_2 laser monitoring system, at least two laser beams at slightly different wavelengths are used [14]. The wavelengths are adjusted to be as much as possible on or off an absorption band of the molecule under investigation. Comparison of the two signals collected by the receiver provides a measure of the integrated concentration of the molecular species along the transmission path. In some applications more than ten different wavelengths are required because of multiple molecular absorption. In Fig. 11, four transmitted CO_2 laser beams are shown, with wavelengths λ_1, λ_2, λ_3, and λ_4. The more wavelengths used, the more efficiently one can separate pollutants from interferences [12]. Each measurement wavelength should also be selected so that the interference from the other gases is a minimum.

We assume three absorbing gases which give four equations and four wavelengths as outlined below. Then for a homogeneous medium $(\int_{z_1}^{z_2} - k(\nu,z)dz = -k(\nu)L)$,

$$A_1 = \ln \frac{P_t^{1}(\lambda_1)}{P_r^{1}(\lambda_1)} - B = -L\left[\alpha_{11}(\lambda_1)p_1 + \alpha_{12}(\lambda_1)p_2 + \alpha_{13}(\lambda_1)p_3\right] \tag{15}$$

$$A_2 = \ln \frac{P_t^{2}(\lambda_2)}{P_r^{2}(\lambda_2)} - B = -L\left[\alpha_{21}(\lambda_2)p_1 + \alpha_{22}(\lambda_2)p_2 + \alpha_{23}(\lambda_2)p_3\right] \tag{16}$$

$$A_3 = \ln \frac{P_t^{3}(\lambda_3)}{P_r^{3}(\lambda_3)} - B = -L\left[\alpha_{31}(\lambda_3)p_1 + \alpha_{32}(\lambda_3)p_2 + \alpha_{33}(\lambda_3)p_3\right] \quad (17)$$

$$A_4 = \text{Background } (\lambda_4) = B \quad (18)$$

The absorbance A_i and the concentration p_j can be written in general notation,

$$A_i = L \sum_{j=1}^{m} \alpha_{ij}\, p_j \quad (19)$$

$$p_j = \frac{1}{L} \sum_{i=1}^{m} \alpha_{ji}^{-1} \left[\ln \frac{P_t^{j}(\lambda_j)}{P_r^{j}(\lambda_j)} - B \right] \quad (20)$$

α_{ji}^{-1} is an element in the inverse absorption coefficient matrix. For a given measurement problem the computer should be programmed to select the optimum wavelength. If we consider two laser wavelengths, where one laser line may be on the absorption line and the other on the background interference, and if the laser transmitted powers at the two wavelengths are assumed equal and the background absorption is assumed small, then the minimum detectable gas concentration can be found from the equations above,

$$p_1 = \frac{1}{2\alpha_{11}(\lambda_1)L_{ret}} \; \frac{P_r(\lambda_2) - P_r(\lambda_1)}{P_r(\lambda_2)} . \quad (21)$$

In the case of detector noise limitation only,

$$P_r(\lambda_2) - P_r(\lambda_1) \simeq \text{NEP} = \frac{\sqrt{2\Delta f\, A_D}}{D^*} , \quad (22)$$

where for large distances with diffraction

$$P_r(\lambda_2) = \left[\frac{A_t\, A_r\, A_{ret}^2\, rT_a^2(\lambda_2)}{\lambda_2^4 L_{ret}^4} \right] P_t . \quad (23)$$

In practical cases it has been found that the measurements are in general limited by other factors than the detector noise, such as atmospheric turbulence and optical noise in the system. Therefore, in experimental situations it has been found that

$$\frac{P_r(\lambda_2) - P_r(\lambda_1)}{P_r(\lambda_2)} = 0.01 = 2\alpha_{11}(\lambda_1)L_{ret}\ p_{1\,min}\,, \tag{24}$$

and, as an example, for ethylene with $\alpha_{11} = 30\ cm^{-1}atm^{-1}$, $p_{1\,min} =$ $= 0.5$ ppb with $L_{ret} = 3.2$ km.

In the above equations A_r, A_D, A_{ret}, and A_t are the areas of the receiver optics, diode, retroreflector, and transmitter optics, respectively; T_a^2 = the two-way background transmission at λ_2; r = the reflection coefficient at the retroreflector; and L_{ret} is the distance from the transmitter to the retroreflector. The NEP stands for the noise equivalent power of the detector, and the other detector parameters are defined in the conventional way.

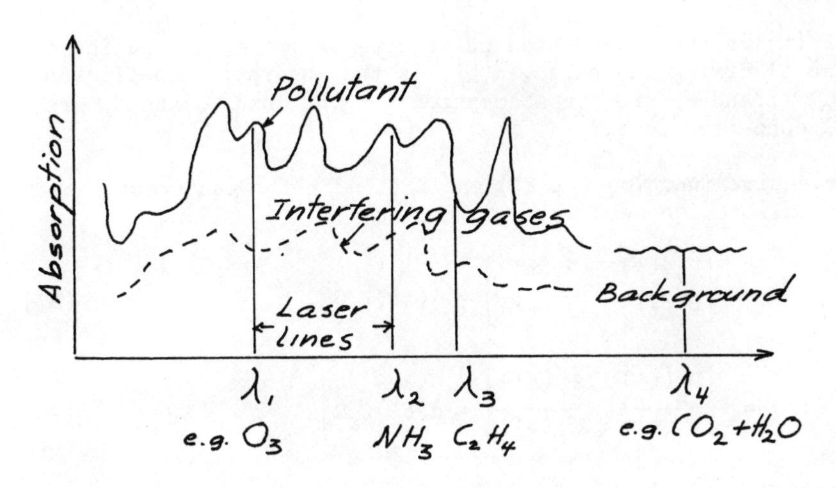

Fig. 11. Measurement of three absorbing gases and background interference.

Derivative - line scan system

The experimental technique used for the diode laser system is a direct line scan with a current pulse through the diode, and the derivative of the absorption line recording is taken by the signal processing electronics [15, 1]. Because of the fast pulses used, the derivative is insensitive to turbulence effects, and by normalizing the derivative signal to the original signal, one cancels the influence of long-term atmospheric effects in the first approximation. Other special features in our particular measurement system are digital processing and correlation technique.

Laser pulses are alternately directed through an absorption cell (X), a reference cell (N) with a known concentration of the gas to be investigated, and an empty cell (0). The pulses through the cells are detected and the derivatives taken electronically and the correlation with the computer.

If only the attenuation caused by the unknown gas, e.g. SO_2, is considered, the received laser pulse power shape $P_X(t)$ from the measurement path X and the corresponding $P_N(t)$ from the calibration cell N become

$$P_X(t') = P_0(t')\exp[-\alpha(t')\ell_X C_X] \tag{25}$$

$$P_N(t') = P_0(t')\exp[-\alpha(t')\ell_N C_N], \tag{26}$$

where $P_0(t')$ is the transmitted pulse power as a function of the time t' during the pulse, $\alpha(t')$ is the absorption coefficient for SO_2, ℓ_X and ℓ_N are the absorption lengths, and C_X and C_N are the SO_2 concentrations.

The derivative functions $D_X(t')$ and $D_N(t')$ in the measurement path and the calibration cell are obtained as

$$D_X(t') = -\left[\frac{\dot{P}_X(t')}{P_X(t')} - \frac{\dot{P}_0(t')}{P_0(t')}\right] = \dot{\alpha}(t')\ell_X C_X \tag{27}$$

$$D_N(t') = -\left[\frac{\dot{P}_N(t')}{P_N(t')} - \frac{\dot{P}_0(t')}{P_0(t')}\right] = \dot{\alpha}(t')\ell_N C_N \tag{28}$$

The dots indicate time derivatives.

The relation between the correlation product $(D_X{*}D_N)$ of $D_X(t')$ and $D_N(t')$ and the autocorrelation product $(D_N{*}D_N)$ of $D_N(t')$ gives the unknown concentration C_X as

$$C_X = \frac{(D_X{*}D_N)}{(D_N{*}D_N)} \cdot \frac{\ell_N}{\ell_X} C_N \tag{29}$$

By analogy with the switched on-off absorption line technique discussed earlier, the line scan on-off the absorption line within a short time interval would seem to yield approximately the same minimum detectable concentration as in Eq. (21). The continuous tuning range of the diode laser is assumed at least as large as the linewidth of the spectral line to be measured. The digital processing and correlation technique used will give improvement in the

$\Delta P_r/P_r$ ratio from the 1% limit quoted for the CO_2 system to a possible 0.1% or less.

7. LASER SPECTROSCOPIC SYSTEMS

A computer-controlled CO_2 laser long-path absorption system [14]

This section describes a CO_2 laser long-path absorption system that has been used in automatic monitoring of ethylene and vinyl chloride within or near a petrochemical industry (Fig. 12). The system is controlled by an HP 21MX minicomputer that can set two lasers to 54 different laser lines and trim the cavity lengths. A mechanical scanner aligns the two laser beams coaxially and switches them sequentially to a transmitting mirror in the center of a receiving Newtonian telescope. Two HgCdTe detectors are used to monitor the transmitted and received laser powers. The detector signals are sampled synchronously with the scanner by two 12-bit A/D converters, one for each detector. In this way the four values $P_t(\lambda_1)$, $P_t(\lambda_2)$, $P_r(\lambda_1)$, and $P_r(\lambda_2)$, corresponding to transmitted and received powers at the two laser wavelengths are obtained. The sampling rate is 110 Hz, and N = 100-500 samples are averaged when the ratio R is calculated according to

$$R = \frac{\sum_{i=1}^{N} (P_r(\lambda_2)/P_t(\lambda_2))_i}{\sum_{i=1}^{N} (P_r(\lambda_1)/P_t(\lambda_1))_i} \qquad (30)$$

This method compensates partially for the scintillation and emphasizes large values of the ratios in Eq.(30) (when the scintillation losses are small). All calculations, including error analysis, are made in real time.

When one pollutant is predominant, its concentration can be calculated immediately by means of the ratio R and the absorption coefficients for the two laser lines. If many different pollutants are present, several line pairs have to be used to form a system of linear equations from which the different concentrations can be solved by the computer.

Field experiments have been performed in the vicinity of several petrochemical factories where ethylene and vinyl chloride are produced. A scan over the area using path lengths of 500- 2 800 m showed ethylene concentrations of from 10 to 400 ppb. The concentrations of vinyl chloride were generally small (< 50 ppb). The uncertainty of the vinyl chloride concentration is largely due to the fact that it was not possible to get a representative measurement over the prime source of vinyl chloride. Measurements close to a

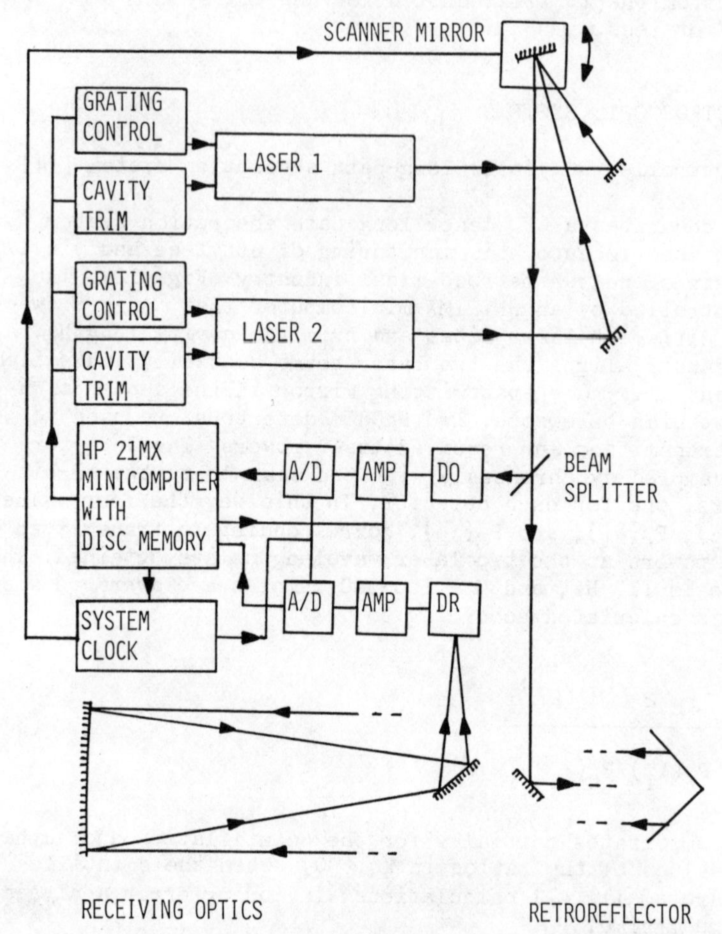

Fig. 12. An automated CO_2 field system for ethylene monitoring.

steam cracker showed ethylene concentrations in the 1-50 ppm region using path lengths of 60-200 m.

The maximum systematic errors of the calculated concentrations were continuously estimated and were typically 1-10%. The main contributors to these errors are limited resolution in the A/D converters and noncoaxial alignment of the two laser beams.

The field experiments performed showed that automated CO_2 laser systems can successfully be used in monitoring ethylene for industrial as well as environmental applications.

Laser spectroscopy in mineral prospecting [2]

The great global demand for minerals is leading to many new exploration studies to locate new ores. Attention has been given recently to surface gases associated with metallic ore deposits. Since the most sensitive techniques for detection and measurement of very low gas concentrations are optical ones, it is very natural to apply laser spectroscopy to mineral prospecting.

Some of the more significant vapors that can be detected, the type of deposits from which they emanate, and a few of the optical sources useful for field detection are listed in Table VII.

Table VII. Some gases indicating ore deposits and optical sources to be used in measurements.

Vapor	Deposits	Optical Source (field use)
SO_2	Sulfide	Semic. laser
Hg	Zn-Pb-Cu sulfides	Mercury lamp Dye laser
H_2S	Sulfide	Dye laser Semic. laser
CO_2 (or CO_2/O_2)	Sulfide, gold	CO_2 laser
NO, NO_2	Nitrate deposits	Semic. laser

Among ore types in which SO_2 anomalies can be measured are base metal sulfides. porphyry copper deposits, and pyrites. Mercury vapor occurs in close association with many types of ore materials such as ZnS and HgS deposits. Furthermore, hydrogen sulfide can be present in the oxidized zone of an ore deposit. Also, the oxidation of sulfide ore will result in the natural gas mixture becoming richer in CO_2 and poorer in O_2. Thus, in the soil gases above a sulfide zone, the ratio of carbon dioxide will be increased from regional background.

We have chosen to focus on SO_2 detection using a PbSe diode laser, and the laboratory experiment performed consisted of measurement of SO_2 above rocks containing metal sulfides, of which pyrite (FeS_2) is the most abundant. Other minerals in the rocks include pyrrhotite (Fe_7S_2-FeS), zinc sulfide, and lead sulfide. The rocks were kept in a glass container to which two teflon tubes were attached; one tube introduced fresh air into the container, and the other led the gases to the measurement cell.

A lead selenide (PbSe) diode injection laser was used as the spectrometer source. The diode was operated at a temperature of 77 K, a pulse width of 30 μs, a duty cycle of $1.6 \cdot 10^{-3}$, and a pulse current of 7 A (which is just above threshold current). In every pulse the laser frequency spectrum varied from 7.326 μm (1365 cm^{-1}) to 7.278 μm (1374 cm^{-1}). Only 13 μs of the pulse was utilized in the experiment.

The optical setup is shown in Fig. 13 and is briefly explained as follows.

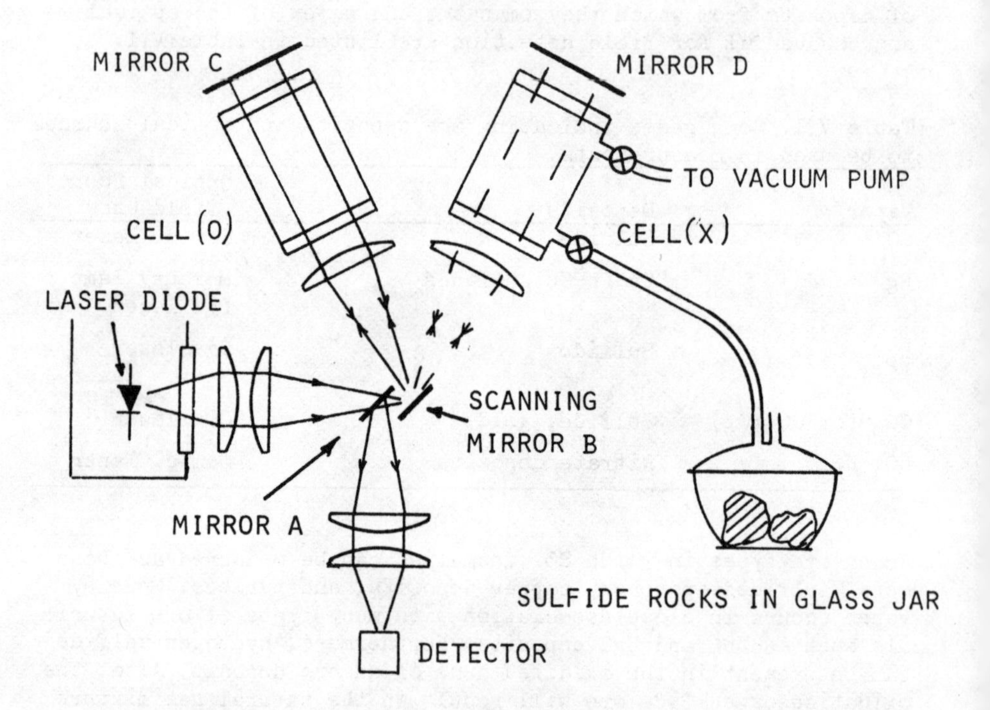

Fig. 13. The experimental optical arrangement. The lenses and windows are made of BaF_2. The length of the cells is 0.2 m.

The beam from the pulsed diode laser is passing just above the mirror (A) and is focused on the scanning mirror (B), which alternately directs the beam through the cells (X) and (O). By the two plane mirrors (C and D) the beam is reflected and then again focused on the scanning mirror (B) but this time 3 mm underneath the first focus spot. This lower orientation of the beam makes the reflected radiation first hit the mirror (A) and then the detector. Gas with an unknown concentration C_X of SO_2 from the glass jar is connected to the cell (X). The ratio between the signals from the two cells is detected and processed by the processing electronics.

The laser frequency is tuned during the pulse, and the time derivative of the amplitude of the detected signal gives an approximation of the derivative of the transmission spectrum. By this method weak absorption lines are accentuated.

With the present laboratory setup (ℓ_X = 0.4 m, laser power ≈ 0.2 mW, detector noise = $1.6 \cdot 10^{-12}$ W$\sqrt{Hz}$, and with a mean value of C_X/C_N from eighty calculations), the lower detectable SO_2 concentration determined by the detector noise is around 70 ppb. A measurement of the SO_2 concentration from a container with sulfide rocks in the laboratory gave a concentration of around 10 ppm. The experiment also shows the importance of making the cells identical. The experimental limitations indicate that we are far, maybe two orders of magnitude, away from the detector noise limit.

A diode laser system for long-path automatic monitoring of SO_2 [16]

An automatic, long-path SO_2 monitoring system has been investigated and tested. The interference by water vapor in the atmosphere is simulated by means of computed absorption spectra of SO_2 and water vapor. The system is a further development of the *in situ* monitoring system reported in the preceding section.

A schematic illustration of the optical system is shown in Fig. 14. A $Pb_{1-x}Sn_xSe$ diode laser with a wide tunable range (*cf*. single heterostructure laser in Table IV) is located in a liquid nitrogen cooled dewar. By pulse operating the laser, we obtained a small wavelength tuning in the 9 μm region.

The laser beam is focused on a scanning mirror and usually directed, *via* a transmitting lens, through the atmospheric path. At the end of the path the beam is reflected by the retroreflector and finally directed by the beam splitter BS to the PbSnTe detector.

For calibration purposes the laser beam is regularly directed by the scanning mirror to the calibration cell, containing a known concentration C_N of SO_2, and to a short reference path for measuring the laser output pulse shape.

A block diagram of the monitoring system is shown in Fig. 15. A minicomputer (HP 21MX) controls the scanning mirror, which directs a certain number of laser pulses through each of the three optical paths. For every second pulse the detector signal is converted from analog to digital, and for the other pulses their time derivatives are digitized. The reason for not making the differentiation in the computer by using a digitized laser pulse is that the amplitude resolution in the converted pulse is too low to resolve the small absorption caused by the SO_2 in the measurement path.

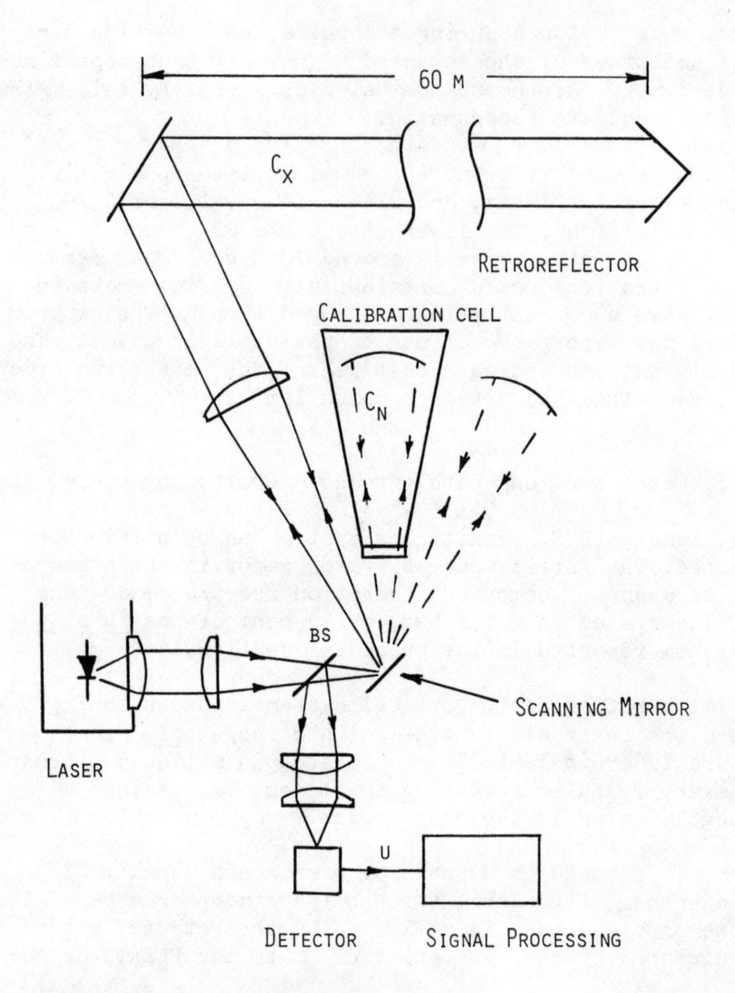

Fig. 14. Optical arrangement. The laser beam is alternately directed through an atmospheric path, a calibration cell, and a reference path by the scanning mirror.

The amplifier is adjusted to use the full dynamic range of the 8-bit A/D converter for each type of pulse. A converted pulse is stored in the buffer memory and transferred into the computer memory in the time "window" between the pulses.

When the desired number of laser pulses and their time derivatives have been recorded, a measurement value of C_X is computed and displayed. Thereafter, a new measurement cycle is automatically started.

This system has been used over a 60 m atmospheric path. The mini-

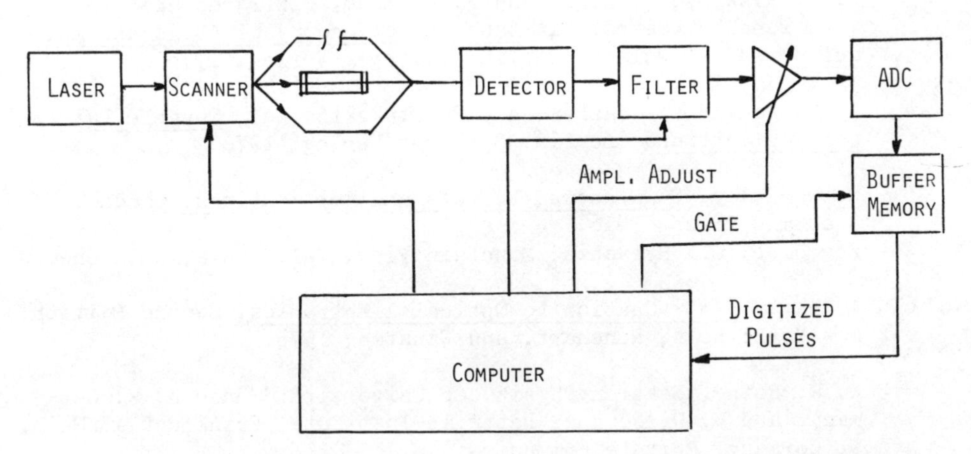

Fig. 15. Block diagram of the monitoring system. The measurements are controlled by the minicomputer and the digitized laser pulses are transferred via a buffer memory to the computer, which automatically computes the SO_2 concentration.

mum detectable SO_2 concentration is at present around 100 ppb.

8. CONCLUSIONS

The fundamental principles of infrared laser instrumentation for environmental monitoring have been reviewed. The technology has advanced so rapidly in the last few years that commercial instrumentation should soon be available. However, there are especially three problem areas left. The first one is improved tunability (wider spectral range) of infrared lasers. Another one is better techniques for minimization of interference effects from background gases, and the final one is compensation technique for overcoming the limitations imposed by atmospheric turbulence. Because of the advantages offered by optical techniques, research in those particular areas seems worthwhile.

REFERENCES

1. E. D. Hinkley, R. T. Ku, and P. L. Kelly, "Techniques for Detection of Molecular Pollutants by Absorption of Laser Radiation," in Laser Monitoring of the Atmosphere, ed. E. D. Hinkley. Berlin: Springer-Verlag, 1976.

2. S. T. Eng and E. Max, "Tunable Laser Spectroscopy in Mineral Prospecting," in Tunable Lasers and Applications, ed. A. Mooradian, T. Jaeger, and P. Stokseth. Berlin: Springer-Verlag, 1976.

3. E. D. Hinkley, K. Nill, and F. A. Blum, "Infrared Spectroscopy with Tunable Lasers," in Laser Spectroscopy of Atoms and Molecules, ed. H. Walter. Berlin: Springer-Verlag, 1976.

4. A. Mooradian, T. Jaeger, and P. Stokseth, ed., Tunable Lasers and Applications. Berlin: Springer-Verlag, 1976.

5. W. W. Duley, CO_2 Lasers, Effects and Applications. London: Academic Press, 1976.
See also: GTE Sylvania, Mountain View, California, Data Sheets.

6. A. Yariv, Introduction to Optical Electronics, Second Edition. New York: Holt, Rinehart, and Winston, 1976.

7. K. W. Rothe, Sektion Physik der Universität München, W. Germany, and W. Wiesemann, Battelle-Institute, Frankfurt am Main, W. Germany. Private communication.

8. T. C. Harman, "Narrow-Gap Semiconductor Lasers," Proc. Conf. on the Physics of Semimetals and Narrow Band Gap Semiconductors, ed. D. L. Carter and R. T. Bates. Oxford: Pergamon, 1971.

9. K. J. Linden, J. F. Butler, and K. W. Nill, "Single Heterostructure Tunable Diode Lasers Formed by Compositional Interdiffusion," IEEE Int. Electron Devices Meeting - Technical Digest, pp.132-135, 1976.
See also: Laser Analytics, Inc., Mass. Data Sheets.

10. J. N. Walpole, A. R. Calawa, T. C. Harman, and S. H. Groves, "Double-Heterostructure PbSnTe Lasers Grown by Molecular-Beam Epitaxy with CW Operation up to 114 K," Appl. Phys. Letter, Vol. 28, No. 9, May 1976.

11. J. Johansson, B. Marthinsson, and S. T. Eng, "Computer Automation and Error Analysis of a CO_2-Laser Long-Path Absorption System for Air Pollution Monitoring," Conf. on Precision Electromagnetic Measurements (CTEM) - Technical Digest, Ottawa, Canada, June 1978.

12. S. E. Craig, D. R. Morgan, D. L. Roberts, L. R. Snowman, "Development of a Gas Laser System to Measure Trace Gases by Long Path Absorption Techniques," Final Report (No: EPA-650/2-74-046-A) prepared for the Environmental Protection Agency, Research Triangle Park, N.C., June 1974.

13. R. A. McClatchey, W. S. Benedict et al., "AFCRL Atmospheric Absorption Line Parameters Compilation," Technical Report (No. AFCRL-TR-73-0096) from the Air Force Cambridge Research Laboratories, Bedford, Mass., 1973.

14. B. Marthinsson, J. Johansson, E. Max, and S. T. Eng, "A Computer-Controlled CO_2 Laser Long-Path Absorption System," Conf. on Laser and Electrooptical Systems (CLEO) - Technical Digest, San Diego, February 1978.

15. E. Max, S. T. Eng, "A PbSe Diode Laser Spectrometer to be Used in Air Pollution Monitoring and Mineral Prospecting," Optical and Quantum Electronics, Vol. 9, pp. 411-418, 1977.

16. E. Max, S. T. Eng, "A Diode Laser System for Long Path SO_2 Automatic Monitoring," Submitted for publication.

ADDITIONAL REFERENCES FOR FURTHER STUDY

17. R. L. Byer, "Remote Air Pollution Measurements," Optical and Quantum Electronics, Vol. 7, pp. 147-177, 1975.

18. R. T. Menzies, "Laser Heterodyne Detection Techniques," in Laser Monitoring of the Atmosphere, ed. E. E. Hinkley, Berlin: Springer-Verlag, 1976.

19. P. L. Hanst, "Optical Measurement of Atmospheric Pollutants: Accomplishments and Problems," Optical and Quantum Electronics, Vol. 8, pp. 87-93, 1976.

20. W. W. Rothe and H. Walter, "Remote Sensing Using Tunable Lasers," in Tunable Lasers and Applications, ed. A. Mooradian, T. Jaeger, P. Stokseth. Berlin: Springer-Vorlag, 1976.

21. P. L. Kelly, R. A. McClatchey, R. K. Long, A. Snelson, "Molecular Absorption of Infrared Laser Radiation in the Natural Atmosphere," Optical and Quantum Electronics, Vol. 8, pp.117-144, 1976.

REMOTE SENSING IN THE ATMOSPHERE BY MEANS OF FIXED FREQUENCY IR LASER

K. W. Rothe

Sektion Physik der Universität München,
Garching, West Germany

ABSTRACT. The application of fixed frequency IR lasers for pollution monitoring and process control is discussed and demonstrated. Field measurements of HF emission from an aluminum plant by means of a cw chemical HF laser are reported. In addition ambient air measurements of water vapour distribution in the vicinity of a cooling tower as well as range resolved measurements of the ethylene concentration over a refinery have been performed using a pulsed TEA CO_2 laser in connection with the differential absorption technique.

INTRODUCTION

The urgent need for pollution monitoring and control has become obvious in recent years. Unfortunately there is an essential drawback inherent in all chemical methods, as only point monitoring measurements can be performed, which may possibly not reflect the current pollution situation. The various laser techniques, however, allow to determine the mean value of pollution concentration not only over a certain pathlength but also over a certain area or volume. In addition to this high power pulsed lasers and the differential absorption technique can be used to obtain a complete range resolved map of the pollution distribution over large urban or industrial areas. This technique is also of some interest to factories, as leakages can be detected in the production process which might help to save energy and may be raw materials.

T. Lund (ed.), Surveillance of Environmental Pollution and Resources by Electromagnetic Waves, 145–153.

Considering the different field measurements which have been performed in recent years, it turns out that apart from NO_2 /1,2,3/ and SO_2 /4,5/ nearly all investigations have been done in the infrared spectral region where almost any constituent shows its individual absorption band. In addition, from all available IR lasers only the semiconductor diode laser (SDL) /6,7/ and the fixed frequency IR laser (FFIR) /8,9/ are usable for field application. All other types are still lacking in reliability and ease of handling. As up to now the SDL still awaits improvements in lifetime and output energy we decided to start measurements using fixed frequency infrared lasers, which can be handled easily and which are able to operate at rather high output energy levels. Furthermore, there are no problems associated with frequency control as the laser lines are fixed.

There is a variety of laser applications for remote pollution monitoring and process control. Therefore one single laser device is not able to meet all different requirements in an optimum way, especially if economical considerations are taken into account. This paper describes two types of applications with the properly selected laser and detection technique for each:
(a) emission control in the exhaust of a factory using a compact cw laser and long path absorption to give the relevant mean value of the total emission /10/, and
(b) range resolved measurements of atmospheric constituents using a high power pulsed TEA laser and the differential absorption technique to get a detailed map of the pollution situation over a larger area /9/.

LONG PATH ABSORPTION MEASUREMENTS WITH A CW HF LASER

The control of the HF emission e.g. from an aluminum plant is a serious problem. The sensitivity of chemical methods for the detection of HF is rather poor, so that even for HF concentrations at the maximum allowed emission level sampling times up to one hour are necessary. Thus there is no hope for chemical methods to detect possible short term peaks in the HF emission.

The most sensitive method for detection of HF is to use the HF laser itself. The active medium is produced from a chemical reaction in excited vibrational states, thus having transitions to the ground state which of course coincide with the HF absorption lines in an optimum way. Details of the laser and the evaluation scheme are given in /10/.

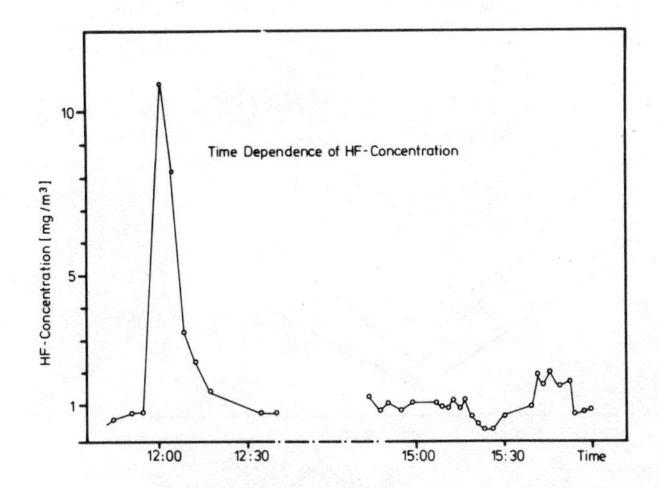

Fig. 1: Short term variation of HF concentration

For the measurements of the HF emission from an aluminum plant the laser beam was directed along the roof of the building with the furnaces, perpendicular to the exhaust stream. Typical results are shown in Fig. 1. The HF concentration is plotted as function of time. The short term peak in the emission is caused by some activities at the furnaces underneath (anode effect, crust breaking). This quick response demonstrates the usefulness of such measurements for the control of production processes.

At the end of the HF campaign simultaneous measurements have been performed for comparison of the laser method with two nearly indentical chemical methods. For this purpose an absorption cell was used in order to ensure that all methods were probing the same object. The air was circulated through the cell by fans at either side of it. To establish a sufficient high concentration level for the chemical measurements a flexible pipe was attached to the cell to suck in air at quite different locations in the vicinity of the furnaces. As the sampling time of the chemical methods was between 15 and 30 minutes, only the mean value of the individual laser measurements over the corresponding time interval could be used for a comparison. The results shown in Fig. 2 clearly indicate a significant disagreement between the different methods. The reason for this, especially for the discrepancies between the

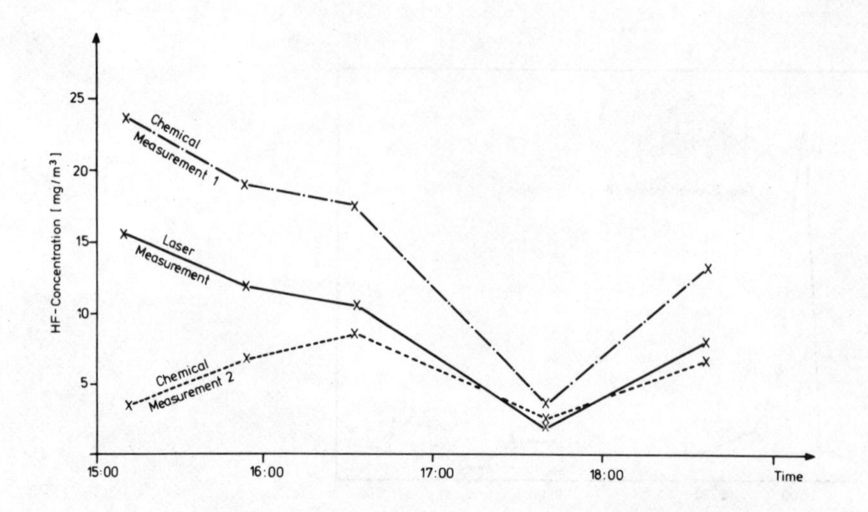

Fig. 2: Results of comparative HF measurements

chemical methods, is up to now subject to speculation and has to be clarified by further investigations and comparisons.

RANGE RESOLVED MEASUREMENTS WITH A PULSED CO_2 LASER

Apart from the measurement of spatially averaged concentration values discussed in the last chapter there is also a need for detailed mapping of different pollutants over larger areas. With such a range resolved picture of the pollutant concentration it is possible to detect and localize emission sources over large industrial areas.

Such requirements are met by LIDAR systems equipped with a high power pulsed laser where Mie-scattering is used as a distributed reflector. The technique of differential absorption has already been described in detail in many papers /1,2,11/ and its applicability has been proven e.g. in the case of NO_2 /1,2,3/, SO_2 /4,5/, and water vapour /12,9/. In this section a mobile LIDAR system is described which up to now has been used to measure the three dimensional distribution of water vapour in the vicinity of a cooling tower and the ethylene distribution over a refinery.

The LIDAR system is equipped with a pulsed TEA CO_2 laser having a beam diameter of 4 cm. By means of a beam-splitter and a focussing mirror a small fraction of the laser light is directed onto a spectrum analyser to control the emitted laser frequency. The main part of the beam is expanded 1:8 and is emitted into the atmosphere. In this way not only eye-safety is improved but also the divergence of the laser beam is reduced to about 0.4 mrad which is significantly smaller than the field of view of the receiving optics. The latter consists of a spherical mirror of 60 cm diameter and 240 cm focal length which focusses the backscattered light onto a fast IR detector. The signal of the detector is amplified, stored in a transient recorder as function of time, and then transferred into a computer for further data evaluation. The computer also performs wavelength control of the laser as well as the mechanical steering of the whole set-up which can be moved hydraulically both in horizontal and vertical direction.

Field measurements of water vapour

The first field measurement campaign with the LIDAR apparatus just described aimed to demonstrate the feasibility of the system for three dimensional remote mapping of atmospheric constituents. For this purpose measurements on the atmospheric water vapour concentration in the vicinity of the cooling tower of a gas burning power plant were performed. The pair of laser lines we selected were the R(18) and R(20) line of the 00^01-10^00 transition of the CO_2 laser, with absorption coefficients of $0.935 \cdot 10^{-4}$ $(atm \cdot cm)^{-1}$ and $8.65 \cdot 10^{-4}$ $(atm \cdot cm)^{-1}$ respectively /13/.

Some typical examples for the different types of measurements of the water vapour distribution near the cooling tower of the power plant are demonstrated in Fig. 3. On the left hand of the figure the cooling tower and the visible plume consisting of condensed water vapour is shown; the cross marks the direction of the laser beam. On the right hand the corresponding water vapour concentration is plotted as function of the distance. The first type of investigations comprises all measurements which were performed onto the plume itself. In this case a sharp increase of the concentration can be observed starting about 100 m to 200 m before the beginning of the visible plume. There was no possibility to extend the measurement into the visible plume as all light is reflected back from the very beginning of the plume. This distance is indicated by the

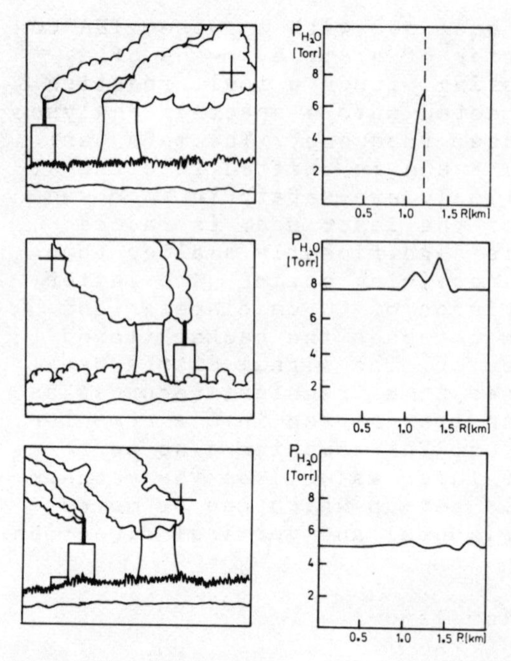

Fig. 3: Different types of water vapour measurements

dotted line in the upper right part of Fig. 3. In the second type of investigations the laser beam was directed by the visible plume on its lee side (centre part of Fig. 3). In this case a distinct increase of the water vapour concentration could be observed even several hundred meters away from the plume, indicating the strong influence of the wind. This behaviour was made plain by the third type of measurements with the laser beam directed at the wind side of the plume (lower part of Fig. 3). In this case no significant increase of the water vapour concentration was observed, even in measurements performed very close to the plume.

A three dimensional profile of the water vapour concentration was obtained by scanning the apparatus horizontally at different altitudes. Fig. 4 shows the area of the power plant, the location of the cooling tower and lines of equal water vapour concentration for one fixed altitude (approx. 200 m). The numbers give the values of the concentration in torr. Such a profile can give detailed information on the distribution of water vapour outside the visible plume, thus allowing propagation studies for various wind directions and other meteorological parameters.

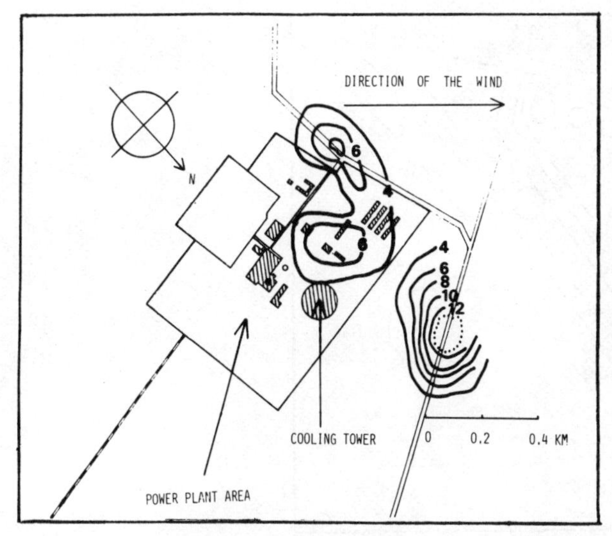

Fig. 4: Water vapour distribution near the cooling tower of a power plant

Field measurements of ethylene

The second field measurement campaign aimed to prove the feasibility of the LIDAR set-up for detection of trace constituents in the atmosphere at the ppb level. To demonstrate the capability of the differential absorption technique also for the detection of leakages in industrial areas we decided to perform measurements of the ethylene distribution over a refinery. Like most hydrocarbons ethylene can be measured near 3 µm, but this is a region of strong interference not only with the other hydrocarbons but also with water vapour and carbon dioxide. In addition ethylene represents only a very small fraction of the total amount of hydrocarbons (typically less than 1 %). To avoid interference with other species we therefore have chosen another spectral region namely with the P(14) line of the 00^01-10^00 transition of the CO_2 laser having an absorption coefficient of 36.1 $(atm \cdot cm)^{-1}$ for ethylene /14/. However, some care has to be taken choosing the appropriate non-absorbing laser line for the reference signal in order to minimize the water vapour interference. Calculation of atmospheric transmission in the spectral region of interest proved the P(16) line not to be the best suited candidate for a reference line as it is influenced by a wing of a water vapour line nearby. Therefore, when using the P(14) and P(16) lines, water vapour absorption has to be taken into account for the

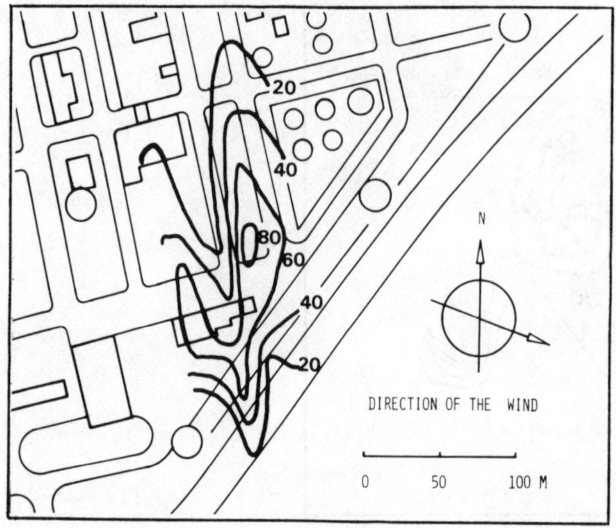

Fig. 5: Ethylene distribution over a refinery

evaluation of the ethylene concentration /15/. In order to avoid this complication we used the P(14)-P(18) laser lines with negligible interference. First topographical measurements confirmed this behaviour and gave a mean ethylene concentration of about 25 ppb over a pathlength of 700 m.

Range resolved measurements of the ethylene distribution were performed using Mie-scattering and the differential absorption technique. The laser was positioned in five different directions at an altitude of about 50 m near the distillation towers. Twelve thousand laser pulses were emitted for each direction. The evaluated ethylene concentration is shown in Fig. 5. Lines of equal ethylene concentration are plotted over the area of the oil refinery. The numbers give the concentration values in ppb. This map shows the mean value of the ethylene concentration over a period of two hours. A distinct increase of the concentration can be seen at the lee side of the distillation towers, indicating them as the emission source for ethylene.

CONCLUSION

Summarizing the present situation it can be said that the laser proved to be a powerful tool for the remote control of pollutants in the atmosphere. The use of fixed frequency IR lasers for this purpose has been demonstrated successfully. Once reliable and operable high pressure gas lasers will become available for field application the detection limit for some species might be reduced considerably. In addition, further measurements should be performed in comparison with chemical methods in order to establish the laser technique as a very helpful instrument in remote and online process control and pollution monitoring.

ACKNOWLEDGEMENT

The financial support of the Bundesministerium für Forschung und Technologie as well as the Minister für Arbeit, Gesundheit und Soziales des Landes Nordrhein-Westfalen is gratefully acknowledged. I would also like to thank J. Wanner for the construction of the cw HF laser and A. Tönnißen and W. Baumer who performed the measurements with the cw HF laser and the pulsed TEA laser, respectively.

REFERENCES

/ 1/ K. W. Rothe, U. Brinkmann, H. Walther, Appl. Phys. 3, 115 (1974)
/ 2/ W. B. Grant, R. D. Hake, Jr., E. M. Liston, R. C. Robins, B. K. Proktor, Jr., Appl. Phys. Lett. 24, 550 (1974)
/ 3/ K. W. Rothe, U. Brinkmann, H. Walther, Appl. Phys. 4, 181 (1974)
/ 4/ J. Kuhl, H. Spitschan, Opt. Commun. 13, 6, (1975)
/ 5/ W. B. Grant, R. D. Hake, Jr., J. Appl. Phys. 46, 3019 (1975)
/ 6/ E. D. Hinkley, Opt. Quantum Electron. 8, 155 (1976)
/ 7/ J. Reid, J. Shewchun, B. K. Garside, E. A. Ballik, Appl. Opt. 17, 300 (1978)
/ 8/ E. R. Murray, Opt. Eng. 16, 284 (1977)
/ 9/ H. Walther, invited paper presented at ECOSA 1, Brighton, U.K., April 1978
/10/ A. Tönnißen, J. Wanner, K. W. Rothe, H. Walther, to be published
/11/ R. L. Byer, Opt. Quantum Electron. 7, 147 (1975)
/12/ E. R. Murray, R. D. Hake, Jr., J. E. van der Laan, J. G. Hawley, Appl. Phys. Lett. 28, 542 (1976)
/13/ M. S. Shumate, R. T. Menzies, J. S. Margolis, L. G. Rosengreen, Appl. Opt. 15, 2480 (1976)
/14/ J. Boscher, BMFT Contract 01 TL 026-AK/RT/WRT 2074
/15/ E.R. Murray, J. E. van der Laan, to be published in Appl. Optics

MILLIMETRE REMOTE SENSING OF THE STRATOSPHERE AND MESOSPHERE

D. L. Croom

SRC., Appleton Laboratory, Ditton Park,
Slough SL3 9Jx, England

ABSTRACT. The purpose of this paper is to emphasise those aspects of atmospheric remote sensing which are of special importance in the millimetre wavelength region of the electromagnetic spectrum; in particular problems of pressure and Doppler broadening, local thermodynamic equilibrium, atmospheric absorption, Zeeman splitting and the optimum observation angle. A simplified discussion of the principles of atmospheric weighting functions is given.

1. INTRODUCTION

Other papers in this volume have discussed the general principles of atmospheric remote sensing and of the invertion of the observed data (Peckham 1978), together with the fundamentals of molecular rotation emission lines (Svanberg 1978), and hence this will not be repeated in detail here. The purpose of this paper is to describe those features which are of particular importance to the millimetre region.

The millimetre emission lines from atmospheric molecules are all due to transitions between rotational levels, or to hyperfine splitting of rotational levels, in the ground vibration state.

The use of observations of these molecular lines in the atmosphere to determine height profiles of constituents or temperature depends on the fact that the lines are pressure-broadened, and that heterodyne millimetre techniques enable resolutions of less than the line-width to be obtained. The slope of the line contains information on the height distribution of the parent constituent (except for O_2, in which case the information relates to the

T. Lund (ed.), Surveillance of Environmental Pollution and Resources by Electromagnetic Waves, 155-169.

vertical temperature structure of the atmosphere). The possibility of using millimetre waves to probe stratospheric temperature and water vapour was first published by Meeks (1961) and Barrett and Chung (1962) respectively, and it depends on the ability of millimetre wavelength receivers to measure the shape of the molecular line.

A valuable survey of the potential of millimetre waves for atmospheric studies was published by Staelin (1969). Although it is now nearly ten years old it still forms a useful introduction to the subject.

2. THE MILLIMETRE WAVELENGTH REGION OF THE ELECTROMAGNETIC SPECTRUM

Because the millimetre region spans the radio and infra-red sections of the electromagnetic spectrum, wavelengths and frequency units commonly used in either section are applied in practice. Approximately the regions covers:

Wavelength 15 - 1 mm (15000 - 1000μ)
Frequency 20 - 300 GHz
Wave Number 0.67 - 10 cm-1

The wavelengths are sufficiently long for the Rayleigh-Jeans approximation to the Planck radiation function to apply (except for high precision observations near 1 mm at effective temperatures of a few degrees K, a condition not normally encountered in the atmosphere).

This leads to the simple black body relationship:

$$P = k\ T\ B \tag{2.1}$$

where P is the emitted black body power, k is Boltzman's constant, T is the black body temperature and B is the spectral bandwidth.

For a partial absorber (gray-body) this can be written as:

$$P = k(\alpha T)\ B \tag{2.2}$$

where α is the absorption coefficient of the medium (= 1 for a black body). More commonly this is in the form:

$$P = k\ T_N\ B \tag{2.3}$$

where $T_N = \alpha T$ is called the effective temperature (or effective black-body temperature, or noise temperature) of the medium.

Because of this linear relationship between power and temperature, and because a convenient method of calibrating millimetre radiometers is to measure the emission from a matched load (i.e. black body) immersed in a boiling liquid (e.g. liquid N_2 at 77K) it is common to refer to signal power as so many degrees K. Only for a perfect black body does T_N equal the kinetic temperature T.

ATMOSPHERIC ABSORPTION AT MILLIMETRE WAVELENGTHS

Figure 1 shows the clear sky zenith absorption in the range 10 - 300 GHz. This absorption is due to rotational transitions of H_2O and to hyperfine splitting of the rotational levels of O_2. The region around 60 GHz consists of about 25 separate O_2 lines which at sea-level are merged into a band due to pressure-broadening. At a height of about 30 km these lines are completely resolved.

Many other atmospheric molecules have lines in this part of the spectrum, but they are not strong enough to appear in Figure 1. They are however potentially capable of being used to study the distribution of some of the molecules in the atmosphere.

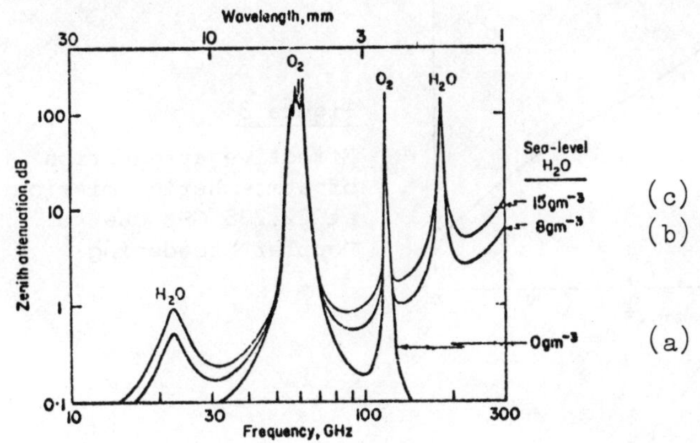

Figure 1 Zenith absorption (in dB) of the atmosphere from 10-300 GHz (a) No H_2O (b) 8 gm^{-3} surface (c) 15 gm^{-3} surface H_2O

One advantage of millimetre waves over infra-red is that many clouds which heavily absorb the infra-red are transparent or partially transparent to millimetre waves.

Millimetre waves are strongly absorbed and scattered by rain, the size of the raindrops being close to the wavelength. Millimetre waves can however be transmitted through light to medium rain over distances of several kilometres.

4. BASIC OBSERVATIONAL TECHNIQUES

According to circumstances and the information required the following observational techniques can be used:

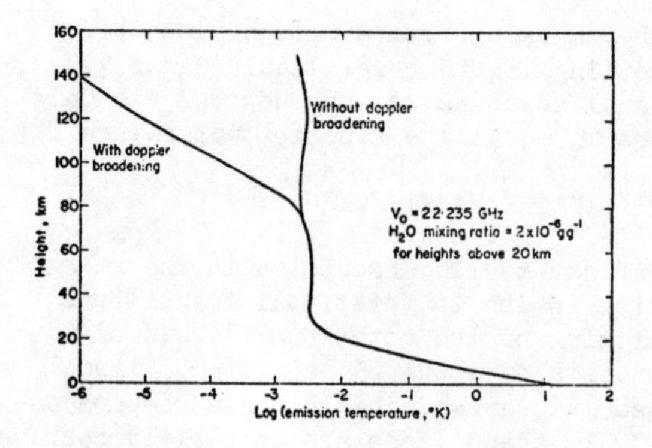

Figure 2

The effect of Doppler broadening on the atmospheric emission temperature at 22.235 GHz

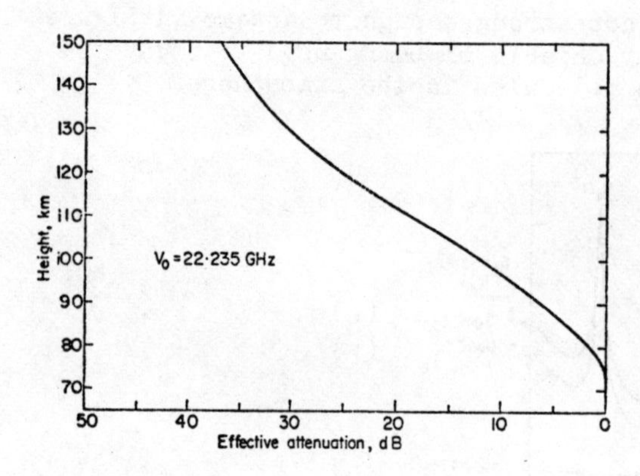

Figure 3

Effective attenuation of atmospheric emission at 22.235 GHz due to Doppler broadening

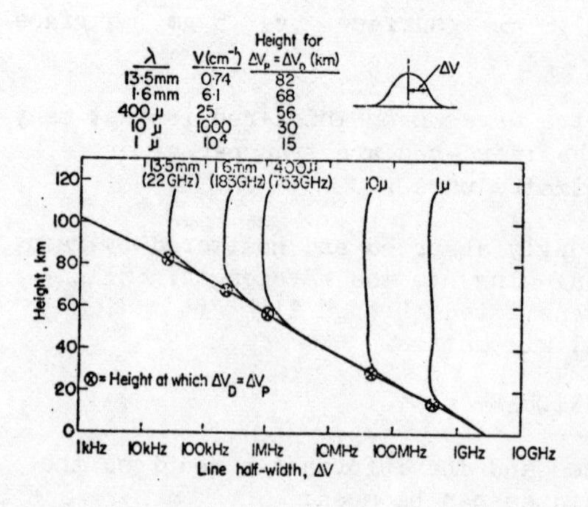

Figure 4

Doppler and pressure line widths as a function of height and frequency

(i) Atmospheric emission from the ground
(ii) Solar absorption from the ground
(iii) Nadir atmospheric emission from space
(iv) Limb emission sounding from space
(v) Limb solar absorption from space (only for spacecraft sunset/sunrise).

5. LOCAL THERMODYNAMIC EQUILIBRIUM (LTE)

Remote sensing techniques based on molecular emission involve the assumption that the molecule under investigation is in local thermodynamic equilibrium. This means that:

(a) The energy level distribution is determined only by thermal collisions and,
(b) spontaneous radiative transitions are negligible. That is:

$$\underset{\text{Time between collisions}}{\tau_C} \ll \underset{\text{Radiative Lifetime}}{\tau_R} \qquad (5.1)$$

τ_C in the atmosphere varies from about 1.5×10^{-10} sec at the surface to 1×10^{-9} sec at 150 km, whereas τ_R is about 10^7 sec and thus inequality (5.1) holds very well.

However the radiative lifetime decreases with ν^3 and for some species LTE does not necessarily hold in the infra-red, especially at the higher altitudes.

6. PRESSURE BROADENING LINE SHAPES

6.1 Lorentz

In the optical/infra-red regions the shape F of the pressure-broadened line was originally determined by Lorentz. Its main part has the form, in the general vicinity of the line:

$$F_p = \frac{1}{\pi}\left[\frac{\Delta\nu_p}{(\nu - \nu_o)^2 + \Delta\nu_p^2} - \frac{\Delta\nu_p}{(\nu + \nu_o)^2 + \Delta\nu_p^2}\right] \qquad (6.1)$$

where ν is the frequency, ν_o is the line centre-frequency and $\Delta\nu_p$ is the line half-power half width.

Lorentz made the assumption that the position x and velocity $\dot{x}$ of the molecules after collision have average values of zero.

6.2 Van Vleck-Weisskopf

The Lorentz shape was found not to fit the wings of lines in the microwave region and Van Vleck and Weisskopf (1945) modified the shape to:

$$F_p = \frac{\nu}{\pi \nu_o} \left[\frac{\Delta\nu_p}{(\nu - \nu_o)^2 + \Delta\nu_p^2} + \frac{\Delta\nu_p}{(\nu + \nu_o)^2 + \Delta\nu_p^2} \right] \tag{6.2}$$

(assuming x and $\dot{x}$ have a Boltzmann distribution).
These modifications are not important in the optical region for which Lorentz derived his line shape.

6.3 Gross (or Zhevakin-Naumov)

Further study revealed some residual discrepancies and Gross (1955) and Zhevakin and Naumov (1963) revised the shape to:

$$F_p = \frac{\nu_o \nu}{\pi} \frac{4 \Delta\nu_p}{(\nu_o^2 - \nu^2)^2 + 4 \nu^2 \Delta\nu_p^2} \tag{6.3}$$

(assuming x unchanged and $\dot{x}$ having a Maxwellian distribution)

At present this is the best available theoretical line shape, but it is still not an exact fit at many line widths from the centre frequency.

Various empirical variations on the above formulae have been published, but they have no theoretical justification for the modifications.

7. DOPPLER (TEMPERATURE) BROADENING

In the upper atmosphere Doppler broadening resulting from the thermal motions of the molecules becomes important and eventually dominates over pressure broadening. When this occurs the line shape no longer contains identifiable information on the height profile.

The shape of the Doppler-broadened line is given by:

$$F_D = \left(\frac{\ell n\ 2}{\pi} \right)^{\frac{1}{2}} \frac{1}{\Delta\nu_D} \exp \left[- \ell n\ 2 \left\{ \frac{(\nu - \nu_o)}{\Delta\nu_D} \right\}^2 \right] \tag{7.1}$$

where $\Delta\nu_D$, the half-power Doppler half-width of the line is given by:

$$\Delta\nu_D = \frac{3.581}{10^7} \left(\frac{T}{M} \right)^{\frac{1}{2}} \nu_o \tag{7.2}$$

where M is the molecular weight, and T is the temperature.

The effect of Doppler broadening can be seen in Figure 2

which shows emission temperature as a function of height of the 22.235 GHz H_2O line for the cases where Doppler broadening is taken into account, and for where it is ignored.

The effect of Doppler broadening can be shown in the form of an effective attenuation as a function of height (Figure 3).

Figure 4 compares the Doppler and pressure line widths for a number of wavelengths from the millimetre to optical regions.

It can been seen that because of the ν_o term in (7.2) the height at which Doppler broadening dominates pressure broadening decreases as one moves from the microwave to infra-red regions. The table in Figure 4 shows the heights at which $\Delta\nu_D = \Delta\nu_p$. In the Infra-red this can occur as low as 30 km, whereas in the micro-wave region it can be over 80 km.

8. ZEEMAN SPLITTING

Unlike most atmospheric molecules O_2 does not have a permanent electric dipole. However the O_2 molecule is in the $^3\Sigma$ electronic ground state, and has a permanent magnetic dipole produced by the unpaired spins of two electrons (i.e. is paramagnetic). This gives rise to a series of over forty transitions centred around 60 GHz, together with an isolated line at 118.75 GHz. These are not transitions between rotational states K, but hyperfine splitting of these K levels by interaction with spin, each K level being split into three J levels. (The first rotational transition line of O_2 is at 368 GHz).

As an added complication each J level is split into further M levels in the presence of a magnetic field (Zeeman splitting). This is illustrated in Figure 5 (Henry 1950, Croom 1971). The resulting components are polarised and their number depends on the rotational states involved and on orientation with respect to the earth's magnetic field. Figure 5 illustrates the simplest cases. More complex examples have been given by Lenoir (1968).

The overall splitting is given by:

$$\nu = \nu_o \pm (1.4 \text{ H MHz}) \tag{8.1}$$

Since the earth's magnetic field is about 0.5 gauss, for the atmosphere this becomes:

$$\nu = \nu_o \pm 0.7 \text{ MHz} \tag{8.2}$$

which means that Zeeman splitting becomes noticeable above about 40-50 km, and will be dependent on the exact intensity and orientation of the earth's magnetic field.

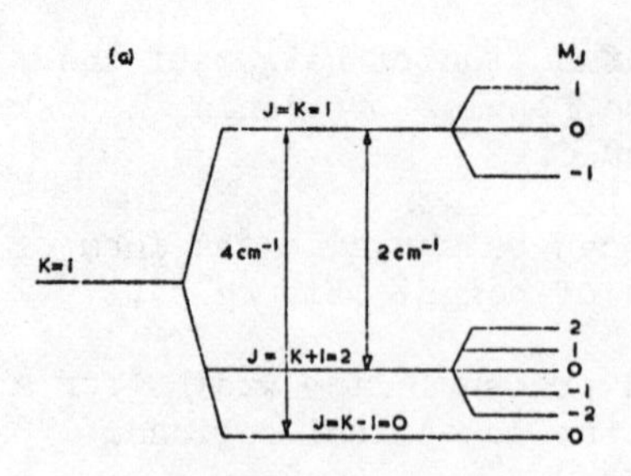

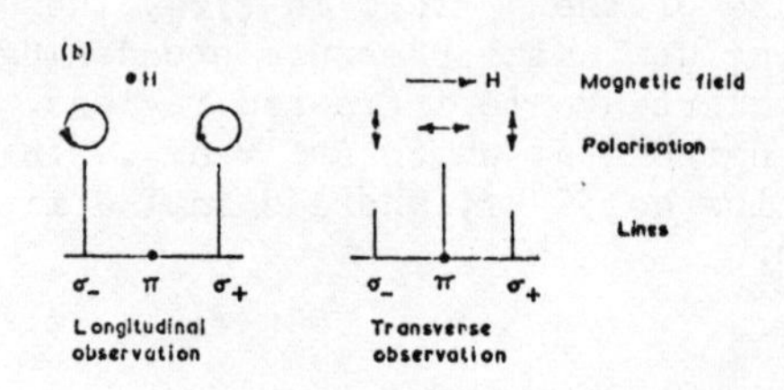

Figure 5

Zeeman splitting of O_2 energy levels (after Henry 1950)

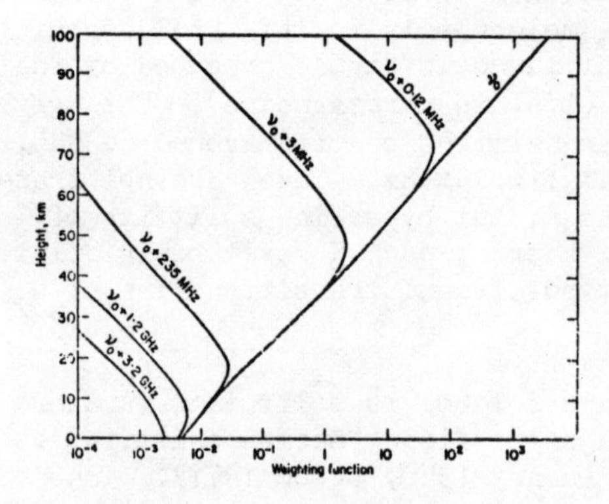

Figure 6

Un-normalised weighting functions for H_2O (22.235 GHz line)

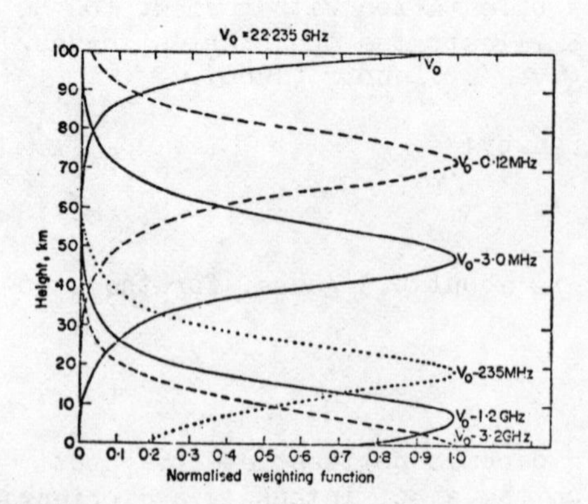

Figure 7

Normalised weighting functions for H_2O (22.235 GHz line)

9. WEIGHTING FUNCTIONS

The observed antenna temperature T_A when the line is observed in absorption against the sun is:

$$T_A = T_S \, e^{-\int_0^h \alpha \sec\theta \, dh} + \int_0^H \alpha . T . \sec\theta . e^{-\int_0^h \alpha \sec\theta \, dh} \, dh \qquad (9.1)$$

where T_S is the solar temperature (6000-20,000°K), H is the effective height of the atmosphere and θ is the look-angle. For emission observations the first term becomes almost zero (a very small residual term remains due to the 3 K black body temperature of the universe, i.e. $T_S = 3$). To a first order approximation it can be equated to zero.

Similarly the first term in the solar absorption case dominates the second term (approximately in the ratio of the solar to atmospheric temperatures), and hence we can write:

$$T_A \approx T_S \, e^{-\int_0^H \alpha \sec\theta \, dh} \qquad (9.2a)$$

or $$T_A \approx T_S \, e^{-\tau} \qquad (9.2b)$$

where $$\tau = \int_0^H \alpha \sec\theta \, dh \qquad (9.3)$$

is total attenuation (often called "optical depth")

However the absorption coefficient α is proportional to the number density N of the molecule and hence we can relate the measured τ to the required N by:

$$\tau = \int_0^H N \left(\frac{\alpha \sec\theta}{N} \right) dh \qquad (9.4)$$

or $$\tau = \int_0^H N \, W \, dh \qquad (9.5)$$

where $$W = \frac{\alpha}{N} \sec\theta \qquad (9.6)$$

is known as the weighting function, and is basically the absorption coefficient per unit density.

The behaviour or W as a function of height clearly depends on the behaviour or α as a function of height.

10. THE ABSORPTION COEFFICIENT (α)

The full expression for the absorption coefficient (Van Vleck and Weisskopf, 1945) is complex. However for the purpose of illustrating the remote sensing technique it can be highly simplified. If, for the moment, we assume the temperature terms to be independent of height (temperature varies by a few tens of percent in the 0-100 km height range, whereas pressure varies by several orders of magnitude) then we can write (very approximately):

$$\alpha = \text{const. } N \left[\frac{\Delta\nu}{(\nu - \nu_o)^2 + \Delta\nu^2} \right] \tag{10.1}$$

where $\Delta\nu_p = \text{const. } P$ (10.2)

At the line centre we have $\nu = \nu_o$, and

$$\alpha = \text{const. } \frac{N}{P} = \text{const. } M \tag{10.3}$$

where M is the mixing ratio.

Whereas at frequencies away from the centre such that $(\nu-\nu_o) >> \Delta\nu$, we have

$$\alpha = \text{const. } N\, P \tag{10.4}$$

Incorporating these expressions in (9.4), and considering observations at a fixed zenith angle θ, we have (again very approximately):

(a) $\underline{\nu = \nu_o}$

$$W = \text{const. } P^{-1}$$

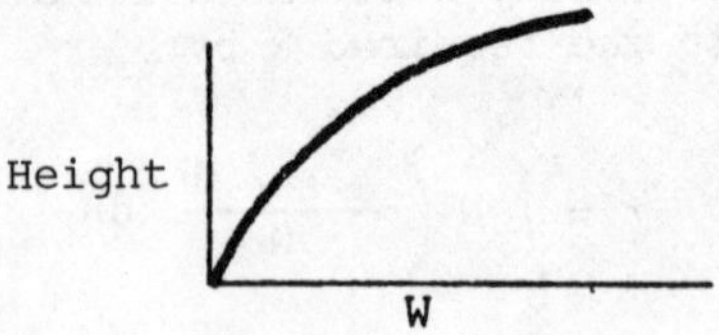

Height

W

(b) $\underline{(\nu - \nu_o) >> \Delta\nu}$

$$W = \text{const. } P$$

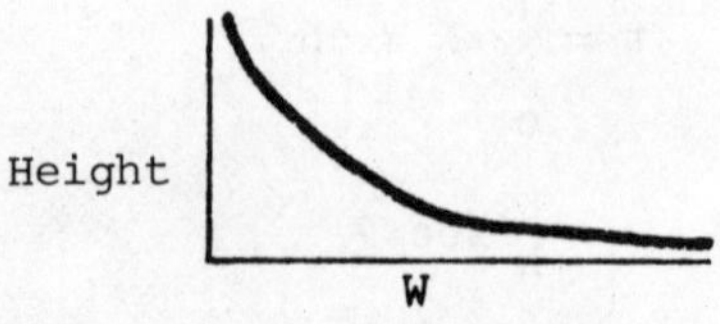

Height

W

(c) The general case of ν intermediate between (a) and (b)

$$W = \text{const.} \frac{S\,P}{(\nu - \nu_o)^2 + S^2 P^2}$$

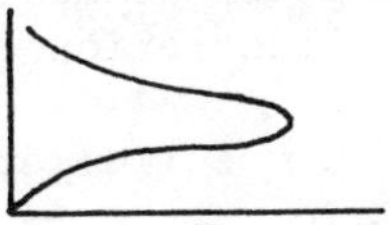

Height

W

where S is also approximately a constant.

(c) gives the shape of the generalised weighting function.

Figure 6 shows the un-normalised weighting functions for the 22.235 GHz H_2O line and Figure 7 the same weighting functions with their peaks normalised to unity (a common form in which weighting functions as shown). In Figure 6 W is shown for frequencies above ν_o and in Figure 7 for frequencies below ν_o. However the difference between them is very small.

Figures 8 and 9, which are taken from Croom (1965), illustrate the weighting function effect in another way.

Figure 8 shows the calculated spectra at the centre of the 22.235 GHz H_2O lines for illustrative layers of stratospheric water vapour each of the same density and thickness but centred at different heights.

For Figure 9 the integrated amount of water vapour has been kept the same as for Figure 8, but the H_2O has been distributed over layers of different thickness centred at one height.

In each case the effect of height distributions on the frequency spectra can be seen.

11. THE SPECIAL CASE OF MOLECULAR OXYGEN

For oxygen we have a special situation in that:

$$N = \text{const. } P \quad (11.1)$$

Hence at $\nu = \nu_o$ we have:

$$\alpha = \text{const.} \quad (11.2)$$

and at $(\nu - \nu_o) >> \Delta\nu$ we have

$$\alpha = \text{const. } P^2 \quad (11.3)$$

The weighting functions still have the same form as in section 10, but since the general behaviour of P with height is known, the effect of P can be largely eliminated and we are left with second order effects due to the dependence of temperature with height (the earlier assumption we made to ignore temperature effects in order to simplify the illustration of the technique now has to be discarded), i.e.

$$\alpha = f \text{ (Temperature profile)} \tag{11.4}$$

Hence in the case of O_2 the measurements of spectral shape can be used to infer the temperature profile (Meeks and Lilley, 1963).

12. DATA INVERTION

The techniques involved in inverting Equations (9.1) or (9.2) to obtain the required height profile are basically those described in this volume by Peckham (1978) and hence are not repeated here. The actual technique to be applied depends on the specific problem to be solved and the amount of a priori information. A valuable detailed survey of the many techniques has been given by Rodgers (1976).

13. OPTIMUM OBSERVATION ANGLE

The signal emitted at zenith angle θ from a layer at height h is α Tsec θ, and this is attenuated in passing through the intervening atmosphere by an amount $e^{-\int_0^h \alpha \sec\theta \, dh}$. The signal at the ground from that one layer is therefore

$$T_A = \alpha T \sec\theta \, e^{-\int_0^h \alpha \sec\theta \, dh} \tag{13.1}$$

For small zenith angles αT is the dominant term and therefore the signal from the layer will increase as θ increases. However at large zenith angles the exponential term becomes dominant, and hence between these extremes there will be an optimum angle of observation for maximum signal given by:

$$\theta_{opt} = \text{Cos}^{-1} (\ell n \, \alpha_o) \tag{13.2}$$

where α_o is the total zenith fractional absorption, or

$$\theta_{opt} = \text{Cos}^{-1} (0.2303 \, \alpha_o) \tag{13.3}$$

where α_o is in db.

Figure 10 shows how the signal from stratospheric water vapour at 22.235 GHz varies with zenith angle, and Figure 11 shows how the optimum zenith angle depends on the total tropospheric absorption. Note that the half power width in Figure 10 is about 25°, and that the exact look-angle is therefore not critical.

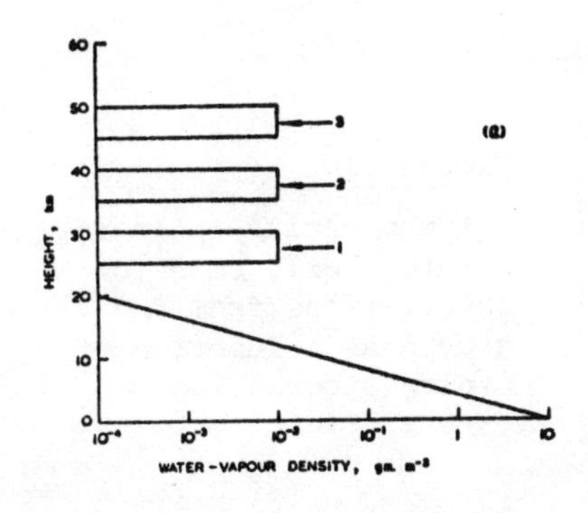

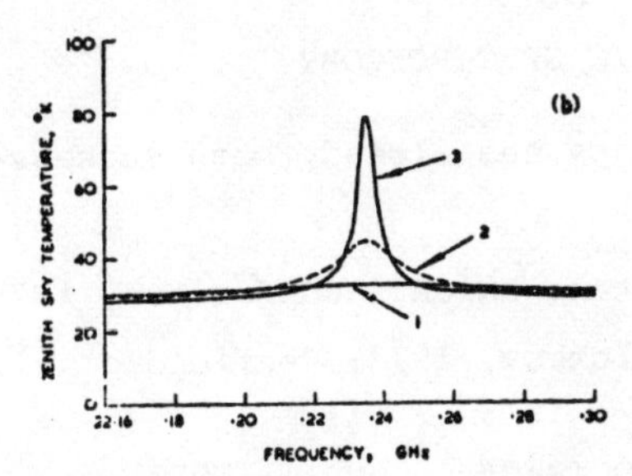

Figure 8 Effect of layer height on atmospheric emission at the centre of the 22.235 GHz H_2O line

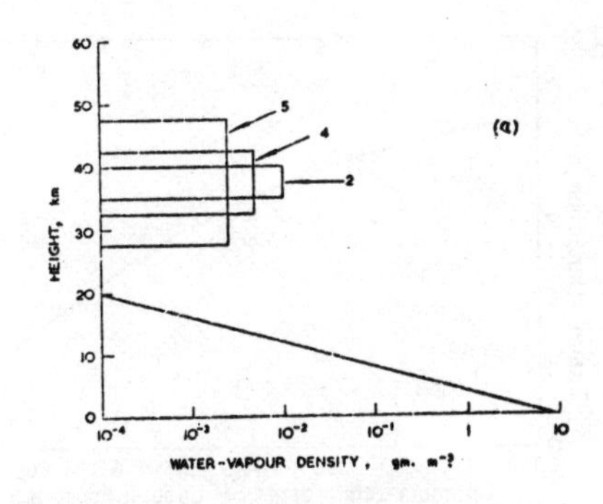

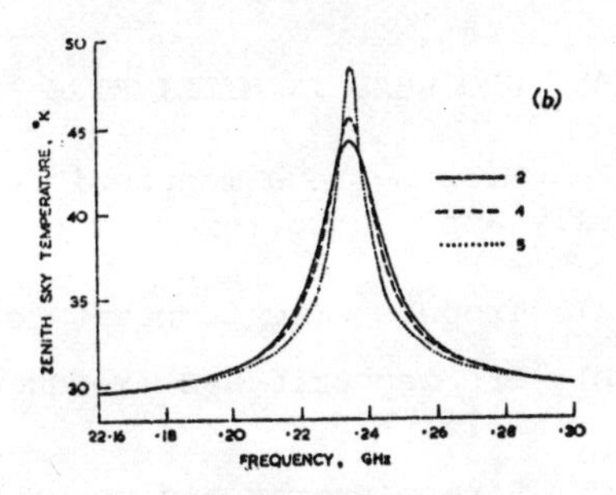

Figure 9 Effect of layer thickness (for same total H_2O) on atmospheric emission at the centre of the 22.235 GHz H_2O line

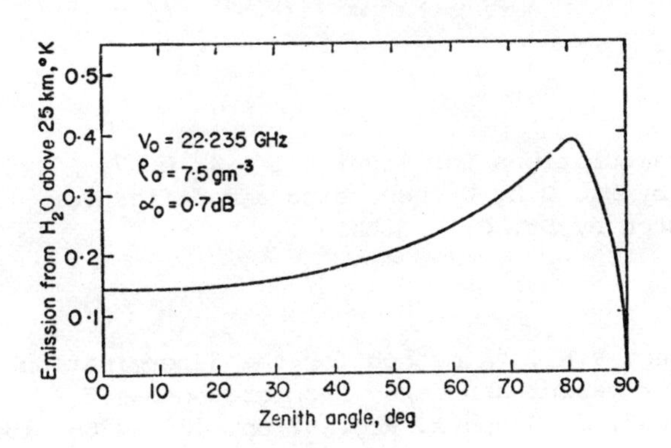

Figure 10 Optimum observation angle for stratospheric H_2O emission observed from the ground (22.235 GHz)

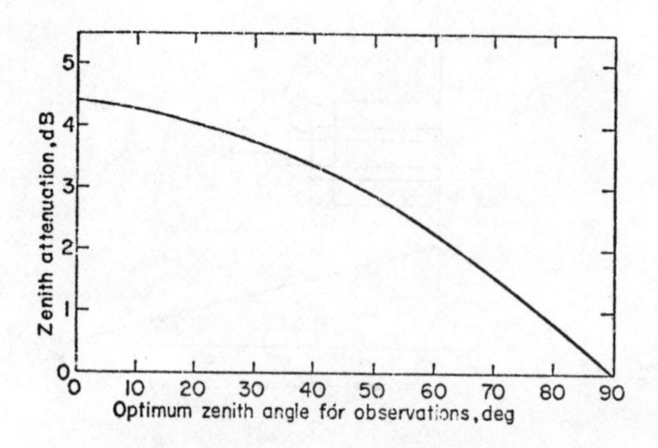

Figure 11

Optimum zenith angle for stratospheric emission observations from the ground as a function of zenith attenuation

14. PROGRESS IN MILLIMETRE ATMOSPHERIC SPECTROSCOPY

Millimetre atmospheric spectroscopy has already been successfully applied to remote sensing of:

(a) Tropospheric temperature (Westwater, Snider and Carlson, 1975)

(b) Tropospheric H_2O (Yershow and Pelchkov, 1977; Penfield et al. 1976)

(c) Stratospheric and mesospheric H_2O (Radford et al. 1977)

(d) Stratospheric and mesospheric O_3 (Shimabukuro, Smith and Wilson, 1977)

(e) Stratospheric temperature (Waters, 1973)

The references listed are representative of many workers in this field. Other molecules that are potentially capable of being studied include:

CO, $C\ell O$, NO, N_2O.

ACKNOWLEDGEMENTS. The calculations for Figures 1 - 4, 6, 7, 10, and 11 were carried out by Mr. C.L. Wrench using a modified version of a programme written by Dr. C.J. Gibbins.

REFERENCES

1. Barret, A.H. and Chung, V.K., "A method for the determination of high altitude water vapour abundance from ground-based microwave observations", J. Geophys. Res. 67 pp. 4259-4266, 1962.
2. Croom, D.L., "Stratospheric thermal emission and absorption near the 22.235 GHz (1.35 cm) rotational line of water vapour", J. Atmos. Terr. Phys. 27, pp. 217-233, 1965.
3. Croom, D.L., "The 2.53 mm molecular rotation line of atmospheric O_2", Planet. Spa. Sci. 19, pp. 777-789, 1971.
4. Gross, E.P., "Shape of collision-broadened spectral lines", Phys. Res. 97, pp. 395-403, 1955.

5. Henry, A.F., "The Zeeman effect in oxygen", Phys. Rev. 80, pp. 396-401, 1950.
6. Lenoir, W.B., "Microwave spectrum of molecular oxygen in the mesosphere", J. Geophys. Res. (Space Physics), 73, pp. 361-376, 1968.
7. Meeks, M.L., "Atmospheric emission and opacity at millimetre wavelengths due to oxygen", J. Geophys. Res. 66, pp. 3749-3757, 1961.
8. Meeks, M.L. and Lilley, A.E., "The microwave spectrum of oxygen in the earth's atmosphere", 68, pp. 1683-1703, 1963.
9. Peckham, G.E., "Remote measurements of atmospheric parameters", Proc. NATO Advanced Study Inst. on Surveillance of Environmental Pollution and Resources by Electromagnetic Waves, Reidel Publ. Co., 1978.
10. Penfield, H., Litvak, M.M., Gottlieb, C.A. and Lilley, A.E., "Mesospheric ozone measured from ground-based millimetre wave observations", J. Geophys. Res. 81, pp. 6115-6120, 1976.
11. Radford, H.E., Litvak, M.M., Gottlieb, C.A., Gottlieb, E.W., Rosenthal, S.K., and Lilley, A.E., "Mesospheric water vapour measured from ground-based microwave observations", J. Geophys. Res. 82, pp. 472-478, 1977.
12. Rodgers, C.D., "Retrieval of atmospheric temperature and composition from remote measurements of thermal radiation", Rev. Geophys. Spa. Sci. 14, pp. 609-624, 1976.
13. Shimabukuro, F.I., Smith, P.L. and Wilson, W.J., "Estimation of the daytime and night-time distribution of atmospheric ozone from ground-based millimetre wavelength measurements", J. Appl. Met. 16, pp. 929-934, 1977.
14. Staelin, D.H., "Passive remote sensing at microwave wavelengths", Proc IEEE, 57, pp. 427-439, 1969.
15. Svanberg, S., "Fundamentals of atmospheric spectroscopy", Proc. NATO Advanced Study Inst. on Surveillance of Environmental Pollution and Resources by Electromagnetic Waves, Reidel Publ. Co., 1978.
16. Van Vleck, J.H. and Weisskopf, V.F., "On the shape of collision-broadened lines", Rev. Mod. Phys. 17, pp. 227-236, 1945.
17. Waters, J.W., "Ground-based measurement of millimetre-wavelength emission by upper stratospheric O_2", Nature, 242, pp. 506-508, 1973.
18. Westwater, E.R., Snider, J.B. and Carlson, A.V., "Experimental determination of temperature profiles by ground-based microwave radiometry", J. Appl. Met. 14, pp. 524-539, 1975.
19. Yershow, A.T., and Plechkov, V.M., "Error estimation for radiometric measurements of atmopsheric moisture content according to Gate-74 expeditionary data", Izvestiya, Atmospheric and Oceanic Physics, 13, pp. 71-73, 1977.
20. Zhevakin, S.A. and Naumov, A.P., "Absorption of electromagnetic waves by water vapour in the region 2 cm - 10 microns". Izvestiya Vuzov (Radiofizika) 6, p. 674, 1963.

ACTIVE HETERODYNE SPECTROMETRY WITH AN AIRBORNE CO_2 LASER SYSTEM*

W. Wiesemann, W.Englisch, J. Boscher, G. Schäfer

Battelle-Institut e.V., Frankfurt am Main, West Germany

ABSTRACT. This paper describes a remote sensing system which is based on active differential absorption measurements in the IR spectral region. As a first step towards a spaceborne system an airborne model equipped with two tunable cw CO_2 lasers has been developed and succesfully tried at flight altitudes from 300 m to 1200 m. By extrapolating the measured data one can conclude that the measuring range may well be extended to Spacelab altitudes. The system can be used not only for analysis of atmospheric trace gases but also for petrographic mapping of the uncovered ground by differential reflection spectroscopy in the reststrahlen bands. Fundamental spectroscopic data have been provided by measuring the pressure-dependent absorption cross-sections of relevant molecular gases for a set of discrete laser wavelengths.

1. MEASURING PRINCIPLE

The long-path differential absorption (DAS) technique utilizes the absorption of laser light as the measurable parameter to infer molecular concentration [1,2]. Transmitting and receiving parts of the measuring apparatus are co-located aboard a high-flying platform (e.g. air-

*This work has been supported by the German Ministry for Research and Technology represented by DFVLR-BPT

T. Lund (ed.), Surveillance of Environmental Pollution and Resources by Electromagnetic Waves, 171-182.

craft or spacecraft). Two laser beams with slightly different wavelengths λ_1, λ_2 are colinearly directed to the earth's surface where they are diffusely reflected. Part of the backscattered laser light is collected by the receiving telescope and detected and analyzed by a sensitive optical heterodyne receiver.

The received laser power P depends on the spectral reflectivity $\rho(\lambda_i)$ of the earths' surface and the atmospheric extinction $T(\lambda_i)$ along the beam path:

$$P = K\, P_o(\lambda_i)\, \rho(\lambda_i) T^2(\lambda_i) \tag{1}$$

where K is a wavelength-independent system constant, $P_o(\lambda_i)$ is the transmitted laser power. It is useful to split up $T(\lambda)$ into two terms:

$$T(\lambda) = \tilde{T}\, T_a(\lambda). \tag{2}$$

$\tilde{T}$ changes only slowly with wavelength and describes the "normal" transmission (due to background absorption and aerosol scattering); $T_a(\lambda)$ is due to selective absorption by molecules and is a rapidly varying function of wavelength.
The wavelengths λ_1, λ_2 of the two laser beams should be chosen in such a way that λ_1 is strongly absorbed by the molecule under investigation while λ_2 suffers, ideally, no selective absorption:

$$T_a(\lambda_1) = \exp\left\{-\int_o^R \sigma_a(\lambda_1, R')\, N_a(R')\,dR'\right\} \tag{3}$$

$$T_a(\lambda_2) = 1 \text{ (ideally)}$$

where $\sigma_a(\lambda_1, R')$ is the wavelength-dependent and pressure-dependent absorption cross-section of the molecule, $N_a(R)$ is the number density of absorbing molecules, and R is the beam path length.

Since λ_1 and λ_2 are spectrally separated by a small wavelength increment $|\lambda_1 - \lambda_2| \ll \lambda$, the reflectivity and the "normal" atmospheric transmission are essentially identical for each; that means $\rho(\lambda_1) \approx \rho(\lambda_2)$ and $\tilde{T}(\lambda_1) \approx \tilde{T}(\lambda_2)$. To the extent that this is valid, the mean number density of absorbing molecules can be calculated from the normalized ratio of the two received signals if the absorption cross-sections and the path length are known. Combining Eqs. 1 - 3 results in

$$\ln T_a^2 = \ln \left\{ \frac{P(\lambda_1)}{P(\lambda_2)} \cdot \frac{P_o(\lambda_2)}{P_o(\lambda_1)} \right\} = -2 \bar{\sigma}_a \bar{N}_a R \quad (4)$$

where $\bar{\sigma}_a$ and $\bar{N}_a$ are mean values.

The mean relative concentration is readily obtained by normalizing N_a to the overall number density N_o of the atmosphere:

$$\xi = N_a/N_o = -\ln T_a^2/(2\sigma_a N_o R). \quad (5)$$

The minimum detectable concentration ξ_{min} is mainly determined by the achievable signal-to-noise ration (S/N). Assuming that a signal change of 5% is measurable (conservative assumption) the limit of sensitivity is given by

$$\xi_{min} = 10^{-21}/\bar{\sigma}_a R \text{ (R in cm, } \bar{\sigma}_a \text{ in cm}^2\text{)}. \quad (6)$$

2. SPECTROSCOPIC MEASUREMENTS

Characteristic molecular absorption spectra occur over a broad portion of the electromagnetic spectrum from the microwave region to the vacuum ultraviolet. Most promising is the middle infrared regime (2.5 µm - 25 µm), where absorption is due to vibrational-rotational bands. This is the so-called fingerprint region of the molecules. Furthermore, molecular lasers (e.g. CO and CO_2 lasers) which can be spectrally tuned to a large number of discrete vibrational-rotational lines with precisely known wavelengths are capable of emission in this region. In nearly all cases it is possible to find two laser lines which are on and off resonance, respectively, with a molecular absorption line, thus fulfilling the requirements for differential absorption.

For vertical sensing of the atmosphere the pressure dependence of σ_a must be taken into consideration. Furthermore, it is necessary to known the possible interference from other gases in order to arrive at accurate measurements of one species. As the required spectroscopic data are not available in the literature, we have measured in the laboratory for some of the most interesting gases the pressure and temperature dependence of the absorption cross-section at lines of the "normal" or "isotopic" CO_2 laser. Up to now a sample of about 3000 data has

Gas	Signal			Reference			Sensitivity (ppb·km)
		Wave-length (μm)	Absorption coefficient (atm^{-1} cm^{-1})	Line	Wave-length (μm)	Absorption coefficient (atm^{-1} cm^{-1})	
C_2H_4	P(14)	10.532	36.1	P(20)	10.591	2.15	8
NH_3	R'(30) R (8) R (8)	9.220 10.334 10.334	77 22.7 22.7	R'(16) P (10) R (10)	9.294 10.495 10.318	13.8 0.18 0.52	4 12 12
C_6H_6	P'(30)	9.639	2.01	P'(26)	9.604	0.40	160
CF_2Cl_2	P(30) P(36)	10.696 10.764	18.4(23.8)[a] 46.5(83.2)[a]	P(22) P(22)	10.611 10.611	0.60 0.60	15(11)[a] 6(3)[a]
O_3 [c]	P'(12)	9.488	12.5(28.6)[b]	P'(24)	9.586	0.88	22(9)[b]
SO_2	$\tilde{R}$'(42)	9.018	0.45	R'(28)	9.229	0.09[d]	710
SF_6	P(12) P(16)	10.513 10.551	186 819	P(10) P(12)	10.495 10.513	9.5 186	1.5 0.4
C_2H_3Cl	P(22)	10.611	8.69	P(10)	10.494	1.14	34
H_2O [e]	R(20)	10.247	$8.4 \cdot 10^{-4}$	R(18)	10.260	$0.9 \cdot 10^{-4}$	$342 \cdot 10^3$ ≙ 1 % [h]

Branches: R,P : 00°1 – 10°0 / R'P': 00°1 – 02°0 — transitions of $^{12}C^{16}O_2$ laser

$\tilde{R}$,$\tilde{P}$: 00°1 – 10°0 / $\tilde{R}'\tilde{P}'$: 00°1 – 02°0 — transitions of $^{12}C^{18}O_2$ laser

a values in brackets refer to pure gas of 0.2 mm Hg
b values in brackets refer to total pressure of 75 mm Hg
c Menzies (1976) d Comera et al. (1977) e Shumate et al. (1976)
f Craig et al. (1974) g estimates based on known spectra
h relative humidity (298K)

Table 1. List of measurable gases (not complete)

Gas	Sensitivity Equivalent Background Concentration (SEB) of Interfering Gases (ppm · km)								
	C_2H_4	NH_3	C_6H_6	CF_2Cl_2	O_3 [c]	SO_2	C_2H_3Cl	H_2O [e] ×10^3	CO_2 [f]
C_2H_4	-	0.38	>13	>1.3	>5 [g]	>50	0.33	90	50000
NH_3	2.1 0.11 0.57	- - -	>13 >13 >13	0.028 >1.3 >1.3	>3 [g] >5 [g] >5 [g]	4.8 >50 >50	0.6 1.2 0.32	90 130 20	140 8500 850
C_6H_6	0.24	6.8	-	>1.3	0.64	>50	2.3	>260	370
CF_2Cl_2	26 1.1	0.31 0.032	>13 >13	- -	>5 [g] >5 [g]	>50 >50	0.04 0.05	50	270 160
O_3 [c]	1.3	0.51	0.83	>1.3	-	>50	0.27	26	580
SO_2	>2 [g]		>5 [g]	0.014	>3 [g]	-	>4 [g]		≫
SF_6	0.18 0.36	0.47 1.4	>13 >13	>1.3 >1.3	>5 [g] >5 [g]	>50	1.2 2.0	90 14	990 1300
C_2H_3Cl	0.14	4.3	>13	>0.4	>5 [g]	>50	-	130	1600
H_2O [e]	0.56	20	>13	>1.3	>5 [g]	>50	1.1	-	1700
	0.2	6		0.15	25	2		22 ≙ 70% [h]	330
	Background Concentration (ppb)								

Branches: R,P : 00°1 - 10°0; R'P': 00°1 - 02°0 — transitions of $^{12}C^{16}O_2$ laser

$\tilde{R},\tilde{P}$: 00°1 - 10°0; $\tilde{R}'\tilde{P}'$: 00°1 - 02°0 — transitions of $^{12}C^{18}O_2$ laser

a values in brackets refer to pure gas of 0.2 mm Hg
b values in brackets refer to total pressure of 75 mm Hg
c Menzies (1976) d Comera et al. (1977) e Shumate et al. (1976)
f Craig et al. (1974) g estimates based on known spectra
h relative humidity (298K)

Table 1 (continued). Influence of interference

been collected [3]. Some preliminary results and conclusions are compiled in Table 1. It indicates the most suitable measuring and reference lines, which have been selected according to a compromise between high sensitivity and low interference from other gases. The interference is characterized by the term "sensitivity equivalent background" (SEB). The SEB determines the concentration of an interfering gas which produces just the same amount of absorption as is produced by the minimum detectable concentration of the gas to be measured. Table 1 shows that the given gases can be measured with high sensitivity and low interference. It should be noted that the list of measurable gases is by no means complete. Cross-section measurements of further interesting gases will performed additionally.

3. THE MEASURING SYSTEM

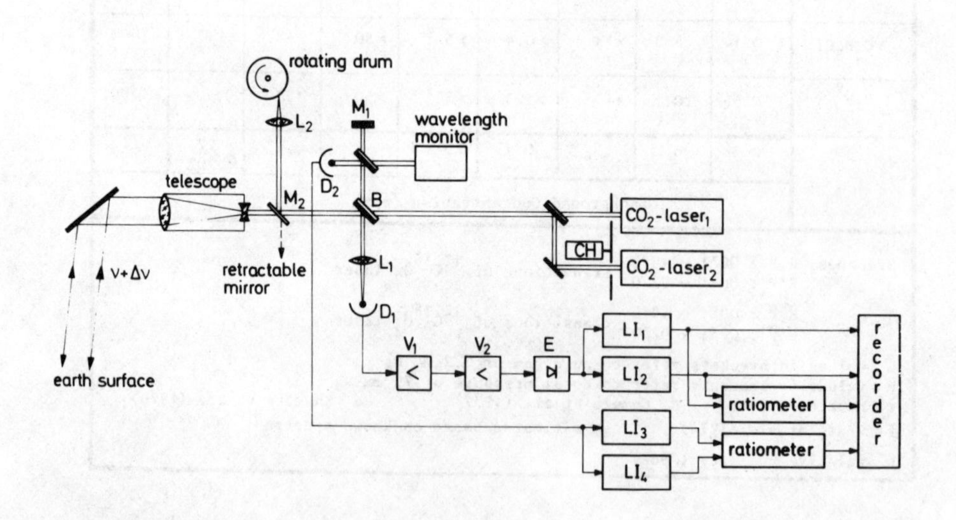

Fig. 1. Differential absorption laser spectrometer with heterodyne detection; optical beam path and signal processing

Fig. 1 shows a simplified block diagram of a cw differential absorption heterodyne system. The optical system is basically a Michelson interferometer. One arm is formed by the beam path from beam splitter B to the earth's surface, and the other by the path from beam splitter B to the plane mirror M_1. The optical signal collected by telescope is focussed on to an optical mixer (fast photodiode D_1) where it is coherently superimposed with the local oscillator radiation provided by mirror M_1. The mixer generates an electrical heterodyne signal with an intermediate frequency. This frequency is equal to the frequency shift $\Delta\nu$ which the laser light suffers during reflection at the earth's surface owing to the Doppler effect. $\Delta\nu$ depends on the direction of the laser beam and is proportional to the spacecraft's velocity component in the direction of the beam. It should be pointed out that only one telescope is used for both transmission and reception. This means not only that the matching conditions for optimum coherent detection are met but also that weight, cost and adjustment problems are significantly reduced.

The two laser beams are chopped at different frequencies, which allows further processing of the signals by means of the lock-in technique with the two spectral channels being separated electronically. The electronics consist of a low-noise impedance machting preamplifier V_1 followed by a band pass amplifier V_2, an envelope detector E, and the lock-in amplifiers LI_1, LI_2. The laser output power is monitored by the detector D_2 and the lock-ins LI_3, LI_4. Logarithmic ratiometers are used to compare and normalize the phase-sensitively detected signals. The equipment also includes the control systems for monitoring the selected laser lines and a rotating drum which simulates the moving earth's surface for in-flight adjustment and calibration purposes.

The measuring system described is actually an active sensor. Its performance, i.e. range, sensitivity, and measuring accuracy is mainly dependent on the achievable signal-to-noise ratio (S/N). Using the well-known range equation and assuming typical values for the different parameters (Table 2) it is possible to calculate the achievable S/N for various altitudes. Fig.2 shows results for some altitudes of spaceborne platforms and different atmospheric conditions (clear, hazy, light clouds). The left scale shows the S/N in the reference channel while the right scale indicates the achievable measuring accuracy $\Delta\xi/\xi$ (cf. Eq. 5), assuming that the "selective" transmission due to the absorbing gas is

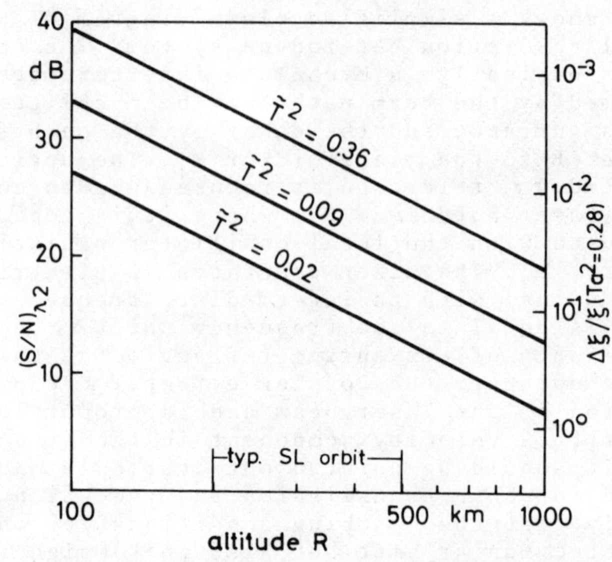

Fig. 2. Signal-to-noise ratio (S/N) and measuring accuracy ($\Delta\xi/\xi$) versus altitude for three different atmospheric conditions (clear, hazy, light clouds).

$T_a^2 = 0.28$. From these results one can conclude that DAS measurements of high accuracy can be performed from space platforms using a system which is well within the present state of the art.

Table 2. System parameters.

Parameter	Value
quantum energy	$1.875 \cdot 10^{-20}$ Ws
detector efficiency	0.6
bandwidth of intermediate frequency	$2 \cdot 10^{5}\ s^{-1}$
post integration time	$2.9 \cdot 10^{-2}$ s
emitted laser power	20 W
telescope diameter	100 cm
mean reflectivity (earth surface)	0.1
optical system transmission	0.15
atmospheric transmission T^2	(0.36 (clear) (0.09 (hazy) (0.02 (light clouds)

4. IN-FLIGHT TRIAL OF AN AIRBORNE SYSTEM

As a first step towards a spaceborne remote sensing system we developed an airborne system and tested it aboard the Dornier 28 Skyservant. The basic design features are the same as outlined in section 3. However, the system is equipped with low-power lasers ($P_o \approx 2W$) and a small telescope only 7 cm in diamter. The two lasers, the mechano-optical subsystem, the heterodyne detector, and part of the electronics are housed in a rigid frame (dimensions 1.5x0.5x0.45 m^3 which are presently being reduced to 1.2x0.5x0.5 m^3) that is shock-mounted on the seat rails of the aircraft. Fig. 3 shows the system together with the electronics rack after integration into the aircraft.

The objectives of the in-flight trial [4] were

- to prove the feasibility of the Doppler heterodyne detection technique;
- to determine the signal to noise ratio;
- to perform some preliminary measurements.

The system has been tried at flight altitudes between 200 m and 1200 m. The heterodyne signal was always clearly above the noise level. The measured S/N at R = 500 m was > 40 dB with the noise level being determined by some electromagnetic interference.

In order to test the concept of differential absorption a cloud of SF_6 was traced which had been generated by a perforated pipeline. This gas does not naturally occur in the atmosphere. Fig. 4 shows the rectified signals for the 9.46 µm channel (upper trace, no absorption) for two subsequent flyovers. The regions of interaction with the SF_6 cloud are characterized by a deep minimum occurring only in the 10.51 µm channel. The amount of absorption corresponds well to the estimated optical transmission of the SF_6 cloud.

The spectral reflectivity of rocks and minerals depends characteristically on their acidity and changes drastically in the 8 - 12 µm spectral region (reststrahlen bands). Measurements of the signatures and the characteristic changes by differential reflection spectroscopy should therefore provide valuable evidence of the mineralogical composition of the earth's surface. R.Steinmann pointed out that the DAS system (which is actually an active reflection spectrometer) should be useful not only for analysis of atmospheric trace gases but also

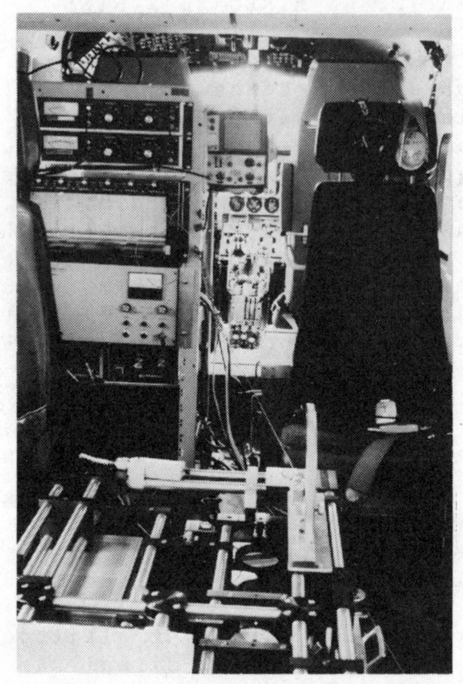

Fig. 3. Differential absorption laser spectrometer installed on DFVLR/Dornier Skyservant

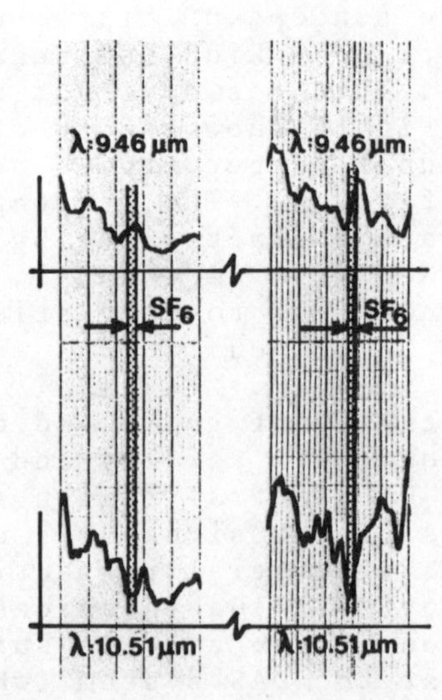

Fig. 4. Rectified heterodyne signals received in the two specified spectral channels while tracing a SF_6 cloud

for petrographic mapping of the uncovered ground by appropriate selection of the laser lines. It was therefore a second objective of the flight tests to detect as an example the strong increase in reflectivity of quartz in the 9 μm region by airborne differential reflection spectroscopy. For the experiment (Fig. 5) the laser system was flown over the open-pit kaolin mines situated near Amberg in Bavaria, because kaolin is rich in crystalline quartz ($\approx$ 70 %). The upper trace (λ = 9,32μm) of Fig. 5 exhibits a dramatic increase in signal amplitude over the lower trace (λ = 10,72μm). The reflectivity within this spectral band is estimated to be increased by a factor of about eight, that is in fair agreement with known spectral data of quartz.

Although these experiments are more qualitative than quantitative, they suggest that active differential reflection spectroscopy might become a valuable tool to measure petrographic signature of the uncovered ground. It might help to obtain evidence of unknown resources of the earth especially in regions (e.g. deserts) where normal exploration is very difficult.

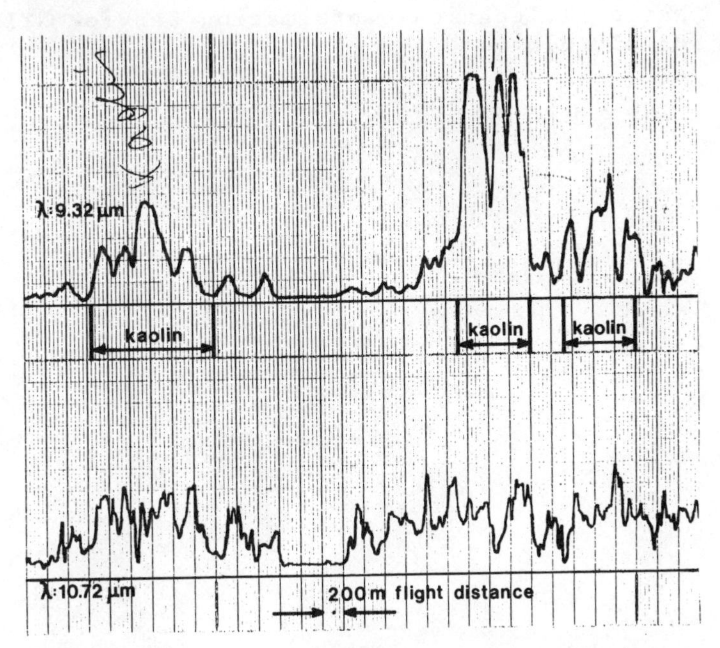

Fig. 5. Rectified heterodyne signals in the two specified spectral channels while crossing rubble dumps of kaolin mines

REFERENCES

1. H. Kildal, R.L. Byer, Proc.IEEE 59, 1644 (1971)
2. E.D. Hinkley (Ed), Laser Monitoring of the Atmosphere, Springer Verlag, Berlin-Heidelbg.-NewYork1976
3. J. Boscher, Entwicklung eines flugzeuggetragenen Meßsystems zur Erkundung der Erdoberfläche und Atmosphäre, Analytische Untersuchungen, Vol.2, Report for DFVLR-BPT, Köln-Porz, Contract No. 01 TL 026-AK/RT/WRT 2074
4. W. Wiesemann, R. Beck, W. Englisch, K. Gürs, Appl. Phys. 15, 257 (1978)
5. R.T. Menzies, Appl.Opt. 15, 2597 (1976)
6. J. Comera, A. Mayer, C. Roussel, R. Pierret, R.Faivre H. Charpentier, C. Jaussaud, Rapp. MNMR/17 March 1977, Commissariat à l'énergie atomique, Centre d'études nucléaires de Grenoble
7. M.S. Shumate, R.T. Menzies, J.S. Margolis, L.G. Rosengren, Appl. Opt. 15, 2480 (1976).
8. S.E. Craig, D.R. Morgan, D.L. Roberts. L.R. Snowman National Technical Information Service (NTIS) PB-236 678

HOLOGRAPHIC REAL TIME SEEING THROUGH MOVING SCATTERING MEDIA

H. Schmalfuß

Physikalisches Institut der Universität Erlangen-Nürnberg, Erwin-Rommel-Strasse 1, 8520 Erlangen, West Germany

ABSTRACT. We developped a method for seeing through fog or any other scattering medium. This method uses holography, which allows us to reject the scattered light that is Doppler shifted due to the interaction with the moving scatterers. By means of a TV method we are able to reconstruct the holographic image in real time.

1. Introduction

At first we will point out shortly the main differences between photography and holography. A general object can be described by a complex amplitude distribution. If you take a photograph from this incoherently or coherently illuminated object you store only the square of the amplitude and you lose the phase information. On the other hand, if you record a hologram from the same object, this object must be coherently illuminated and in the hologram plane one must add a reference wave during the exposure. Now, the wanted information of amplitude and phase is coded in a system of interference fringes, created by the object wave and the reference wave. For the storage of these fringes, it is necessary that they are not moving during the exposure time. This demand is not fulfilled if the two waves are detuned, i.e. the wavelengths of the two waves are not the same. In a sense you can say the hologram plate acts like a heterodyne receiver. Furthermore, as you know, light interacting with moving obstacles picks up a Doppler frequency shift. Now, if you make use of these two facts you will get a good tool for the seeing through moving scattering media, e.g. fog. At first Spitz /1/ and Stetson /2/, later on Lohmann and Shuman /3/, /4/showed that holography through fog works well.

T. Lund (ed.), Surveillance of Environmental Pollution and Resources by Electromagnetic Waves, 183–189.

The next step was the development of a real time set up /5/, which we now discuss in more detail.

2. Condition for the light source /6/ and the fog velocity

At first we will make some assumptions. In the following, we use only image plane holography, because this type of hologram you can reconstruct in real time by electronic means. The reference wave will be produced by amplitude division. Then we suppose that the object can be thought as a composition of plane waves and that the rules of linear superposition are essentially fulfilled. Next we assume that the temporal frequency amplitude distribution S of the light source is given by :

$$\sqrt{S(\nu)} = \mathrm{rect}\left(\frac{\nu - \nu_0}{\Delta\nu}\right)$$

and that the time average of the interaction of the phase terms $\langle \exp(2\pi i\,[\varphi(\nu,t) - \varphi(\nu',t)]) \rangle$ creates a Dirac delta function $\delta(\nu-\nu')$. Furthermore we imagine that the fog droplet is moving in the Fourier plane of the imaging system. From that droplet we get in the image plane a Doppler shifted plane wave, tilted by $\delta\alpha$ and shifted by $\Delta\nu_D$.
Here in the hologram plane three types of light are interacting: the undisturbed object wave, the reference wave and the detuned scattered wave.
After the exposure time τ we have a look at two of the resulting signals:

a) undisturbed object wave and reference wave

$$E(x) \sim 1 + \cos(2\pi\mu_0 x) \cdot \mathrm{sinc}(\Delta\mu \cdot x)$$

b) scattered light and reference wave

$$E(x) \sim 1 + \cos(2\pi\mu_0 x)\ \mathrm{sinc}(\Delta\mu x)\,\mathrm{sinc}(\Delta\nu_D \cdot \tau)$$

where

$$\mu_0 = \frac{\sin\alpha}{c}\nu_0,\ \Delta\mu = \frac{\sin\alpha}{c}\Delta\nu,\ \alpha = \text{angle between object and reference wave}$$

We recognize, that the term under a) is the good one and that the term under b) is the bad one. Therefore we have to take care that the $\mathrm{sinc}(\Delta\mu x)$, which depends on the spectral linewidth of the lightsource, becomes broad enough, and, on the other hand, that the $\mathrm{sinc}(\Delta\nu_D \cdot \tau)$, which depends on the Dopplershift, is essentially zero. From these demands we get the following conditions:

$$1 \geqslant \mathrm{sinc}(\Delta\mu x) \geqslant 0 \Rightarrow 0 \leqslant |\Delta\mu x| \leqslant 1$$

$$\mathrm{sinc}(\Delta\nu_D \tau) \approx 0 \Rightarrow \Delta\nu_D \tau \geqslant 1$$

The meaning of these conditions is given for the example of a real experiment:

Data: $\tau = 1/25s$, $\Delta\nu = 10^9 Hz$, $\alpha = 1°$, $\lambda = 514{,}5nm$, $c = 3.10^8 m/s$

$$\Rightarrow \quad |x| \leqslant \frac{c}{\sin\alpha \cdot \Delta\nu} = 17{,}2m$$

$$\Rightarrow \quad \Delta\nu_D \geqslant \frac{1}{\tau} = 25Hz$$

We still should mention, that the possible image width here is about 17 meters, but in our TV case we only use 20mm. That means that in general the spectral purity of the light source causes no trouble.
The given condition for the necessary Doppler shift leads us with a couple of further assumptions to a condition for the fog velocity. If we assume the velocity v perpendicular to the optical axis and we define a angle γ as the angle between the direction of the scattered light and the optical axis then the Doppler shift is given by :

$$\Delta\nu_D = \frac{1}{\lambda} \cdot v \cdot \sin\gamma$$

Furthermore we make use of the fact that the spatial bandwidth of the imaging system is limited. If δx is the resolution limit of the system and f the focal length then we can define a minimal angle $\min(\sin\gamma) = \delta x/f$. Now we put this fact and the definition of the aperture width $B = \lambda f/\delta x$ in our expression and get a simple condition for the fog velocity:

$$v \cdot \tau \geqslant B$$

3. Experimental results and set ups

The general set up is shown in fig.1

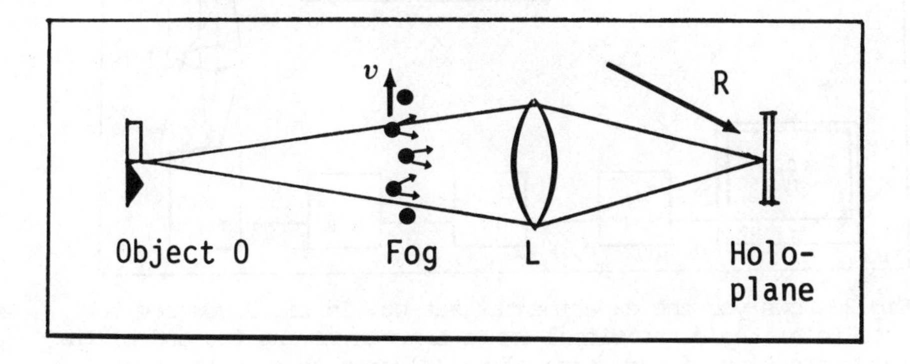

Here we get a hologram exposure:

$$E \sim |O + R + S|^2 = |O|^2 + |S|^2 + |R|^2 + O^*R + OR^* ,$$

where $R = \exp(2\pi i\mu x)$. Now we have to discriminate two configurations.
First, the off-axis holography with $\mu \neq 0$.
By converting the spatial signal into a electronic signal by means of a TV camera, we get

$$E \sim \left\{ |O|^2 + |R|^2 + |S|^2 \right\} + 2 \cdot O \cdot \cos(f_T \cdot t)$$

The terms in { }-brackets are low frequency terms, due to the limitations of the imaging system. These terms will be removed by a electronic high pass filter.
By squaring the output, we demodulate the interference term

$$E \sim O^2 + O \cdot \cos(2f_T \cdot t)$$

After the rejection of the appearing highfrequency term by a low pass filter, the monitor signal looks like:

$$I_{Mon} \sim O^2$$

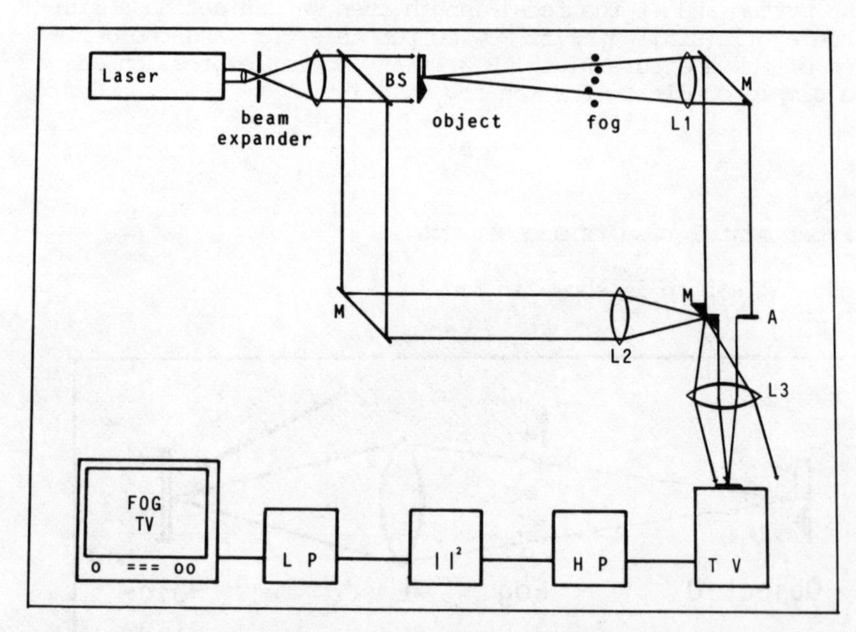

Fig.2 shows you the experimental set up. In fig.3 you see the results a) the conventional image through moving fog and b) the reconstruction of our image plane hologram through the same fog.

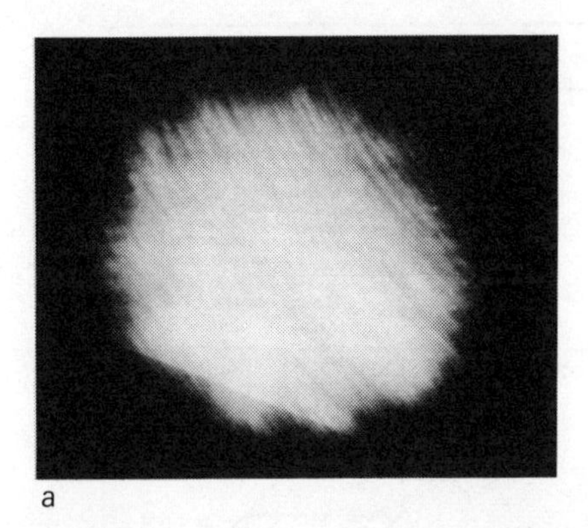

a

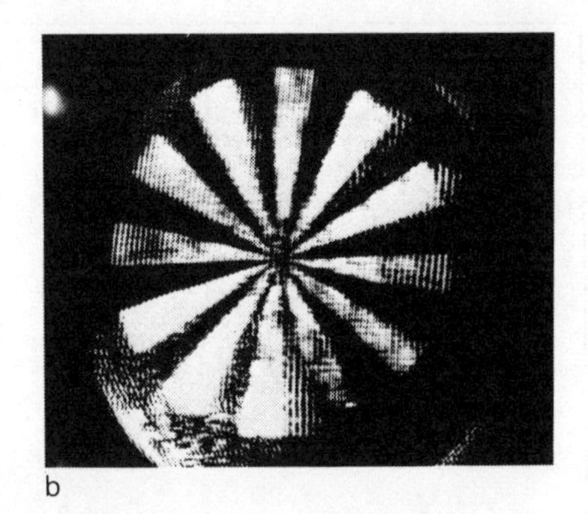

b

Figure 3

Second, we discuss the on-axis hologram with $\mu = 0$./7/
Here we have to take two exposures, one with R1 = +1 and the next with R2 = -1. That means that we have to shift the phase of the reference wave by π .
Therefore we get two videosignals:

$$E_1 \sim |O|^2 + |S|^2 + |R|^2 + 2\mathrm{Re}(O)$$

$$E_2 \sim |O|^2 + |S|^2 + |R|^2 - 2\mathrm{Re}(O)$$

By subtracting these two signals we get after squaring a monitor signal :

$$E_1 - E_2 = 4\mathrm{Re}(O) \quad \Rightarrow \quad I_{Mon} \sim [\mathrm{Re}(O)]^2$$

There are a couple of experimental set ups possible, but here we will only show our practical set up in fig.4.
Here we made use of the fact that the fringes which appear at the two gates of an interferometer are shifted by π . At first the shutter S1 was opened and S2 closed. We stored this image. Next we closed S1 and opened S2. This image was electronically inverted and then added to the first. The result then was given to a normal TV monitor. Fig. 5 shows the results.

4. Conclusions

As you see holography through fog works well. But we are still working to make this idea more applicable and more realistic.

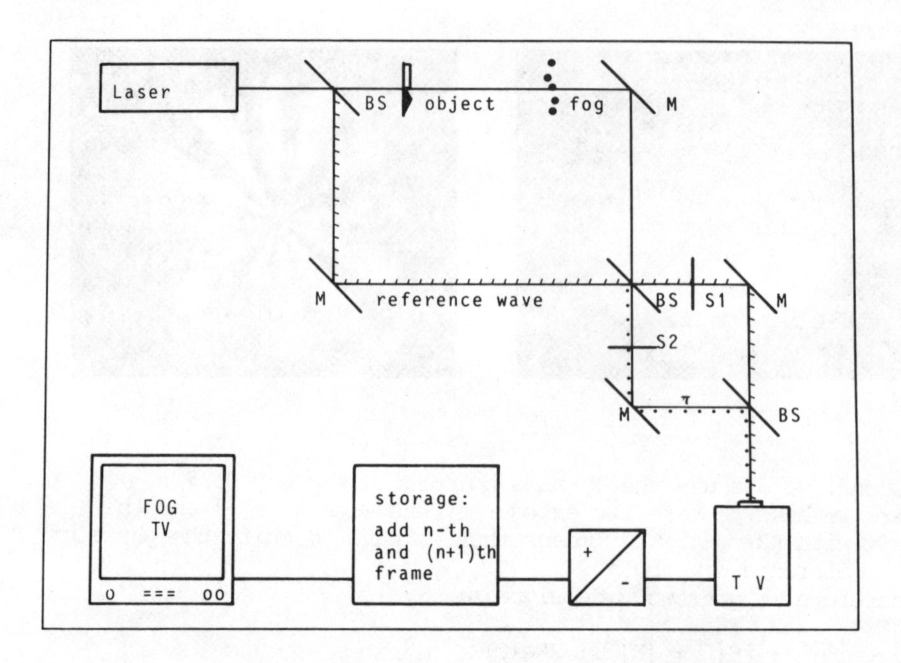

Figure 4

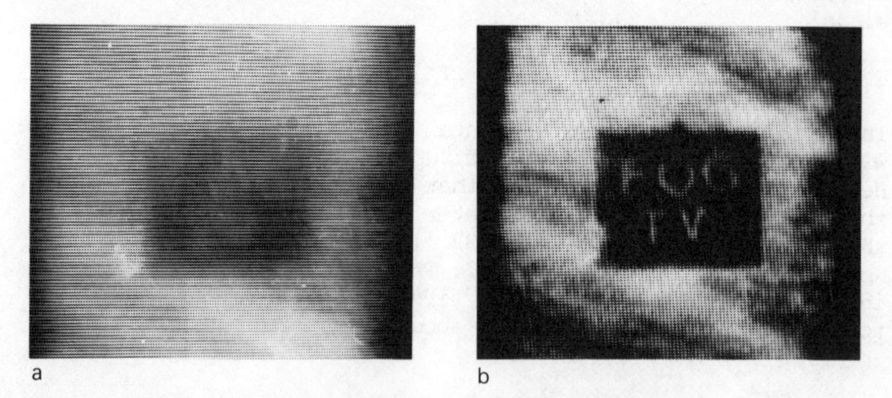

Figure 5: a) the conventional image through fog;
b) the reconstructed on-axis hologram.

References

1. E.Spitz, C.R.Acad.Sc. Paris 264 (1967) 1449
2. K.A.Stetson, Journ.of the Opt.Soc.of Am. 57 (1967) 1060
3. A.W.Lohmann, C.A.Shuman, Image Holography Through Convective Fog, Opt.Comm. 7 (1973) 93
4. C.A.Shuman, Holography Imaging Through Moving Diffusive Media, UCSD - APIS Report 73-2 (1973)
5. H.Schmalfuß, Real Time Seeing Through Moving Fog, Opt. Comm. 17 (1976) 245
6. H.Schmalfuß, Coherence Requirements For Holography Through Fog, To appear in OPTIK (1978)
7. A.W.Lohmann, H.Schmalfuß, Image Restauration by Means of Image Holography (in preparation)

PROBING OF THE ATMOSPHERE

G. Peckham, K.W. Rothe, Chairmen
T.C. Berg, A.R. Birks, H. Bjor, W.G. Burrows, R. Capitini, D. Croom, S. Eng, O. Gylling-Nielsen, R. Heylen, H.K. Myrabo, J. Schlesak, W. Wiesemann

Introduction

The need for effective techniques to monitor various constituents of the atmosphere has become obvious in recent years. In all industrial regions effective control of quite different pollution situations is necessary. In addition the global distribution of many pollutants has become a problem the severity of which possibly is still not realized everywhere. The long range transport of trace constituents requires detailed study - the transportation of sulfur dioxide from the centre of Europe to Scandinavia is only one of numerous examples.

The laser is a powerful and sensitive aid for the investigation and solution of these problems, which - because it enables measurements of pollutants to be made over large areas - is in many applications superior to any chemical method. At the same time the global distribution of atmospheric parameters and constituents can be investigated more appropriately by passive techniques. In the following we discuss selected examples of both these approaches, but first of all some atmospheric effects are considered, which may affect remote sensing methods.

Atmospheric Effects

The ability to remotely probe land, ocean, and atmosphere accurately using electromagnetic waves, is limited

T. Lund (ed.), Surveillance of Environmental Pollution and Resources by Electromagnetic Waves, 191-196.

	Definition, description	Spectral range affected	Size/number distribution	Height occurence	Dynamics	Notes
Aerosols	Dispersed system of particles suspended in a gas	VIS IR	r=0.01μm to 20μm $N \propto r^{-4}$	All heights (maximum near the ground)	Lifetime: minutes to months	Origins: earth surface to cosmic dust.
Hydrometeors	Fog, rain, ice crystals, hail snow, cloud	2GHz to 50GHz	Different size distribution	Troposphere	Weather fronts	Up to 30 dB/km attenuation, phase shifts
Turbulence	Eddies with refr. index inhomogenities	Mainly UV,VIS, IR	Typical volume: cm^3-m^3	Troposphere mainly near the ground	Transport by wind, heating	Apperture size dependent. Scintillation
Refraction effects	Refractive index changes	All wavelenghts		All atmosphere	Diurnal variations	Temperature, pressure, density dependant
Polarisation rotation	Rotation of polarized light in the ionosphere	f<10GHz. rotation ≃0.1 rad at 4GHz		Ionosphere	Variation at geomagnetic activities	Proportional to total electron content and $1/f^2$
Non-thermal atmospheric emission	Aurora, airglow, afterglow	0.2μm-5μm	Emitting area: $1m^2$-$1km^2$	Ionosphere	Brightness variations 1 - 10^4	Atoms, Molecules, Radicals

Table 1: Atmospheric mechanisms affecting EM sensing

by the perturbing effects that atmospheric irregularities may impose on electromagnetic propagation. A summary of the major interfering mechanisms is given in table 1.

More information must be obtained about these mechanisms and to what extent they interfere with the various remote sensing techniques. More specifically there is a need to outline and investigate the man-made processes that may cause abnormal size and height distribution of aerosols which may give rise to abnormal Mie scattering conditions. In addition further study of the spatial distribution and overall dynamics of the non-thermal atmospheric emissions is required.

Passive Sensing

Passive remote sensing may be performed by means of transmission measurements using a natural source of radiation such as the sun, or by means of thermal radiation emitted by some constituent. In the latter case there must be a temperature contrast between the constituent and its background; e.g. it is very difficult to see emission from pollutants in the lower levels of the atmosphere when viewed against the ground at much the same temperature. The following types of measurements either have been proved to be practicable in the field - or appear promising for the future:

(a) Tropospheric water vapour may be mapped by satellite-borne radiometers scanning about the nadir and responding to emitted thermal radiation in an infrared or microwave absorption band. Cloud cover may also be mapped from satellites, and both ground-based and satellite measurements of the liquid water content of clouds - again by means of the thermal emission at microwave frequencies - have been attempted.

(b) The emission from CO_2 at 4.2 μm or 15 μm, and alternatively that from O_2 at either 5 mm or 2.5 mm, may be used to deduce the vertical temperature profile in the atmosphere. The measurements are made from satellites using one of two geometries; either limb scanning or nadir looking. Ground-based measurements are also possible.

(c) Ground-based, airborne or satellite-borne experiments may be used to measure the composition of the stratosphere. Satellite-borne radiometers look towards the limb and see emission from molecules with infrared or microwave absorption bands against a cold space background. Clouds prevent the extension of such measurements to the troposphere. Backscattered sunlight in the ultra-violet has been used to measure the distribution of ozone.

Ground-based measurements of stratospheric composition may be made either by examining the solar absorption spectrum or by observing thermal emission. Measurement of line shape can yield the vertical distribution of the absorber provided that the line is predominantly pressure broadened. For work at high altitudes (greater than 30 km) microwaves are more suitable than infrared wavelengths because of the lesser importance of doppler broadening at microwave frequencies.

Considering future trends in passive sensing, these techniques will become widely used for the measurement of atmospheric temperature on meteorological satellites so as to increase the geographical coverage over and above that provided by the international network of radio-sonde stations. Interest in the effects of pollutants on stratospheric photochemistry is likely to continue to stimulate measurements of atmospheric composition in this region. Limb sounding is a powerful tool here, and developments in microwave radiometer technology will allow the extension of ground-based monitoring. Studies of the techniques of data inversion are most important if the ground-based methods are to be exploited to the full. In some applications, passive sensing of pollutants may offer an alternative to lidar. The equipment would possibly be cheaper, although no range discrimination is obtained from this technique.

Monitoring atmospheric pollutants by means of lasers

With the present state of laser technology only a limited number of systems can be used for reliable and automatic field application. These comprise the tunable dye laser, semiconductor diode laser and fixed frequency infrared laser. Up to now all other types are still lacking in reliability and ease of handling. This is made plain by the fact that in spite of numerous laboratory investigations with different laser systems only a limited number of field measurements have been performed.

Since most pollutant molecules have an absorption spectrum in the infrared region, the wavelengths between 1 and 13 μm are of particular interest. Furthermore, in this region there are atmospheric windows which allow pollutants to be detected at very long ranges. Many constituents, which have been investigated in the laboratory, have sufficiently high absorption coefficients at the available laser wavelengths /1/. Therefore, from the physical point of view, field measurements on these pollutants at ambient levels could be performed given the present state of research and technology. Once reliable and operable high pressure gas lasers become available for field application, the minimum detection limit might be significantly reduced for some species. Until then the use of fixed frequency infrared gas lasers will be the most promising way to perform pollution control.

It is often believed that there is one single laser system which is suitable for all possible applications in the field of pollutant sensing. This is not so; we have identified three distinct classes of application each with its own special practical requirements. The following discussion aims to identify the most suitable laser system for each, when both technical and economic considerations are taken into account.

The first type of application comprises stationary control of emission from factories as well as sensitive pollution control in urban areas. Major requirements for monitoring stations for this purpose are low cost and ease of handling. Furthermore they should operate automatically and should have a very fast response, thus enabling local authorities to react rapidly and factories to optimise the production process. The most suitable system for this stationary type of application will be a low cost cw-laser performing long path absorption measurements. This has already been successfully demonstrated in the case of the HF-emission from an aluminium factory /2/ and the ethylene emission from a petrochemical factory /3/.

The second type of application comprises the detailed mapping of different pollutants over the area of a city or a large industrial area. An instrument like this should be able to give a range resolved picture of the pollutant concentration over an area of several km diameter. Here the need is for periodic observation at a variety of different locations, and so the system has to be mobile. These requirements are met by a van-mounted lidar system equipped with a high power pulsed laser able to use Mie scattering as a distributed reflector and thus giving range resolution. Systems like this do already exist and have proven their applicability e.g. in the case of mapping NO_2 /4/, SO_2 /5/, H_2O /6/, and ethylene /7/.

The third type of application comprises the more global mapping of pollutants over large areas. An apparatus for this purpose should show the mean pollution situation over several square kilometers or even over whole districts. For faster scanning purposes such a system should be air-borne and in addition should be able to map the global distribution of a pollutant at different heights up to 15 km. These requirements are met by combining the long path absorption method with the heterodyne detection technique, which improves the sensitivity relative to the backscattered signal by several orders of magnitude. Using a stabilised CO laser in conjunction with the earth as a topographical reflector the feasibility of these systems has already been demonstrated /8/.

To summarize the present demands and activities, it can be stated that the laser is a very helpful tool for the control of pollutants. Improvements in laser technology will considerably increase its usefulness. In addition, comparative measurements with chemical methods must be performed in order to establish the laser method as an extremely powerful help in pollution monitoring and process control. Standardisation of optical pollution measurements will be of the utmost importance in the near future.

REFERENCES

/1/ J. Boscher, BMFT Contract 01 TL 026-AK/RT/WRT 2074
/2/ A. Tönnißen, J. Wanner, K. W. Rothe, H. Walther, to be published
/3/ S. Eng, private communication
/4/ K. W. Rothe, U. Brinkmann, H. Walther, Appl. Phys. 3, 115 (1974)
/5/ W. B. Grant, R. D. Hake, Jr., J. Appl. Phys. 46, 3019 (1975)
/6/ E. R. Murray, R. D. Hake, Jr., J. E. van der Laan, J. G. Hawley, Appl. Phys. Lett. 28, 542 (1976)
/7/ H. Walther, invited paper presented at ECOSA 1, Brighton, U.K., April 1978
/8/ W. Wiesemann, R. Beck, W. Englisch, K. Gürs, Appl. Phys. 15, 257 (1978)

PART III : THE EARTH'S SURFACE; LAND — SEA

SCATTERING OF ELECTROMAGNETIC WAVES FROM THE OCEAN

G. R. Valenzuela

Ocean Sciences Division, Naval Research Laboratory
Washington, D. C. 20375, U.S.A.

ABSTRACT. The basic principles and analytical methods in scattering theory that treat the interaction of electromagnetic (EM) waves with the ocean surface are reviewed. An exact solution of this boundary-value problem is not feasible at this time, because the ocean surface is a complex dynamical and spatial nonlinear (statistical) rough surface still under investigation. In these circumstances, the high-frequency scattering methods (geometrical and physical optics) and the low-frequency scattering methods (perturbation) yield tractable analytical solutions (models) useful for remote sensing applications of the ocean surface. The state of the art and recent advances in radio-oceanography are described.

1. INTRODUCTION

Advances in remote sensing of the ocean surface from satellite, aircraft and coastal installations offer some new opportunities and challenges in basic research to oceanographers (geophysicists) and in technological development to radio scientists.

To take advantage of these powerful techniques and learn about ocean parameters (tides, meso-scale eddies, geostrophic currents, wind speed, wave-height, two-dimensional spectrum and apparent ocean temperature), monitoring oil pollution or implementing fast ship navigational control centers, we must not only know about the physical processes involved in the interaction of electromagnetic (EM) radiation with the ocean surface, but we should also be familiar with the analytical scattering techniques in EM theory to extract the information from the probing sensors.

The present state of understanding of this interaction process was attained through a number of theoretical and experimental investigations after Crombie's pioneer experiments [12] with 13.56 MHz. We will not be able to include all the specific contributions in this field; we refer in this respect to the extensive reviews by Valenzuela [71] and Wright [81].

T. Lund (ed.), Surveillance of Environmental Pollution and Resources by Electromagnetic Waves, 199–226.

Classical methods are available in EM theory to treat the scattering from bodies and corrugated rough surfaces of known electrical properties [45, 53, 63, 72]. In principle, these same methods may be applied to statistical rough surfaces when the statistics of the surface roughness are known [9]. However, in dealing with a geophysical surface (in our case, the weakly-nonlinear ocean surface), only the first few moments of the surface displacement of the distribution are known. In this case, an exact solution of this boundary value problem is clearly out of question. Nevertheless, the high-frequency scattering methods (geometrical and physical optics) and the low-frequency scattering method (perturbation) or a combination of the two methods provide tractable analytical results (i.e., the specular-point, Bragg and composite-surface models). Figure 1 is a summary of the scattering mechanisms and models to be covered in this work. At present, these models are widely used in remote sensing applications of ocean parameters at HF (3-30 MHz) and microwave frequencies; they also serve as "limiting" cases to check the validity of other more general scattering results as they become available.

The ocean surface in the linear approximation is homogeneous, stationary and Gaussian. At higher-order of wave slope, energy and momentum are continuously transferred between all water waves. This transfer destroys the homogeneous and stationary properties of the ocean and modifies the Gaussian statistics. Accordingly a thorough knowledge of the dynamics of ocean waves is crucial in establishing with certainty the oceanic parameters that may be probed (derived) from the EM sensors, and in attempting to apply the optimal surveillance methods described by Gjessing [19].

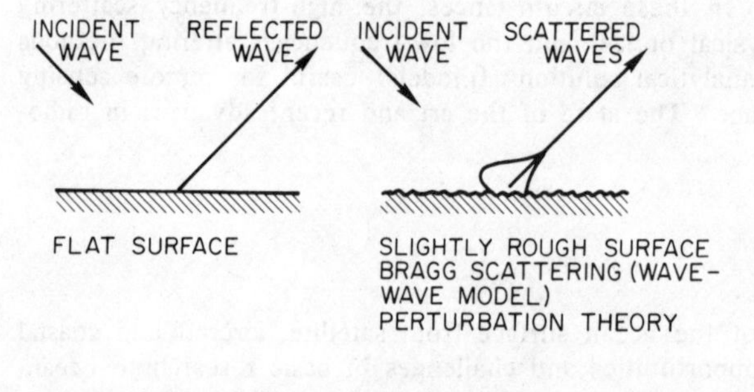

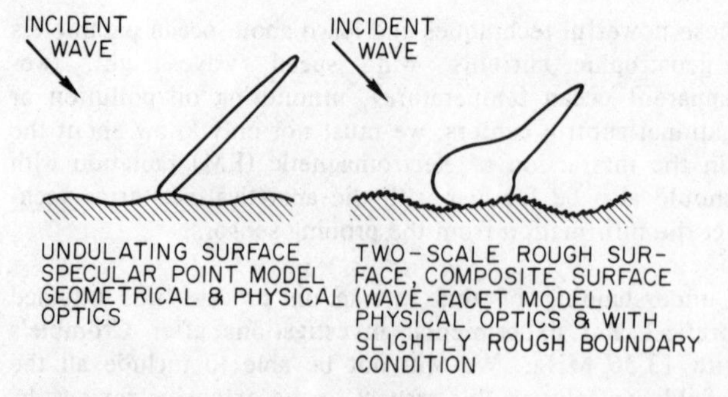

Fig. 1. Summary of EM scattering models for the ocean.

Therefore, when dealing with the EM scattering from the dynamical ocean surface, it is senseless to use rigorous EM scattering methods to solve this boundary-value problem while characterizing the ocean surface as a static "sandpaper" type of rough surface. For realistic solutions we should be prepared to include in the analysis the input from the wind, the energy transfer by nonlinear resonant interactions, and the dissipation by viscosity (or wave-breaking), processes which control the propagation and development of ocean waves.

In fact, a great deal of the success of the EM scattering models (Bragg, composite-surface and specular-point) is based on their simplicity and compatibility with dynamical oceanographic studies allowing the inclusion of higher-order effects in the ocean as they become known.

2. OCEAN WAVES

First, we give a brief background on the dynamics and kinematics of ocean waves. For a treatment in depth of ocean waves, refer to Kinsman [35] and Phillips [47]. However, to deal with the dynamics of an ocean wave spectrum we must use the more powerful techniques of theoretical physics and statistical mechanics used by Hasselmann [25] to formulate a weak-interaction theory for water waves.

2.1 Basic hydrodynamics

Water waves are governed by Navier-Stokes nonlinear differential equations for incompressible fluids which express the balance between inertial, convective, and restoring forces of the wave motion. In the Eulerian specification, one deals with the wave motion as a function of position x and time t, while in the Lagrangian description one deals with the fluid motion (particle trajectory) as a function of position and time. The restoring force for surface waves longer than 1.7 cm (gravity waves) is the gravitational acceleration g, while the restoring force for capillary waves which are shorter than 1.7 cm is surface tension. In addition to the dynamical equation, appropriate boundary conditions apply at the air-water interface and at the water-bottom (kinematical conditions).

In the linear approximation (very small slopes), the surface displacement $\zeta(x, t)$ of the waves is sinusoidal.

$$\zeta(x, t) = A \cos(Kx - \Omega t), \tag{2.1}$$

where K is the wavenumber and Ω is the angular frequency. Ω and K are related by the dispersion relation of water waves. At this order, the fluid describes closed circular orbits and the wave motion travels at the classical phase speed

$$C_{og} = \Omega/K = (g/K)^{1/2} \tanh^{1/2}(KD) \quad \text{(gravity waves)}, \tag{2.2}$$

and

$$C_{oc} = (TK)^{1/2} \quad \text{(capillary waves)}, \tag{2.3}$$

where D is the water depth and $T \approx 73$ cm^3/sec^2 is the surface tension-to-water density ratio. The minimum speed of water waves is about 23 cm/sec for 1.7 cm waves.

As the amplitude "A" of the wave increases in relation to its wavelength, finite amplitude effects modify the phase speed and shape of the water waves. The phase speed of gravity waves increases

$$C_g = C_{og}\left[1 + \frac{1}{2} A^2K^2 + ...\right] \text{ (deep water, } D = \infty) \tag{2.4}$$

$$C_c = C_{oc}\left[1 - \frac{1}{16} A^2K^2 + ...\right], \tag{2.5}$$

and the shape of finite amplitude gravity waves tends toward a trochoid, while the shape for a finite amplitude capillary wave tends toward an inverted trochoid (Crapper wave) [35]. For gravity waves KA/π is less than 0.143, while for capillary waves this quantity may be as large as 0.73.

Mean currents $\mathbf{U}$ advect the surface waves, modifying their phase speed and angular frequency

$$\mathbf{C} = C_o + \mathbf{U} \text{ and } \Omega = \Omega_o + \mathbf{K} \cdot \mathbf{U} \tag{2.6}$$

where C_o and Ω_o are the values in the absence of the current. For rotational motion (with vorticity), (2.2)-(2.5) are more complex. The phase speed and the growth rates of shear waves are affected by the mean flow in the air and in water [51, 59, 69].

The fluid motion is also modified as the wave grows; at second-order the fluid no longer describes closed orbits. A wave drift enhances the horizontal fluid velocity $u(x, z, t)$ (z is the vertical coordinate, positive into the air).

$$u(x, z, t) = \Omega A e^{Kz} \cos \chi + \Omega A^2 K e^{2Kz} \tag{2.7}$$

$$w(x, z, t) = \Omega A e^{Kz} \sin \chi, \tag{2.8}$$

where $\chi = Kx - \Omega t$ and $w(x, z, t)$ is the vertical component of fluid velocity. The constant mean flow, with x and t, in (2.7) is called the Stokes drift.

Keep in mind that the magnitude of the orbital velocity (ΩA) is small compared with the phase speed of the wave. For waves of finite amplitude, the orbital velocity may be as large as 10 percent of the phase speed. The ideal trochoidal wave shape that gravity waves in the ocean resemble have vorticity and no drift (Gerstner wave), while a Stokes wave has drift but no vorticity. Accordingly, ocean waves are a transient between these two limiting waves.

2.2 Ocean spectrum

On the ocean, however, the wind is responsible for the generation of waves. Accordingly, the ocean is composed of an infinite number of wind-waves. The Fourier transforms of the spatial and temporal correlation functions of the surface displacement $\zeta(\mathbf{x}, t)$ yield the two-dimensional wavenumber spectrum $F(\mathbf{K})$ and the frequency spectrum $S(\Omega)$ of the ocean surface. Spectra transform as probability density functions. Hence,

$$F(\mathbf{K})\, d\mathbf{K} = S(\Omega)\, d\Omega, \tag{2.9}$$

where Ω and $K = |\mathbf{K}|$ are related by the dispersion relation

$$\Omega = (gK + TK^3)^{1/2} + \mathbf{K} \cdot \mathbf{U}, \tag{2.10}$$

and $\mathbf{U}$ is a mean current.

The mean-square height of the ocean is then obtained by integrating over $\mathbf{K}$ or Ω

$$\overline{\zeta}^2 = \iint_{-\infty}^{\infty} F(\mathbf{K})\, d\mathbf{K} = \int_o^{\infty} S(\Omega)\, d\Omega. \tag{2.11}$$

For fully-developed seas the Pierson-Moskowitz spectrum has been shown to apply

$$S(\Omega) = 0.0084\, g^2 \Omega^{-5} \exp\left[-0.74 \left(\frac{\Omega_m}{\Omega}\right)^4\right], \tag{2.12}$$

where $\Omega_m = g/W$ is the angular frequency of the dominant wave of the ocean and W is the wind speed at a height of 19.5 m. Fully-developed seas are only encountered when the wind field blows over a sufficiently long fetch (spatial distance over water) and for a sufficiently long time that the phase speed of the dominant wave in the ocean becomes equal to the wind speed.

Phillips [47], using dimensional analysis and assuming that wave-breaking is the main mechanism controlling the maximum amplitude of ocean waves, postulated than in the "equilibrium range" (waves longer than the dominant wave in the ocean and shorter than capillary waves), the frequency spectrum should approach the spectral law

$$S(\Omega) \sim \alpha g^2 \Omega^{-5}, \tag{2.13}$$

where α is an empirically determined constant (≈ 0.01).

Recently in the JONSWAP experiments [29], it was determined that the development of the gravity wave spectrum with time and fetch is controlled by three competing processes: the input from the wind, the energy transfer by nonlinear resonant interactions, and dissipation by viscosity and wave-breaking. Under these conditions, the spectrum of gravity waves is well approximated by

$$S(f) = \alpha g^2 (2\pi)^{-4} f^{-5} \exp\left[\frac{-5}{4}\left(\frac{f_m}{f}\right)^4\right] \gamma^{\exp\left[\frac{-(f-f_m)^2}{2\,\sigma^2\, f_m^2}\right]}, \tag{2.14}$$

where α decreases with fetch, and for a fully-developed spectrum α approaches 0.01 and $\gamma = 1$. In particular, for short fetches it was found that nonlinear resonant interactions play an important role in the development of the ocean spectrum.

2.3 Wave generation, growth and propagation

According to Phillips [47], ocean waves are generated through linear resonant coupling of turbulent pressure fluctuations in the air with the water surface. In the initial stage of growth, this mechanism predicts a linear growth of the spectrum with time. Therefore,

$$\frac{\partial F(\mathbf{K}, t)}{\partial t} = \text{constant}, \tag{2.15}$$

where the constant is proportional to the spectrum of turbulent fluctuations in the air. As the wind increases, the first waves that are generated by this mechanism are 1.7 cm waves, those of minimum phase speed and travel in the direction of wind. For higher winds, waves of phase speed C are generated at an angle ψ with respect to the wind, where

$$\psi = \cos^{-1}(C/U_c), \tag{2.16}$$

and U_c is the convection velocity of the air pressure fluctuations near the water surface.

Recently, a nonresonant mechanism has been observed to be responsible for the generation of 16.5 cm waves in a wave-tank [50]. For light winds, up-wind and down-wind waves of equal magnitude appear initially. As the wind increases, the magnitude of the down-wind wave predominates.

Exponential growth rates are predicted by Miles [41] instability mechanism of a phase shifted, wave-induced pressure in the air acting on the water wave. Energy and momentum are transferred from the "critical" layer (the height at which the mean wind speed is equal to the phase speed of the wave) of the mean flow in the air. Hence, according to this mechanism,

$$\frac{\partial F(\mathbf{K},\ t)}{\partial t} = \beta F(\mathbf{K},\ t), \tag{2.17}$$

where β is the growth rate which is a function of the mean wind profile parameters. Exponential growth is also obtained by a perturbation of the boundary layers in the air and water for short-gravity waves [69] (viscid mechanism).

In addition to these direct transfer mechanisms between the air and water, there are indirect transfer mechanisms: the modulated stress and direct input to forced waves which reappears at other waves [31, 70].

Negative growth, or damping, is obtained from viscous losses and wave-breaking. For viscous damping,

$$\frac{\partial F(\mathbf{K},\ t)}{\partial t} = -4\nu K^2 F(\mathbf{K},\ t), \tag{2.18}$$

where ν is the kinematic viscosity (0.01 cm^2/sec for clean water). According to Hasselmann [30], damping by wave-breaking is a quasi-linear process

$$\frac{\partial F(\mathbf{K},\ t)}{\partial t} = -\eta \Omega^2 F(\mathbf{K},t), \tag{2.19}$$

where η is an empirical parameter.

The development and propagation of a system of water waves (spectrum) under the influence of processes which are weak in the mean may be treated by the Boltzmann transport equation [25].

$$\frac{DF}{Dt} = \frac{\partial F}{\partial t} + \frac{\partial \Omega}{\partial K_i}\frac{\partial F}{\partial x_i} - \frac{\partial \Omega}{\partial x_i}\frac{\partial F}{\partial K_i} = S_t(\mathbf{K}) \tag{2.20}$$

K_i, x_i are the components of $\mathbf{K}$, $\mathbf{x}$ respectively and $S_t(\mathbf{K})$ is the source function representing the net transfer of energy to (or from) the spectrum at the wavenumber $\mathbf{K}$ due to all interaction processes affecting the component $\mathbf{K}$. The net source function is

$$S_t = S_{in} + S_{nl} + S_{ds}, \tag{2.21}$$

where S_{in} represents the input from the wind, S_{nl} is the energy transfer due to nonlinear resonant interactions and S_{ds} is the dissipation due to viscosity and wave-breaking. The energy transfer by nonlinear resonant interactions for gravity waves is of third-order [29] and for gravity-capillary waves is of second-order [67]. When straining effects by varying current are present, we must include an additional source in (2.21). This is

$$S_{st} = \left[\left(\frac{\partial \Omega}{\partial K}\right)\left(\frac{K}{\Omega}\right) - 1\right]\frac{\partial \mathbf{U}}{\partial x}\, F(\mathbf{K}). \tag{2.22}$$

For example, the modulation of short-gravity waves by longer gravity waves was investigated with the "transport" equation [34].

3. CLASSICAL METHODS IN EM SCATTERING

Classical EM theory deals with two kinds of scattering problems: the "direct" and the "inverse". In the "direct" scattering problem, from a given current distribution on a body (surface) one desires to find the scattered fields at some distance from the body (or surface). On the other hand, in the "inverse" problem given the scattered fields, one wishes to find the current distribution (or body) that produced the fields. Both of these problems are of interest to remote sensing. In this work, we will concentrate on the "direct" scattering problem, where one ultimately desires the scattered power and cross-section far away from the body or surface.

3.1 Classical approach in direct scattering

The EM fields $\mathbf{E}$, $\mathbf{H}$ (electric and magnetic) at an observation point of radius vector $\mathbf{r}$ inside a volume region enclosed by a simply connected surface S may be determined from the following integral equations [72]

$$\mathbf{E}(\mathbf{r}) = \mathbf{E}_i(\mathbf{r}) + \nabla \times \oint(\mathbf{n} \times \mathbf{E})\, \phi ds + (i\omega\epsilon_o)^{-1}\nabla \times \nabla \times \oint(\mathbf{n} \times \mathbf{H})\, \phi ds \tag{3.1}$$

$$\mathbf{H}(\mathbf{r}) = \mathbf{H}_i(\mathbf{r}) + \nabla \times \oint (\mathbf{n} \times \mathbf{H})\, \phi ds - (i\omega\mu_o)^{-1}\, \nabla \times \nabla \times \oint (\mathbf{n} \times \mathbf{E})\, \phi ds. \tag{3.2}$$

Where $\mathbf{n}$ is the unit normal to the surface (positive toward free space), ω is the angular frequency of the incident EM radiation and μ_o, ϵ_o are the magnetic permeability and the electric permittivity of free space, $\phi(\mathbf{r}, \mathbf{r}') = \exp(-ik|\mathbf{r} - \mathbf{r}'|/4\pi|\mathbf{r} - \mathbf{r}'|$ is the Green's function of free space ($\mathbf{r}'$ is the radius vector of a point on the surface, k is the propagation constant of the EM radiation in free space, ∇ is the nabla differential operator with respect to $\mathbf{r}$ and the time dependence is $e^{i\omega t}$.

The physical significance of (3.1) and (3.2) are evident when they are expressed in terms of the effective electric $\mathbf{K}_e = \mathbf{n} \times \mathbf{H}$ and magnetic $\mathbf{K}_m = -\mathbf{n} \times \mathbf{E}$ surface currents. The scattered fields $\mathbf{E}_s = \mathbf{E} - \mathbf{E}_i$ and $\mathbf{H}_s = \mathbf{H} - \mathbf{H}_i$ are the result of the effective surface currents established on the surface via the boundary conditions. The scattering configuration in regard to the ocean surface is illustrated in Figure 2. In this work, we will be mostly interested in backscattering, the fields scattered back in the same direction of the incident EM radiation.

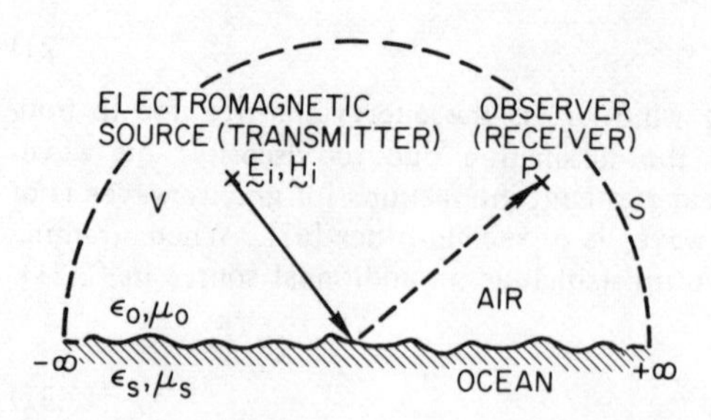

Fig. 2. Geometrical configuration for scattering from the ocean.

Usually one is interested in the scattered fields far away from the surface. If A_{ill} is the area illuminated on the ocean, at a mean distance R_o from the surface such that $R_o >> A_{ill}/\lambda$ (The Far-Field), where λ is the wavelength of the EM radiation. For the Far-Field (3.1) and (3.2) simplify to

$$\mathbf{E}_s \sim \frac{ike^{-ikR_o}}{4\pi R_o} \int [\mathbf{r}_2 \times \mathbf{K}_m + Z_o \mathbf{r}_2 \times \mathbf{r}_2 \times \mathbf{K}_e] \, e^{i\mathbf{k}_2 \cdot \mathbf{r}'} \, dS \tag{3.3}$$

$$\mathbf{H}_s \sim \frac{-ike^{-ikR_o}}{4\pi R_o} \int [\mathbf{r}_2 \times \mathbf{K}_e - Z_o^{-1} \mathbf{r}_2 \times \mathbf{r}_2 \times \mathbf{K}_m] \, e^{i\mathbf{k}_2 \cdot \mathbf{r}'} \, dS, \tag{3.4}$$

where $\mathbf{k}_2$ is the propagation constant in the scattered direction, $\mathbf{r}_2 = \mathbf{k}_2/|\mathbf{k}_2|$, Z_o is the intrinsic impedance of free space (120 π ohms) and the integrals are over the illuminated area on the surface. Accordingly, now the scattering problem is reduced to finding the effective surface currents (fields) on the surface.

3.2 Boundary conditions, iteration, tangent plane and perturbation

In scattering of EM waves from deterministic bodies or corrugated surfaces of arbitrary dielectric properties, the surface fields (effective currents) may be determined from a pair of coupled integral equations resulting from (3.1) and (3.2) when the observation point is moved onto the surface ($\mathbf{r} \to \mathbf{r}''$). The contribution to the integrals as $\mathbf{r}'' \to \mathbf{r}'$ is explicitly evaluated [43, 45].

For perfectly conductive surfaces $\mathbf{n} \times \mathbf{E}$ and $\mathbf{n} \cdot \mathbf{H}$ vanish on the surface. For this case, the integral equations decouple and $\mathbf{H}$ on the surface may be determined from the integral equation

$$\frac{1}{2} \mathbf{n} \times \mathbf{H}(\mathbf{r}'') = \mathbf{n} \times \mathbf{H}_i(\mathbf{r}'') + \mathbf{n} \times \left[\nabla \times \int_{\bar{S}} (\mathbf{n} \times \mathbf{H}) \, \phi \, dS\right], \tag{3.5}$$

where $\bar{S}$ is the punctured surface with $\mathbf{r}'' \neq \mathbf{r}'$. For deterministic bodies and corrugated surfaces, (3.5) may always be solved numerically by matrix inversion methods. However, this is not too practical for extended surfaces, since the surface must be sampled at increments small compared to the wavelength of the EM radiation.

Approximate methods are also available to solve (3.5). For example, in a solution by iteration $\mathbf{H}$ is expanded in the series

$$\mathbf{H} = \mathbf{H}^{(0)} + \mathbf{H}^{(1)} + \mathbf{H}^{(2)} + \dots, \tag{3.6}$$

with the various iterations for the field being given by

$$\mathbf{n} \times \mathbf{H}^{(0)} = 2\mathbf{n} \times \mathbf{H}_i, \dots, \mathbf{n} \times \mathbf{H}^{(n)} = 2\mathbf{n} \times \left[\nabla \times \int_{\bar{S}} (\mathbf{n} \times \mathbf{H}^{(n-1)})\, \phi\, dS\right], \dots . \tag{3.7}$$

As a matter of fact in Physical Optics, a standard method in scattering, the surface fields are approximated by the "tangent plane" value (the zeroth order term in the iteration expansion) and the scattered fields are obtained from (3.3) and (3.4).

The conditions for the validity of Physical Optics (or Kirchhoff method) are [61]

$$\frac{1}{2}\, k\rho \cos^3\theta >> 1 \quad \text{and} \quad 2k\rho \cos\theta >> 1, \tag{3.8}$$

where θ is the local angle of incidence and ρ is the local radius of curvature of the surface. For surfaces of large conductivity, the magnetic fields are still determined from (3.7) and the electric fields are approximated with an impedance boundary condition [43]

$$\mathbf{K}_m = -Z_s(\mathbf{n} \times \mathbf{K}_e), \tag{3.9}$$

where $Z_s << Z_o$ is the inpedance of the surface.

Another method also very useful in scattering is perturbation. The method of perturbation has been invaluable in theoretical physics in extending the applicability of theories that apply to some well known problem to other cases which are slight variations [44]. In scattering from slightly rough surfaces (the amplitude of the surface roughness is small compared with the wavelength of the EM radiation), the surface fields may be obtained from the fields when the roughness is absent applying the boundary conditions at the perturbed rough surface [42, 52].

The boundary-value problem becomes more complex when dealing with the scattering of EM waves from statistical rough surfaces. Now the surface fields and scattered fields are statistical dependent on the temporal and spatial statistics of the rough surface. If the statistics of the surface are completely known, then one could use the recently developed Green's function method [15, 61] which offers a diagrammatic formulation for finding the scattered fields from the incident fields on the surface. One could also implement a numerical Monte-Carlo simulation for the scattered fields.

Finally, we are interested in obtaining scattered power P_s in a give direction

$$P_s = \frac{1}{2}\, \mathcal{Re} < \mathbf{E}_s \times \mathbf{H}_s^* >, \tag{3.10}$$

where the brackets denote the ensemble average and $\mathbf{H}_s^*$ is the complex conjugate value of $\mathbf{H}_s$. Another quantity used in scattering problems is the mean cross-section σ of the rough surface

$$\sigma = \lim_{R_o \to \infty} \{4\pi R_o^2 < |\mathbf{E}_s|^2 > / |\mathbf{E}_i|^2\}, \tag{3.11}$$

where R_o is the mean distance to the rough surface.

4. EM SCATTERING FROM THE OCEAN

In remote sensing applications where the scattering surface is the ocean, one generally does not know the complete statistics of the surface. Hence, a rigorous solution of this boundary-value problem is not possible. Even then, the high-frequency (geometrical and physical optics) and the low-frequency (perturbation) scattering methods have yielded tractable and useful results. The EM backscatter from the ocean surface is of "specular nature" near normal incidence, "diffuse" away from normal incidence and more complex near grazing incidence as shadowing, diffraction and trapping become significant.

4.1 Specular point model

Specular scattering from a body or a rough surface may be obtained by geometrical or physical optics. Geometrical optics is widely used in optics to calculate the reflection, transmission, refraction and scattering of light beams from dielectric media and metallic reflectors (gratings) [10]. One deals with the ensemble of "rays" as it is affected by the medium and by including the proper amplitude, phase and polarization of each "ray" one obtains by superposition the light at a plane or point in space. Physical optics is useful to include wavelength dependent terms in the geometrical optics solution. As mentioned earlier, in scattering from rough surfaces, in physical optics one uses the "tangent plane" approximation for the surface fields and the with this approximation the scattered fields are obtained with (3.3) and (3.4). In the limiting case of vanishing wavelength of the EM radiation, the physical optics results approach the geometrical optics solutions and become exact. The physical optics method may be used whenever condition (3.8) are satisfied (points of discontinuity in slope are excluded). Therefore, generally the physical optics results apply for high frequency and near the specular direction.

Geometrical optics scattering results are polarization independent for backscatter, and only surface facets normal to the direction of incidence contribute [22]. Using physical optics and the method of stationary phase in the integration Kodis [36] derived an expression for the cross-section of a perfectly conductive rough surface of Guassian statistics

$$\sigma_{\infty} = \pi \overline{\rho_1 \rho_2} <N>, \tag{4.1}$$

where $\overline{\rho_1 \rho_2}$ is the average product of the principal radii of curvature of the surface and $<N>$ is the average number of specular points per unit area. Later Barrick [1] generalized this result to rough surfaces of finite conductivity and obtained bistatic cross-sections. For backscatter, the cross-section per unit area is given by

$$\sigma_{o\infty} = \pi \sec^4 \theta \, p(\zeta_{xs}, \zeta_{ys}) \, |R(o)|^2, \tag{4.2}$$

where $p(\zeta_{xs}, \zeta_{ys})$ is the joint-probability density of slopes of the surface with displacement $\zeta(x, y)$ ($\zeta_x = \partial\zeta/\partial x$ and $\zeta_y = \partial\zeta/\partial y$) evaluated at the specular points, θ is the angle of incidence with respect to the mean surface and $R(o)$ is the Fresnel reflection coefficient for normal incidence. The Fresnel reflection coefficients for vertical and horizontal polarization are: (air-water interface)

$$R_V = \frac{\epsilon \cos\theta - \sqrt{\epsilon - \sin^2\theta}}{\epsilon \cos\theta + \sqrt{\epsilon - \sin^2\theta}} \tag{4.3}$$

$$R_H = \frac{\cos\theta - \sqrt{\epsilon - \sin^2\theta}}{\cos\theta + \sqrt{\epsilon - \sin^2\theta}}, \tag{4.4}$$

respectively. θ is the angle of incidence of the EM radiation with respect to the normal to the surface and ϵ is the relative dielectric constant of the surface. For volume scattering (4.2) becomes [4]

$$\sigma_{o\infty} = \pi \sec^4\theta \, p(\zeta, \zeta_{xs}, \zeta_{ys}) \, |R(o)|^2, \quad \text{(volume)} \tag{4.5}$$

where $p(\zeta, \zeta_{xs}, \zeta_{ys})$ is the joint-probability density of surface height and slopes. Equations (4.1), (4.2) and (4.5) are denoted as the "specular point model". Seltzer [57] generalized the above results including third derivative phase terms in the analysis. With this modification, the scattering results are more complicated and contain a weak frequency dependence.

To apply (4.2) and (4.5) to the ocean, we need the probability density of slopes and the relative dielectric constant of the ocean to compute the Fresnel reflection coefficient at normal incidence. Cox and Munk [11] obtained this distribution and found it obeyed the Gram-Charlier distribution. However, the Gaussian distribution may serve as a first approximation. The relative dielectric constant of sea water is complex $\epsilon = \epsilon' - i\epsilon''$ (for free space $\epsilon = 1$) and was investigated by Saxton and Lane [54]. They found that

$$\epsilon' = \frac{\epsilon_s' - 4.9}{1 + \omega^2\tau^2} + 4.9 \tag{4.6}$$

and

$$\epsilon'' = \frac{(\epsilon_s' - 4.9)\omega\tau}{1 + \omega^2\tau^2} + \frac{4\pi\sigma_i}{\omega}, \tag{4.7}$$

where ϵ_s' is the static dielectric constant, σ_i is the ionic conductivity and τ is the relaxation time. These three quantities depend on temperature and salinity of the sea water. For sea water at 20°C and a salt consentration of 0.5 Normal $\epsilon_s' = 71$, $\tau = 9.2 \times 10^{-12}$ sec and $\sigma_i = 0.4 \times 10^{11}$ e.s.u. Hence, the ocean surface is a lossy conductor in the HF region and a lossy dielectric for centimeter radiation.

Clearly, from conditions (3.8) the probability distribution of slopes to be used in (4.2) for the ocean depends on the frequency of the EM radiation and the mean-square slopes to be used for the distribution should only contain the slopes from the dominant waves in the ocean (undulating surface). It should also be clear that the reflection coefficient to be used in (4.2) and (4.5) in practice is not the Fresnel reflection coefficient, but an effective reflection coefficient dependent on the reflectivity of the facets of the dominant waves in the presence of the higher frequency waves in the ocean (short-gravity, gravity-capillary and capillary waves) [71].

Figure 3 shows the backscatter to a nadir-looking altimeter mounted on a tower [83], as the ocean waves travel past. The largest return comes from the trough of the waves which are flatter and smoother.

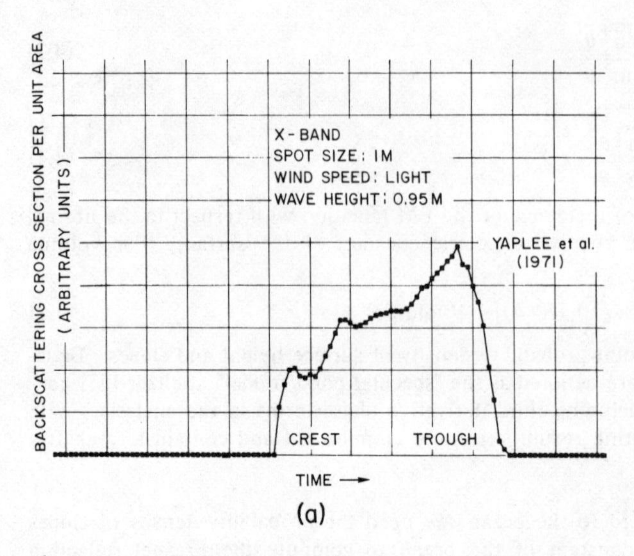

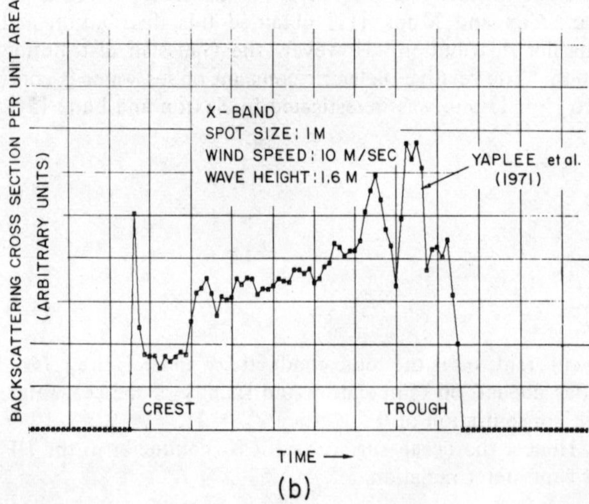

Fig. 3. Ocean backscatter to nadir-looking, tower-mounted altimeter a) light wind b) 10 m/sec wind.

4.2 Resonant (Bragg) model

For EM radiation less than 100 MHz, the ocean surface scatters as a true "diffraction grating". This fact was first observed by Crombie [12] using 13.56 MHz radio waves. The backscatter at grazing incidence is contributed by ocean waves one-half the EM wavelength travelling parallel to the line of radar sight. The doppler shift of the return is identical to the frequency of these ocean waves. Later, in more detailed measurements in a wave-tank, Wright [78] found that this same mechanism is responsible for the backscatter of 3 cm radiation by mechanically generated water waves if the height of the water waves is small compared to the wavelength of the EM radiation. Another characteristic of this scattering mechanism for water surfaces is the large polarization ratio (the return or vertical polarization is much larger than the return for horizontal polarization) away from normal incidence, and toward grazing incidence.

Surprisingly, Rice [52] had already developed a perturbation scattering formulation that could explain this resonant scattering. The scattering formulation applied to slightly rough surfaces (the surface displacement ξ is small compared to the wavelength of the EM radiation and the slopes of the surface are gentle). The polarization ratio predicted by this formulation explains the observations on water surfaces. Peake [46] applied Rice's linear results to vertical polarization and dielectric surfaces. According to this first-order theory, only water waves of wavenumber

$$K = 2k \sin \theta \tag{4.8}$$

contribute to the backscatter, which at grazing incidence ($\theta = 90°$) explains Crombie's observations. For bistatic condition longer water waves contribute.

In first-order, the backscatter cross-sections per unit area are given by (the plane of incidence is in the $y = 0$ plane)

$$\sigma_o^{(1)}(\theta)_{ij} = 16\pi k^4 \cos^4\theta |g_{ij}^{(1)}(\theta)|^2 F(2k\sin\theta, o), \tag{4.9}$$

where k is the wavenumber of the EM radiation, θ is the angle of incidence with respect to the surface normal, $g^{(1)}$ are the first-order scattering coefficients and $F(K_x, K_y)$ is the two-dimensional (Cartesian) wavenumber spectrum of the ocean evaluated at the Bragg resonant condition. The spectrum has been normalized so that $\overline{\xi^2} = \int F(\mathbf{K})\, d\mathbf{K}$. The scattering coefficients are:

$$g_{HH}^{(1)}(\theta) = \frac{(\epsilon - 1)}{\left[\cos\theta + (\epsilon - \sin^2\theta)^{1/2}\right]^2} \quad \text{(horizontal)} \tag{4.10}$$

$$g_{VV}^{(1)}(\theta) = \frac{(\epsilon - 1)\left[\epsilon(1 + \sin^2\theta) - \sin^2\theta\right]}{\left[\epsilon\cos\theta + (\epsilon - \sin^2\theta)^{1/2}\right]^2} \quad \text{(vertical)}. \tag{4.11}$$

As mentioned earlier, the first-order doppler spectrum $\sigma_o^{(1)}(\omega_D, \theta)$ according to this mechanism is a line-spectrum doppler shifted from the carrier frequency of the EM radiation by the frequency of the Bragg resonant water wave (water waves travel at their phase speed) [3]. Accordingly,

$$\sigma_o^{(1)}(\omega_D, \theta) = \sigma_o^{(1)}(\theta)\, \delta(\omega_D \pm \Omega), \tag{4.12}$$

where $\delta(...)$ is the Dirac delta function and $\Omega = \Omega(2k \sin\theta)$ is the radian frequency of the Bragg resonant water wave. At HF frequencies, where sea-water is of high conductivity it is convenient to modify Rice's development and derive the scattering results from the magnetic fields in the air, and with an impedance boundary condition obtain the electric fields at the surface. In this manner, a formulation for the "ground wave" propagation and scattering may be developed [2].

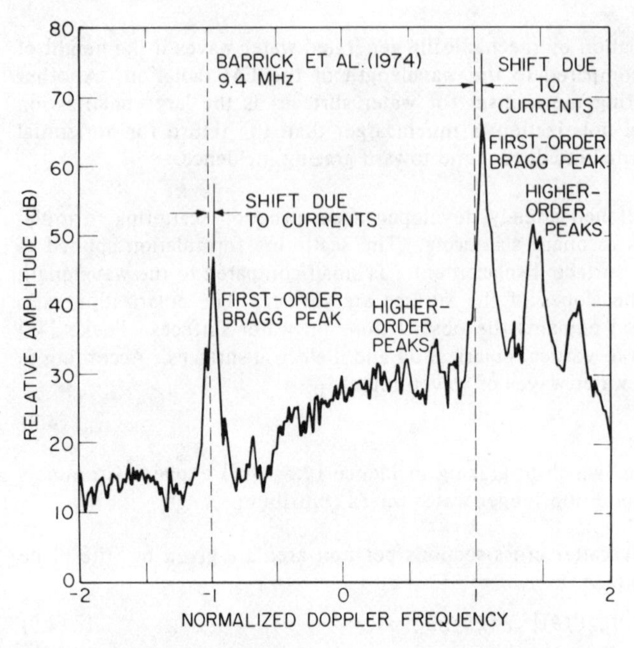

Fig. 4. Typical HF doppler spectrum.

For wind-waves the dispersion relation of short-gravity waves should be modified to include the advection by the wind-induced surface drift which amounts to 3 to 4 per cent of the wind speed and is highly sheared with depth [59, 80]. Finite amplitude effects to the phase speed or irrotational water waves were recently investigated [6, 32, 74]. For the shorter-gravity wind-waves which are rotational waves (with vorticity), no finite amplitude corrections to the phase speed have been obtained to our knowledge [51].

Higher order contributions are also present in the EM scattering, but generally these are smaller than the first-order contributions (at least initially), in particular for frequencies in the HF region or lower. In Figure 4 the doppler spectrum at HF frequencies is shown from work by Barrick et al. [5]. The second-order contributions introduce side-bands on each side of the first-order Bragg lines which are replicas of the low frequency spectrum of the ocean if the Bragg resonant wave is shorter than the dominant wave of the ocean [27]. Second-order doppler contributions were obtained in a more exact manner by Barrick [4] and from a generalized theory including surface tension, the dielectric constant of the surface and viscosity [68]. Originally second-order contribution to the cross-sections were obtained by Valenzuela [65, 66].

In a more explicit manner, the second-order doppler spectrum is the result of contributions by all pair of water waves which add-up to the Bragg resonant condition

$$\sigma_o^{(2)}(\omega_D, \theta) = 64\pi k^4 \cos^4\theta |g^{(2)}(\theta)|^2 \cdot \iint T^{(2)} F(\mathbf{K}_1) F(\mathbf{K}_2) \delta(\omega_D - s_1\Omega_1 - s_2\Omega_2) \, d\mathbf{K}_1, \tag{4.13}$$

where $T^{(2)}$ is the transfer coefficient including EM and hydrodynamic contributions. Ω_1 and Ω_2 are angular frequencies of water waves $\mathbf{K}_1$ and $\mathbf{K}_2$ satisfying the interaction condition

$$2k \sin\theta = s_1\mathbf{K}_1 + s_2\mathbf{K}_2, \qquad \omega_D = s_1\Omega_1 + s_2\Omega_2, \tag{4.14}$$

where s_1 and s_2 are sign indices ($\pm$).

As the ocean becomes rougher with wind-speed or EM frequency, higher-order contribution also become significant to the scattering. A dramatic demonstration of this fact was obtained by Wright [81] in a wave-tank (Figure 5). The complete hierarchy of doppler spectra were recorded starting with very light winds. Initially we have the line-spectrum typical of first-order Bragg scattering and for high winds finally we observe the continuum doppler spectrum typical of the return at microwaves from the ocean (two-scale surface).

4.3 Composite-surface model

The polarization ratio in first-order Bragg resonant backscatter is large for water surfaces, but in the microwave region this ratio decreases with the averaging of the slightly rough cross-section by the slopes of the dominant wave in the ocean (two-scale scattering) (Figure 6). In comparison the polarization ratio for land is smaller than for water (Figure 7) [13]. Grasping the physical significance of the reduction of the polarization ratio with sea roughness, Wright was led to the formulation of the composite-surface scattering model [79]. This scattering model was also developed independently by the Soviet scientists [8, 58].

The composite-surface model has been quite successful for predicting the ocean backscatter. The cross-sections for a slightly rough surface, equations (4.9), are averaged over the distribution of slopes of the dominant waves in the ocean. In this averaging procedure, proper reference must be kept of the polarization in relation to the local plane of incidence ("tilts" normal to the plane of incidence require special attention of the polarization of the local fields) [66, 79].

At normal incidence, the backscatter is of specular nature. Therefore, in first approximation, the cross-section for a composite-surface model may be obtained by adding the cross-sections from the specular-point model (4.2) and the cross-section for first-order Bragg resonant scattering (4.9) (this later property averaged over the distribution of slopes of the dominant waves in the ocean) [71]. Accordingly, for frequencies above 100 MHz the cross-section per unit area of the sea is approximately given by

$$\sigma_{os}(\theta)_{ij} = \sigma_{o\infty}(\theta) + \langle\sigma_o^{(1)}(\theta')\rangle_{ij}, \tag{4.15}$$

where the brackets on the second term denote the ensemble average over the distribution of slopes of the dominant waves (in this averaging local polarization effects must be included). θ' is the local angle of incidence. For vertical polarization, $\langle\sigma_o^{(1)}(\theta')\rangle_{VV} \approx \sigma_o^{(1)}(\theta)_{VV}$ which yields

$$\sigma_{os}(\theta)_{VV} \approx \sigma_{o\infty}(\theta) + \sigma_o^{(1)}(\theta)_{VV}. \tag{4.16}$$

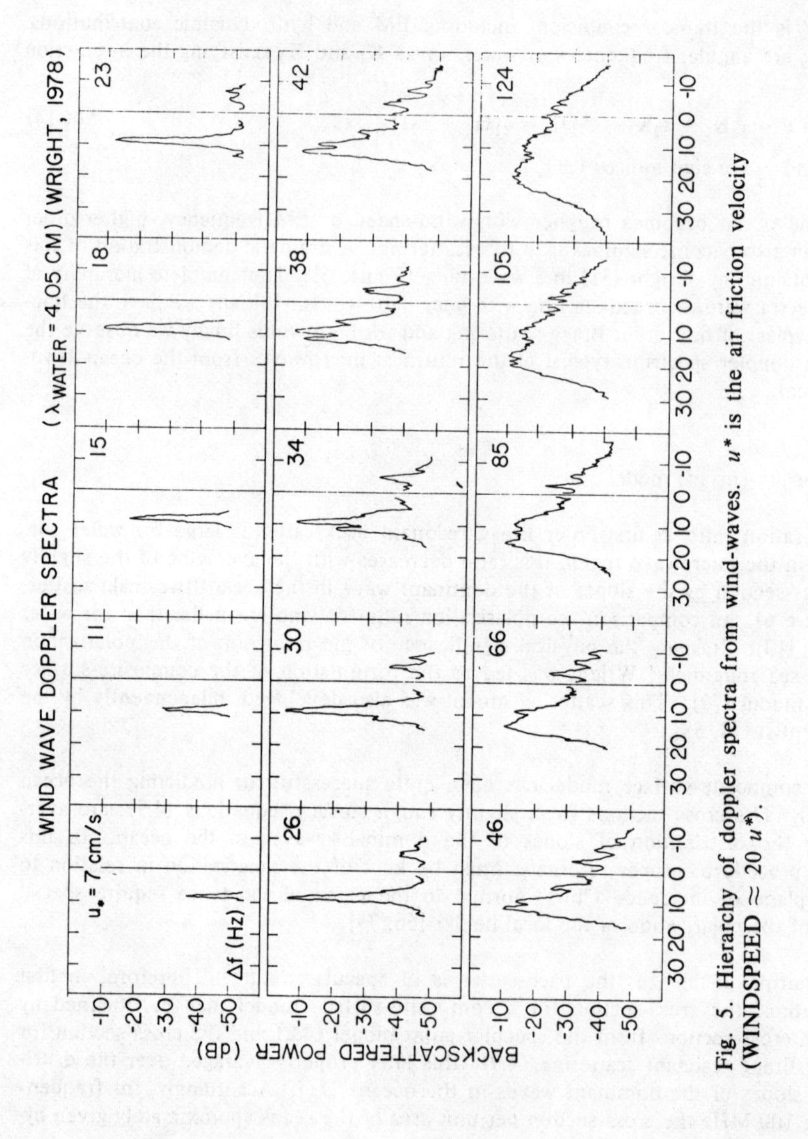

Fig. 5. Hierarchy of doppler spectra from wind-waves. u^* is the air friction velocity (WINDSPEED $\simeq 20\,u^*$).

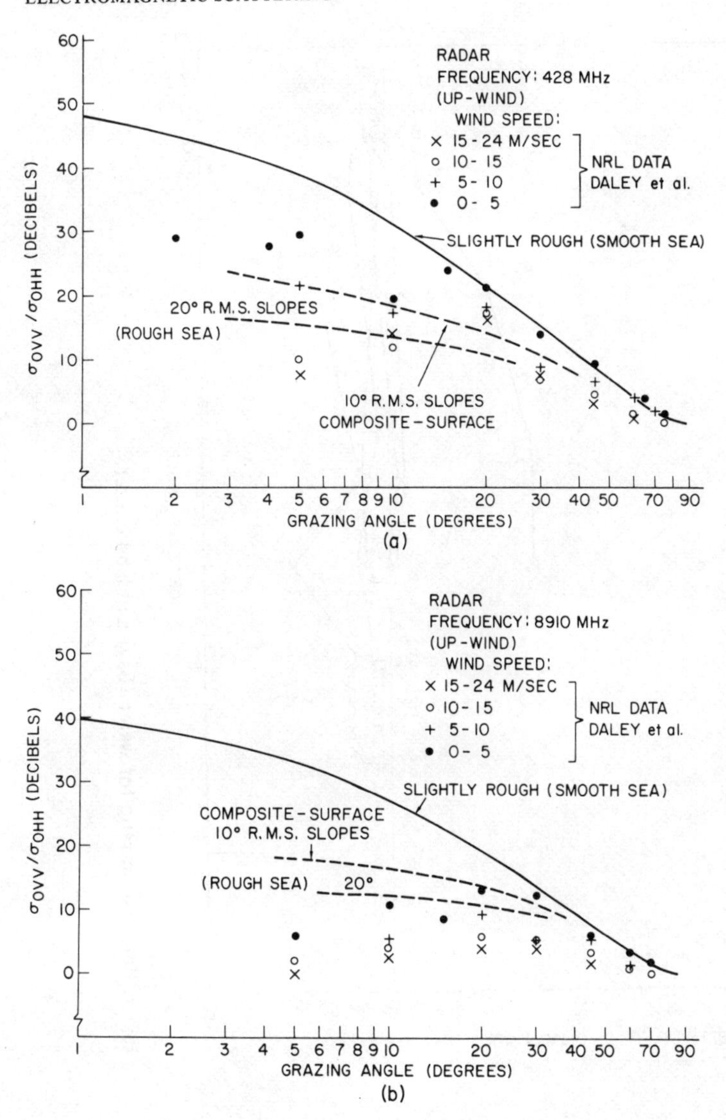

Fig. 6. Polarization ratio for the ocean. a) 428 MHz b) 8910 MHz. References in [71].

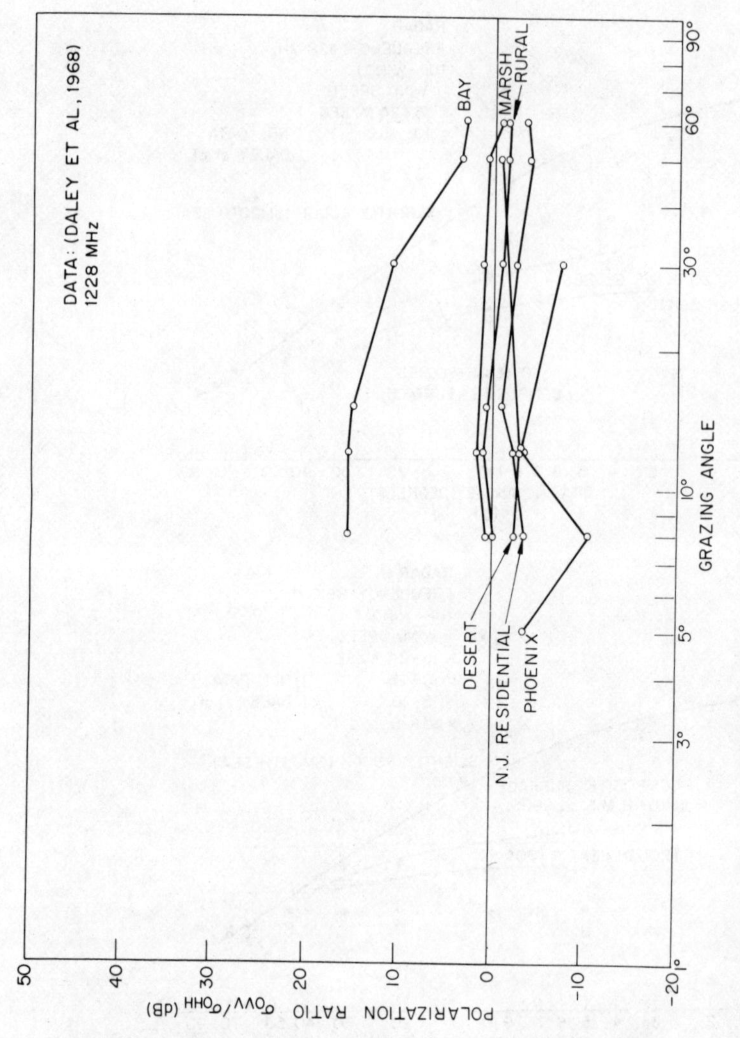

Fig. 7. Polarization ratio for water (Bay) and land clutter.

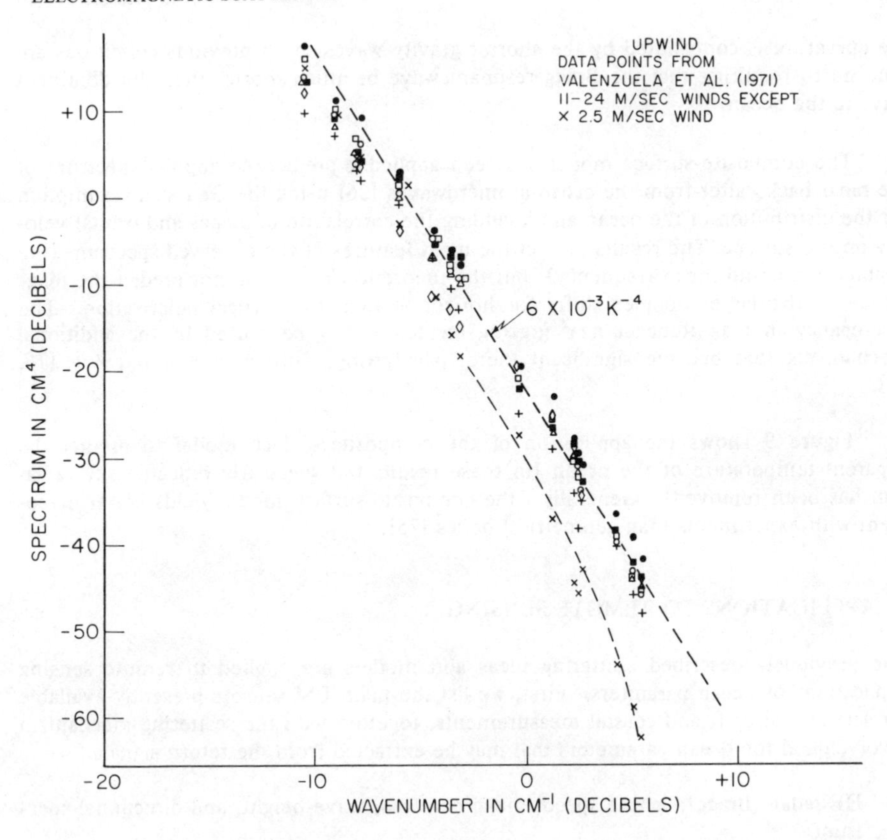

Fig. 8. Ocean wavenumber spectrum by inversion of vertically polarized cross-section, (up-wind) reference in [71].

Using this fact in Figure 8, we have obtained ocean wavenumber spectrum from radar cross-section data for water waves near 1.7 cm for 2.5 m/sec winds is typical for water waves. For the higher winds, the spectrum follows closely the K^{-4} law (Ω^{-5} in angular frequency) for fully-developed seas in the "equilibrium range."

The composite-surface scattering model predicts a depolarized contribution away from normal incidence due to "tilts" normal to the plane of incidence [79]

$$\sigma_{os}(\theta)_{VH} = \sigma_{os}(\theta)_{HV} \approx \frac{4\sin^2\theta}{(1+\sin^2\theta)^2}\,\sigma_p^2 \cdot \sigma_o^{(1)}(\theta)_{VV}, \tag{4.17}$$

where σ_p^2 is the mean-square slopes of the dominant waves of the ocean in a direction perpendicular to the line of sight.

The conditions for the applicability of the composite-surface model are (3.8) for the dominant waves in the ocean and the slightly rough condition (gentle slopes and small amplitude Bragg scatterers in relation to the wavelength of the EM radiation). Actually, for the ocean surface where slope is contributed by the dominant waves and the curvature is contributed by the shorter gravity waves, both previous conditions are reduced to requiring that the Bragg resonant wave be much shorter than the dominant wave in the ocean.

The composite-surface model has been applied to predict the doppler spectrum of the radar backscatter from the ocean at microwaves [26] using the Gaussian assumption for the distribution of the ocean and including the correlation of slopes and orbital velocity on the surface. The results predict the main features of the observed spectrum (the doppler width and mean frequency), but the theoretical results do not predict the magnitude of the larger doppler shifts for horizontal than for vertical polarization. The discrepancy in magnitude at near grazing incidence may be caused by the additional mechanisms that become significant there: shadowing, diffraction and trapping [76, 77].

Figure 9 shows the application of the composite-surface model to predict the apparent temperature of the ocean (in these results the specularly reflected sky radiation has been removed). Generally , the composite-surface model yields better agreement with experiments than geometrical optics [75].

5. APPLICATIONS TO REMOTE SENSING

The previously described scattering ideas and models are applied to remote sensing applications of ocean paramters. First, we list the main EM sensors presently available for satellite, aircraft and coastal measurements, together with the scattering mechanism involved and the ocean parameters that may be extracted from the return signals.

HF radar (Bragg): phase speed, mean currents, wave-height, and directional spectrum.

Altimeter (specular): wave-height, geoid, tides, geostrophic currents and mesoscale eddies.

Scatterometer (composite-surface): wind-speed and wavenumber spectrum

Coherent microwave radar (composite-surface): orbital velocity, gravity-capillary spectrum, modulation of short-gravity waves, and wave-height.

Dual-frequency radar (composite-surface): phase speed of long gravity waves, mean currents, tides, modulation of short-gravity waves.

Radiometer (composite-surface): apparent ocean temperature

SAR, SLAR (composite-surface): direction and wavenumber of dominant waves, and large-scale ocean patterns (current, oil slicks and modulation of short-gravity waves).

In the next sub-sections, we expand somewhat on these sensors and specifically report some recent results on the HF radar, altimeter, dual-frequency radar and the SAR systems.

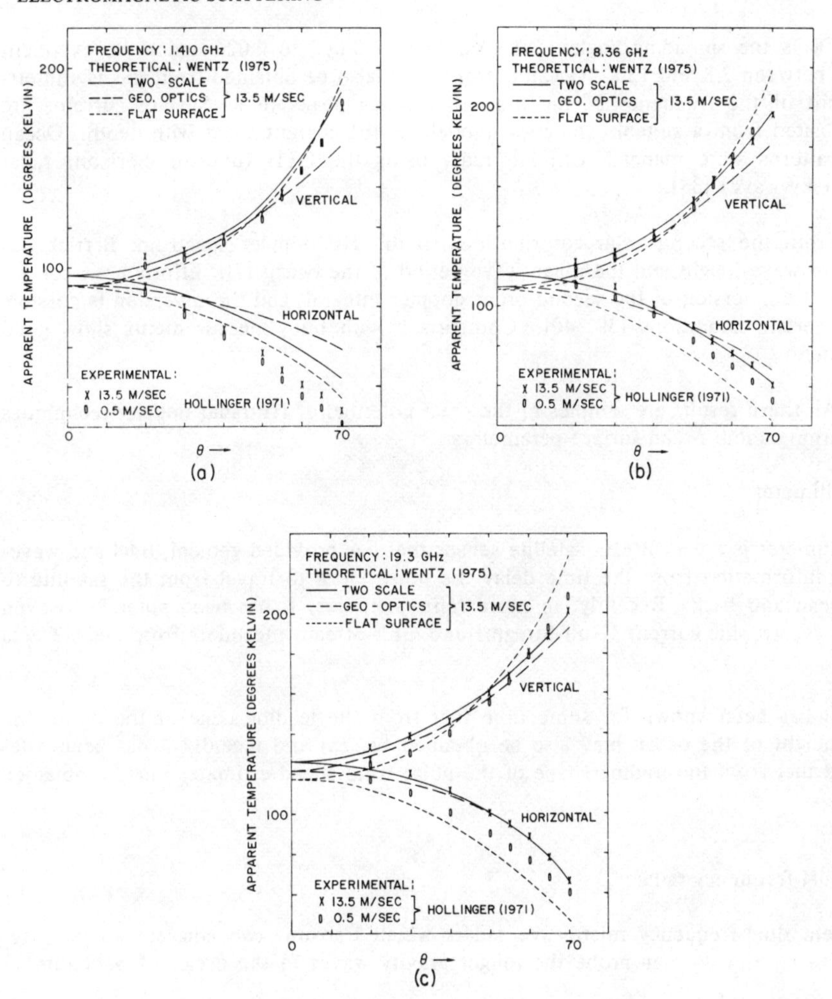

Fig. 9. Application of the composite-surface (two-scale) model to the apparent temperature of the sea. a) 1.41 GHz, b) 8.36 GHz and c) 19.36 GHz. (The specularly reflected sky-radiation has been removed)

5.1 HF radar

From the first-order peaks of the HF doppler spectrum (refer to Figure 4), a number of ocean parameters may be recovered. For example, Tyler et al. [64] obtained the directional distribution of ocean waves, and find that it is well approximated by the expression

$$G(\psi) = a + (1 - a) \cos^r(\psi/2) \quad |\psi| \leqslant \pi, \tag{5.1}$$

where ψ is the spreading angle. "a" ranged from 0.003 to 0.026 and r the exponent varied between 2.8 and 12.8. Ocean currents may also be obtained from the asymmetrical shift of the frequency of the first-order Bragg peaks [62]. Surface currents are extrapolated from a suitable analytical model for the current shear with depth. Ocean wind-patterns were mapped with HF radar using the OTH (over-the-horizon) radar mode (sky-wave) [38].

From the second-order contributions to the HF doppler spectrum, Barrick has obtained wave-height and the mean-wave period of the ocean [7]. Efforts have continued on the inversion of the second-order doppler integral, and the inversion is possible under certain conditions [39, 40]. Comparison with buoy-measurements show good agreement.

All these results are samples of the great potential of HF radar doppler techniques for learning about ocean surface parameters.

5.2 Altimeter

The altimeter is a very useful satellite sensor that has provided geoidal, tidal and wave-height information from the time delay the pulse takes to travel from the satellite to the ocean and back. Recently, in more refined studies, it has been possible to even obtain geostrophic current (Gulf-Stream) and Gulf-Stream meanders from Geos-3 data [33].

It has been known for some time that from the leading edge of the pulse, the wave-height of the ocean may also be obtained [4, 28] and recently it has been established that from the trailing shape of the pulse wind-speed estimates may be obtained [23].

5.3 Dual-frequency radars

Coherent dual-frequency microwave radars which transmit two adjacent angular frequencies ω_1 and ω_2 may probe the longer gravity waves in the ocean of wavenumber [49]

$$K = \frac{2(\omega_1 - \omega_2)}{c} \sin\theta, \tag{5.2}$$

where c is the speed of light and θ is the angle of incidence. If E_1 and E_2 represent the backscattered fields at ω_1 and ω_2, the Fourier transform of the autocorrelation of the product $E_1 E_2^*$ yields a narrow-line doppler spectrum at the angular frequency of the long gravity wave (5.2) resonant to the difference frequency. The magnitude of this line-spectrum is related through a "transfer" function to the energy of the resonant long gravity wave. The "transfer" function depends on the modulation of the short-gravity waves by the longer gravity waves [81]. Initial measurements were performed with a c.w. (continuous wave) system of 3 cm radiation (X-Band) [49]. A pulsed L-Band dual-frequency radar system has been developed [56] and recently Schlude et al. [55] of DFVLR are attempting to develop an airborne version of the system.

The modulation of short-gravity waves by longer gravity waves is actually important to interpret other sensors as well, for example, the SAR imaging radar.

5.4 The SAR system

Great expectations have been placed on the SAR radar system that will operate on SEASAT-A satellite to record on a global basis ocean information. Presently, most SAR measurements have been from airborne systems. For example SAR images may provide information on ocean surface features, direction and period of the dominant waves in the ocean [17, 37, 60]. Figure 10 is an SAR image of the north wall of the Gulf-Stream in which a filament of entrained cold water is visible. The resolution in this image is about 17m × 50m.

According to Bragg scattering the backscattered power to the EM sensors is proportional to the energy of the Bragg resonant waves and oil slicks and other natural surfactants damp water waves shorter than 1 meter. Hence, areas in the ocean which are contaminated give less backscatter and they appear in the SAR images as darker patches [21, 48]. Figure 11 is an example of an oil slick in a SAR image.

The damping rates of short-gravity waves by surface filter may be obtained theoretically [16] and this prediction has been proven experimentally [14]. Garrett [18] has also obtained damping rates of capillary waves by natural ocean slicks. The input from the wind has also been incorporated to the theoretical predictions [20]. However, no estimate has been obtained for the damping of short-gravity waves for a contaminated patch of finite size. To treat this more complex problem one would have to use the "transport" equation [25] and include the input from the wind and the energy transfer by nonlinear resonant interactions.

LARSON ET AL., 1976
1228 MHz
SAR IMAGE OF GULF STREAM
16 x 26 Km AREA

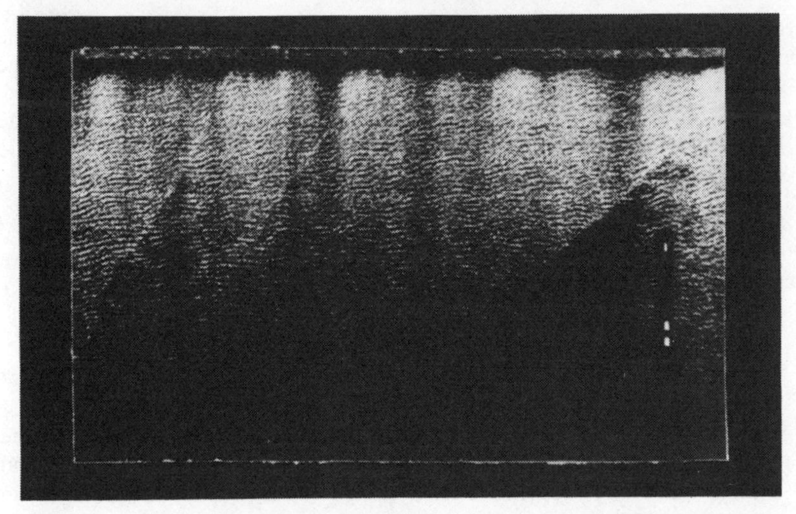

Fig. 10. SAR image of Gulf-Stream at 1228 MHz. (The resolution is 17 m × 50 m.)

Fig. 11. SAR image of oil slick near coast at 1228 MHz. (The resolution is 17 m × 50 m.)

SAR images also have been useful in investigating the age and roughness of sea-ice [21]. Figure 12 is an image of Alaskan frozen tundra and lakes.

There are two main processes which cause distortion in SAR images from the ocean. One is the modulation of short-gravity waves by the dominant waves in the ocean [34], which introduces an additional amplitude weighting to the radar return. The second mechanism is the distortion of the image due to phase errors caused by the advection of the Bragg scatterers by the orbital motion of the dominant waves in the ocean. Investigations still are continuing on these two mechanisms.

At this point, we would like to add a little more detail on the distortions produced by the orbital motion. In a simplified manner, let's describe the instantaneous return signal from one resolution cell in range to the airborne radar moving at the velocity V in the x direction

$$S(t) = \int_{-L/2}^{L/2} A(x, t) \exp\left\{i\left[\phi(x, t) - a\left(t - \frac{x}{V}\right)^2\right]\right\} dx, \tag{5.3}$$

where $A(x, t)$ is the amplitude weighting due to the modulated cross-section of the ocean, $a = kV^2/R$, R is the range to the resolution cell and $\phi(x, t) = 2\int^t \mathbf{k} \cdot \mathbf{U}(x, t)\, dt$. $\mathbf{U}(x, t)$ is the orbital motion of the dominant waves in the ocean. In conventional processing the phase error $-at^2$ is removed by a cylindrical lens (equivalent to the compression of the frequency spectrum of the return) and the

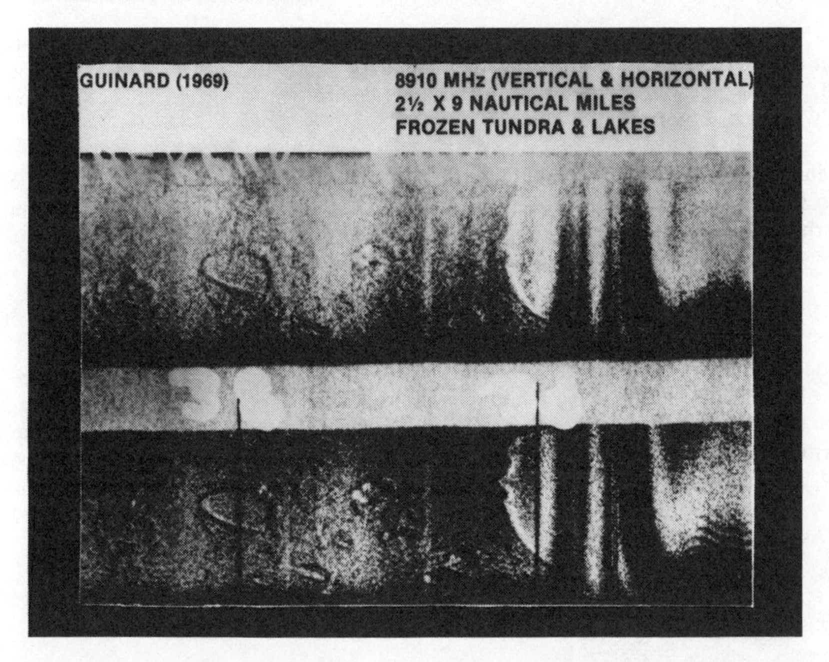

Fig. 12. SAR image of Alaskan frozen tundra and lakes at 8910 MHz. (The resolution is 17 m × 50 m.)

square of the Fourier transform is recorded on film [24]. Here, we process with an arbitrary phase correction $a't^2$ and the Fourier transform of the corrected signal return is

$$m(x') = \int_{-L/2}^{L/2} \int_{-T/2}^{T/2} A(x, t) \exp\left\{ i \left[\phi(x, t) - \frac{ax^2}{V^2} - (a - a')\, t^2 + \frac{2a}{V}(x - x')\, t \right] \right\} dx dt, \tag{5.4}$$

where x' is the spatial coordinate on the film, T is the integration time and L is the azimuth coverage of the radar antenna. Integrals of the form (5.4) when L and T are large are well known in diffraction theory [10]. Contributions to the integral come mainly from the critical points of the integral (stationary phase points). Assuming that $A(x, t)$ is a slowly changing or constant amplitude weighting, the critical points are obtained from

$$\left\{ \begin{array}{l} -\dfrac{ax_o}{V^2} + \dfrac{at_o}{V} + \displaystyle\int \mathbf{k} \cdot \dfrac{\partial \mathbf{U}}{\partial x} \Big|_{x=x_o} dt = 0 \\ -(a - a')\, t_o + \dfrac{a}{V}(x_o - x') + \mathbf{k} \,.\, \mathbf{U}(x_o, t_o) = 0 \end{array} \right\} \tag{5.5}$$

If $a = a'$ and $\mathbf{U} = 0$, (5.4) is a map of the ocean scene. However, since $\mathbf{U} \neq 0$ an image of the ocean surface is a map of the stationary points of (5.4). In (5.5) if a' is changed, various portions of the ocean surface will be brought to focus. This effect was originally suggested by Larson et al. [37].

Hence, for a proper interpretation of SAR images from the ocean surface, one needs to investigate in more detail the solution of (5.5) for a realistic parameterization of the orbital motion of the ocean.

6. CONCLUSIONS

The understanding of the physical processes involved in the interaciton of EM waves and the ocean surface has reached a high level of maturity. This should attract oceanographers and geophysicists to use the powerful techniques of remote sensing of the ocean with EM sensors in the near future, and to obtain ocean parameters with more accuracy and higher resolution than has yet been possible with conventional oceanographic experimental methods. Furthermore, with the launching of SEASAT-A and later SPACELAB, it will be possible to have routine global information on the oceans.

REFERENCES

1. D. E. Barrick, *IEEE Trans.* (4), **AP-16**, 449, 1968.
2. D. E. Barrick, *Radio Sci.* (5) **6**, 517, 1971.
3. D. E. Barrick, *IEEE Trans.* (1), **AP-20**, 2, 1972.
4. D. E. Barrick, Remote Sensing of Sea State by Radar, in: *Remote Sensing of the Troposhere*, ed. by V. Derr, U.S. Govt. Printing Office, Washington, 1972.
5. D. E. Barrick, et al., *IEEE Proc.* (6), **62**, 673, 1974.
6. D. E. Barrick and B. L. Weber, *J. Phys. Ocean*, (1), **7**, 11, 1977.
7. D. E. Barrick, *Radio Sci.* (3), **12**, 415, 1977.
8. F. G. Bass et al., *IEEE Trans.* (5), **AP-16**, 554, 1968.
9. P. Beckmann and A. Spizzichino, *The Scattering of Electromagnetic Waves from Rough Surfaces*, Macmillan, New York, 1963.
10. M. Born and E. Wolf, *Principles of Optics*, Macmillan, New York, 1964.
11. C. Cox and W. Munk, *J. Opt. Soc. Am.* (11), **44**, 838, 1954.
12. D. D. Crombie, *Nature* (4459), **175**, 681, 1955.
13. J. C. Daley et al., *NRL Terrain Clutter Study* (II), NRL Report 6749, 1968.
14. J. T. Davies and R. W. Vose, *Proc. Royal Soc. A.* (1405), **286**, 218, 1965.
15. J. A. DeSanto, *J. Math. Phys.* (3), **15**, 283, 1974.
16. R. Dorrestein, *Proc. Acad. Sci. Amst. B.*, **54**, 260, 1951.
17. C. Elachi and W. E. Brown, *IEEE Trans.* (1), **AP-25**, 84, 1977.
18. W. D. Garrett, *J. Marine Res.* (3), **25**, 279, 1967.
19. D. T. Gjessing, *Remote Surveillance by Electromagnetic Waves for Air-Water-Land*, Ann Arbor Science Publishers, Ann Arbor, 1978.
20. J. C. Gottifredi and G. J. Jameson, *J. Fluid Mech.* (3), **32**, 609, 1968.
21. N. W. Guinard, *Proc. 6th Int. Symp. Rem. Sens. Env.*, 737, 1969.
22. T. Hagfors, *J. Geophys. Res.* (18), **69**, 3779, 1964.
23. D. L. Hammond et al., *IEEE Trans.* (1), **AP-25**, 61, 1977.
24. R. O. Harger, *Synthetic Aperture Radar Systems*, Academic Press, New York, 1970.

25. K. Hasselmann, Weak-Interaction Theory of Ocean Waves, in: *Basic Developments in Fluid Dynamics*, ed. by M. Holt, Academic Press, New York, 1968.
26. K. Hasselmann and M. Schieler, *Proc. VIIIth Naval Hydr. Symp.*, 361, 1970.
27. K. Hasselmann, *Nature* (1), **229**, 16, 1971.
28. K. Hasselmann, *U.S. Nat. Oceanic and Atmospheric Adm. Rep.*, ERL-228, Boulder, 1972.
29. K. Hasselmann et al., *Deutsche Hydrogr. Z.* (12), **S.A8**, 1973.
30. K. Hasselmann, *Bound-Layer Meteor.* (2), **6**, 107, 1974.
31. D. Hasselmann, *J. Fluid Mech.* (3), **85**, 543, 1978.
32. N. E. Huang and C. Tung, *J. Fluid Mech.* (2), **75**, 337, 1976.
33. N. E. Huang et al., *J. Geophys. Res.*, (in press), 1978.
34. W. C. Keller and J. W. Wright, *Radio Sci.* (2), **10**, 139, 1975.
35. B. Kinsman, *Wind Waves*, Prentice-Hall, Englewood Cliffs, 1965.
36. R. Kodis, *IEEE Trans.* (1), **AP-14**, 77, 1966.
37. T. R. Larson et al., *IEEE Trans.* (3), **AP-24**, 393, 1976.
38. A. E. Long and D. B. Trizna, *IEEE Trans.* (5), **AP-21**, 680, 1973.
39. B. Lipa, *Radio Sci.* (3), **12**, 425, 1977.
40. B. Lipa, J. Geophys. Res. (C2), **83**, 959, 1978.
41. J. W. Miles, *J. Fluid Mech.* (2), **3**, 185, 1957.
42. K. M. Mitzner, *J. Math. Phys.* (12), **5**, 1776, 1964.
43. K. M. Mitzner, *Radio Sci.*, (12), **2**, 1459, 1967.
44. P. M. Morse and H. Feshbach, *Methods in Theoretical Physics*, McGraw-Hill, New York, 1953.
45. C. Müller, *Grundeprobleme der Mathematischen Theorie Elektromagnetischer Schwingungen,* Springer-Verlag, Berlin, 1957.
46. W. H. Peake, Theory of Radar Return from Terrain, in: *IRE Nat. Conv. Record* (1), **7**, 1959.
47. O. M. Phillips, *The Dynamics of the Upper Ocean*, Cambridge Univ. Press, Cambridge, 1966.
48. R. O. Pilon and C. G. Purves, *IEEE Trans.* (5), **AES-9**, 630, 1973.
49. W. J. Plant, *IEEE Trans.* (1), **AP-25**, 28, 1977.
50. W. J. Plant and J. W. Wright, *J. Fluid Mech.* (4), **82**, 767, 1977.
51. W. J. Plant and J. W. Wright (submitted to J. Fluid Mech.), 1978.
52. S. O. Rice, *Comm. Pure Appl. Math.* (2/3), **4**, 351, 1951.
53. G. T. Ruck et al., *Radar Cross-Section Handbook*, Plenum Press, New York, 1970.
54. J. A. Saxton and J. A. Lane, *Wireless Eng.* (349), **29**, 269, 1952.
55. F. Schlude et al., *ESA Meeting*, Toulouse, Fr., 1978.
56. D. L. Schuler, *Radio Sci.* (2), **13**, 321,1978.
57. J. E. Seltzer, A Third-Order Specular-Point Theory for Radar Backsctter, in: *Electromagnetic Wave Propagation Involving Irregular Surfaces and Inhomogeneous Media,*, ed. by A. Ince, AGARD-CP-144, 1975.
58. B. Semyonov, *Radio Eng. and Electron. Phys.* (8), **11**, 1179, 1966 (Translation).
59. O. Shemdin, *J. Phys. Ocean.* (2), **2**, 411, 1972.
60. O. Shemdin, et al., *Bound-Layer Meteor.*, **13**, 225, 1978.
61. A. B. Shmelev, *Usp. Fiz. Nauk.*, **106**, 459, 1972.
62. R. H. Stewart and J. W. Joy, *Deep Sea Res.* (12), **21**, 1039, 1974.
63 J. A. Stratton, *Electromagnetic Theory*, McGraw-Hill, New York 1941.
64. G. L. Tyler et al., *Deep Sea Res.*.(12), **21**, 989, 1974.
65. G. R. Valenzuela, *IEEE Trans.* (4), **AP-15** , 552, 1967.
66 G. R. Valenzuela, *Radio Sci.* (11), **3**, 1057, 1968.

67. G. R. Valenzuela and M. B. Laing, *J. Fluid Mech.* (3), **54**, 507, 1972.
68. G. R. Valenzuela, *J. Geophys. Res.* (33), **79**, 5031, 1974.
69. G. R. Valenzuela, *J. Fluid. Mech.* (2), **76**, 229, 1976.
70 G. R. Valenzuela and J. W. Wright, *J. Geophys. Res.* (33), **81**, 5795, 1976.
71. G. R. Valenzuela, *Bound-Layer Meteor.*, **13**, 9, 1978.
72. J. Van Bladel, *Electromagnetic Fields,*, McGraw-Hill, New York, 1964.
73. E. J. Walsh, *Radio Sci.* (8/9), **9**, 711, 1974.
74. B. L. Weber and D. E. Barrick, *J. Phys. Ocean.* (1), **7**, 3, 1977.
75. F. J. Wentz, *J. Geophys. Res.* (24), **80**, 3441, 1975.
76. L. Wetzel, *Radio Sci.*, (5), **12**, 749, 1977.
77. L. Wetzel, *Radio Sci.* (2), **13**, 313, 1978.
78. J. W. Wright, *IEEE Trans.* (6), **AP-14**, 749, 1966.
79. J. W. Wright, *IEEE Trans.* (2), **AP-16**, 217, 1968.
80. J. W. Wright and W. C. Keller, *Phys. Fluids* (3), **14**, 466, 1971.
81. J. W. Wright, *Bound-Layer Meteor.*, **13**, 83, 1978.
82. S. T. Wu and A. K. Fung, *J. Geophys. Res.* (30), **77**, 5917, 1972.
83. B. S. Yaplee et al., *IEEE Trans.* (3), **GE-9**, 170, 1971.

THE RADAR SIGNATURE OF NATURAL SURFACES AND ITS APPLICATION IN ACTIVE MICROWAVE REMOTE SENSING

E.P.W. Attema

Department of Electrical Engineering, Delft University of Technology, Delft, the Netherlands†

1. INTRODUCTION

The development of airborne and spaceborne remote sensing technology has expanded man's ability to observe the surface of the earth and to control its environment. The applications of the technique are in a wide variety of disciplines but, limiting ourselves to natural land surfaces the application areas, presently considered most promising, are geology, hydrology, agriculture and land use inventory.

Most of the research in this area has been focused on optical and infrared sensors as evidenced by LANDSAT. Landsat imagery has provided scientists and engineers with valuable information; its use, however, is limited to cloudfree conditions and in the interpretation process, tonal changes due to temporal sun-angle variations have to be accounted for.

In contrast radar sensors operating in the microwave portion of the electromagnetic spectrum are virtually weather and time-of-day independent. How important this feature is, depends on the region of the world, as can be seen in figure 1 showing mean annual cloud cover percentages.

The technology of imaging radars facilitate microwave viewing by performing a transformation of the microwave spectrum, for which the human eye is not sensitive, to the optical portion of the spectrum.

---

† Delft University of Technology, Postbox 5031, Mekelweg 4, 2600 GA Delft, the Netherlands

T. Lund (ed.), Surveillance of Environmental Pollution and Resources by Electromagnetic Waves, 227–252.

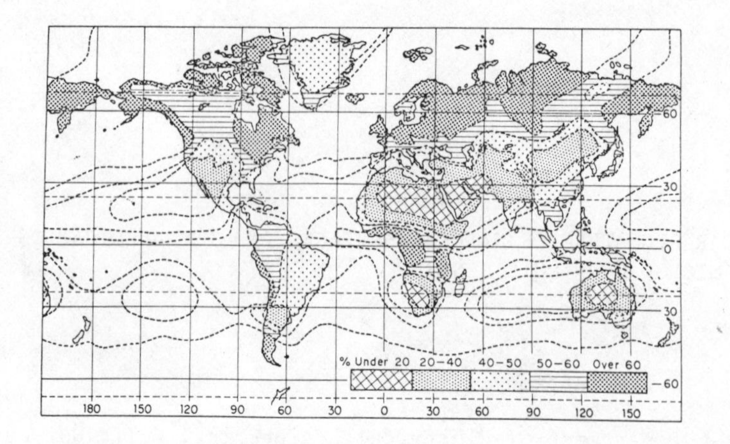

Fig. 1 Mean annual cloud cover in percent of sky covered (From Rumney [1])

Airborne radar systems are presently in use for a variety of applications and although radar lags behind optical sensors in terms of its development as a space platform operational sensor, the feasibility of a spaceborne radar has been demonstrated by skylab's non-imaging radar scatterometer and current satellite projects such as Seasat and the shuttle series include synthetic aperture imaging radars as part of their payloads.

An effective use of active microwave remote sensing requires an understanding of the interaction of the electromagnetic waves with the objects of interest. Because it is the echo signal that carries all the information a radar sensor provides about the object, the characteristics of the scattered signal and their relation with the target parameters such as moisture content, vegetation cover and surface profile are of interest.

These characteristics or what is called the radar signature of the object facilitate the evaluation of the potential applications of radar remote sensing, define the sensor parameters to be chosen for a particular application and finally provide the tools for the interpretation of the images produced.

In the present presentation an example of microwave viewing technology will be discussed, the general characteristics of the radar echo from natural land surfaces especially vegetation and bare soil will be presented, the theoretical and experimental evidence available to date about radar signatures will be reviewed and finally application areas currently under investigation will be identified.

2. MICROWAVE IMAGE FORMATION

Unlike the ordinary imaging radars commonly in use for radio location and navigation, that have a rotating stationary aerial and image presentation on a plane position indicator, imaging radars for remote sensing are usually mounted on a moving platform. One example of this is the side looking radar mounted alongside an aircraft or satellite. The operation of such a radar is as follows. The transmitter radiates a short pulse of microwave energy through the antenna. Echoes from the targets on the earth surface are scattered back and received subsequently after the round trip delay according to the distance to the target and the wave velocity. Since electromagnetic wave propagation velocity very nearly equals the speed of light, the delay times are very short and consequently it takes only 200 μsec to scan over a 30 km swath. The received echoes are displayed in the order of arrival, for instance on a cathode ray tube, in the form of small light dots. The position of which being proportional to distance and the light intensity proportional to the echo strength. Figure 2 shows the geometry and system components of the side looking airborne radar of SLAR.

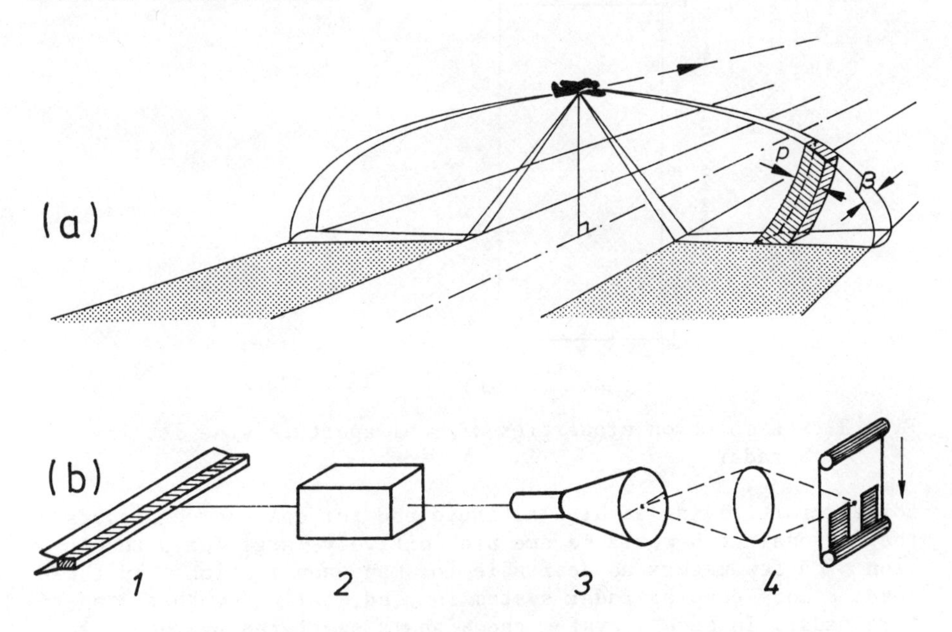

Fig. 2. Side looking airborne radar (SLAR)
2a. Scan configuration
2b. Radar system components
antenna (1); transmitter/receiver (2); image display tube (3); optical recording system (4)

After the reception of the last echo from the far end of the swath, another pulse can be transmitted and the next strip can be mapped. An optical projection of the display on film can produce a hard copy of the received echoes. If the film speed is proportional to the aircraft groundspeed a continuous image of the scene is obtained. Geometrical distortion mainly stems from the fact that radar operation is along the slant range rather than ground range and from platform motion deviations from the ideal flightpath, but corrections can be readily applied to the image from the motion and orientation sensors on board the platform.

Geometrical resolution obtained in the system in cross track direction is determined by the pulse width or the bandwidth of the radarsystem. The along track resolution is determined by the antenna beamwidth. For real aperture systems this beamwidth is given by the ratio of the antennalength and the wavelength. It follows from the spherical wave propagation of electromagnetic waves that the along track resolution linearly decreases with distance. From figure 3 the required antennalength for a given frequency can be readily determined on the basis of the required resolution at a given distance.

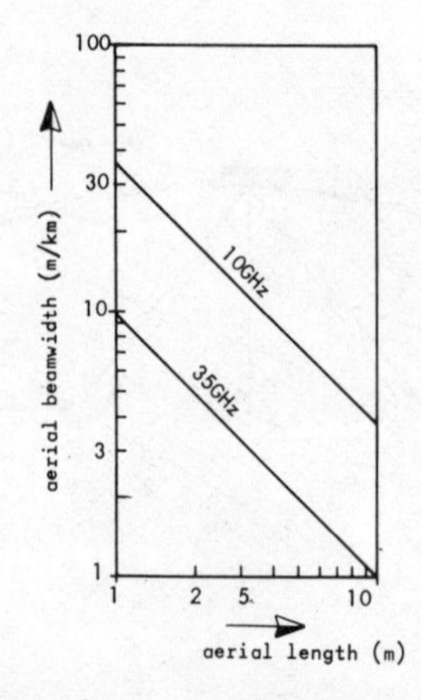

Fig. 3. Resolution properties of real aperture side looking radar

For aircraft flying at high altitude and for spaceborne sensors the antennalength would become prohibitively large for a resolution of a few meters as desirable for many applications. In these cases a more complex radar system is used, called synthetic aperture radar. In such a system the signals scattered by the object from subsequent pulses are coherently processed while the platform is moving along the flight track. In doing so an effective resolution can be obtained equivalent to a huge synthetic antenna-array extending to a distance over which the object is in the field-of-view of the radar. Because the length of the synthetic array increases with target range an along track resolution independent of range can be obtained. The theoretical limit for this resolution is equal to half the length of the azimuth dimension of the antenna actually on board the platform. As an example of this capability: the seasat launched this year will carry an L-band synthetic aperture radar having approximately the same geometrical resolution as the high resolution the thematic mapper planned to be mounted on Landsat-D.

3. RADAR IMAGERY INFORMATION

After having introduced radar sensor technology let us have a look at the first imagery. Figure 4 shows a radar image of the Netherlands (at the bottom) and in comparison a Landsat image of the same area.

Fig. 4. LANDSAT image of the Netherlands (top) and a SLAR image (bottom) in comparison

Geometry looks quite similar but apparently different grey levels are assigned to corresponding areas. Even more strikingly figure 5 shows the differences between an ordinary photograph of an aircraft and the radar image.

Fig. 5. Airplane photograph compared to corresponding radar imagery

Although both images characterise the object, they are fundamentally different. These examples illustrate the problem we are concerned with in the present discussion: what do we see on radar imagery and what information about the target can be obtained.

There is the object-interaction problem, the problem of selecting the proper sensor parameters and finally there is the interpretation problem. With respect to microwave viewing we have to learn to see all over again. Unfortunately in most places of the world the initial approach to the problem has been an "a posteriori"approach: most of the attention being given to the output end of the system in data processing. However, this is only a part of the problem.

In our childhood we did a lot of short range basic research to try and understand the interaction of our eyes with the objects we were interested in, before we started to look out of the window. In remote sensing the situation is even more difficult because then we opened our eyes without any concern about frequency, polarization or sensitivity, but in remote sensing we have to select the proper sensor parameters for a particular application, since there is no computersystem or algorithm that can compensate for the lack of information in an image due to inadequate sensitivity, insufficient resolution or incorrect choice of frequency or polarization. And if we fail to demonstrate the usefulness of the technique to the user, who after all paid for the experiment, he will be very reluctant for some time to consider the introduction of remote sensing techniques again.

The information content of radar images originates from three important radar capabilities.

The first one is the ability to measure distance. This will be possible provided the element of the scene to be detected scatters enough energy back into the receiver and exhibites enough contrast with its surroundings. This capability of radar is the best known and although it will therefore not receive too much emphasis it has many important applications in the area of geology, mineral exploration, landuse planning, water resource survey and the like.

The second is the ability to measure echo strength.The greytone representing the backscattered power intensity and its relation to the target parameters, analysed in much more detail later on, is the key to most applications in agriculture and hydrology, such as crop monitoring and the estimation of soil moisture.

Finally the texture of the image is a possible key to assist in classification. For land applications, however, specific texture differences are partly obscured by the common Rayleigh type signal fluctuations discussed later on.

4. BASIC SIGNATURE STUDIES

The objectives of basic signature studies are:

- identifying potential applications of active microwave remote sensing,
- acquiring a signature databank for the selection of appropriate sensor parameters for a particular application,
- providing tools for data interpretation in an operational remote sensing service.

Accomplishing these objectives effectively requires an understanding of the wave interaction with the object. This insight can be obtained by carefully designed experiments and by theoretical and empirical modelling on the basis of the experimental data. Research of this type has been reported by the Ohio State University [2] and presently most of the research in this area in the USA is carried out by the University of Kansas Remote Sensing Laboratory, where groundbased scatterometer measurements are carried out in a frequency range presently extending from 1 to 36 GHz over various vegetated and bare soils throughout the growing season [3]. Research of the same type started in the Netherlands in 1968 using pulse type radar systems operated from TV-towers. The program was extended in 1973 to mobile short range scatterometry using pulse-type as well as FM-CW scatterometers [4,5].

Presently the program is carried out by an interdisciplinairy research team called ROVE in which biologists, physicists and radar specialists cooperate. In this team participate the Centre for Agrobiological Research (CABO, Wageningen), the Physics Laboratory TNO (The Hague) and the Microwave Laboratory of the Delft University of Technology (Delft).

The groundbased measurements are verified by SLAR testflights carried out by the National Aerospace Laboratory (NLR, Amsterdam). Collaboration with the University of Kansas facilitates the comparison of data and may bring to light possible differences in the results due to different environmental conditions. Fortunately the interest in research of this type is growing in other countries of Europe as well. Presently the European Association of Remote Sensing Laboratories founded in 1977 hopefully will stimulate a coordinated European effort in this field of research.

Experimental programs carried out so far form a quite extensive databank of radarsignatures that contains information on signal characteristics of the radar return in relation with various target properties for each set of sensor parameters included in the program.

5. SIGNAL CHARACTERISTICS

Figure 6 shows an example of the targets we encounter in remote sensing applications.

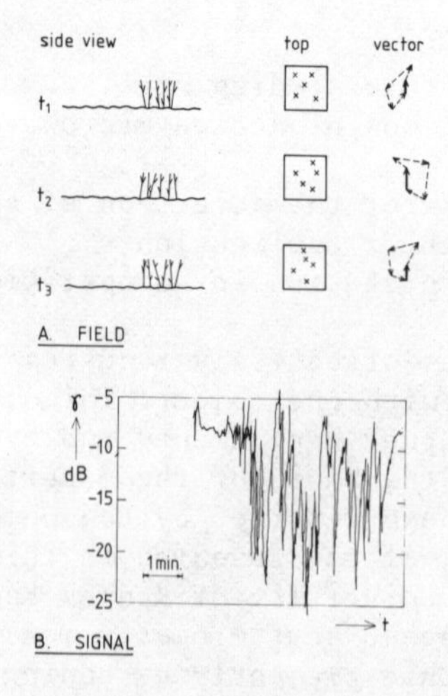

Fig. 6. Radar echo as a vector summation

The complicated structure of targets of this type causes the back scatter from coherent illumination to consist of a vectoral summation of the contributions from individual scatter elements. The random character of the object implies that the instanteneous backscattered power actually is a random variable. The distribution of this variable can be observed experimentally at the output of a stationary radar looking at vegetation moving in the wind. If there is no wind or if we deal with bare soil similar fluctuations can be observed when moving along the field. Signal distribution measurements made for different crop types showed very similar behaviour in good agreement with the Rayleigh distribution to be expected for targets of this type. Based on this evidence, the radarsignature of vegetation and bare soil is expressed as the radar cross section per unit area σ^o which is equal to the average backscattered power normalised with respect to the illuminated area and the incident power density. Meanwhile the signal-fluctuations present a major measurement problem, since the signal spread (5.6 dB for the logarithm of the amplitude) is unacceptably large for almost all applications.

Because of the vector character of the received signal, the more scatter elements the target comprises the better does the amplitude distribution approach the Rayleigh distribution. Therefore, averaging over more scatter elements by enlarging the resolution cell does not reduce the signal fluctuations at all. For higher accuracy we have to average a number of independent measurements of the received power, which is a scalar quantity. To derive at the number of independent measurements required to obtain a certain accuracy for the estimation of the actual average power scattered by a target area, we need to know the distribution that predicts the uncertainty in our estimation of the true average power when we use the K-samples average as an estimator. This distribution, that was derived at using the Rayleigh distribution, has been indicated in figure 7.

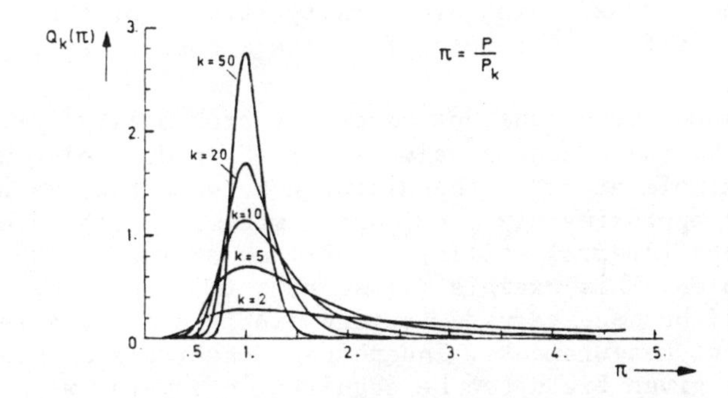

Fig. 7. The distribution of the average intensity as estimated from the K-samples average

The distribution has been plotted as a function of the normalized variable π with k as a parameter. Apparently the convergence of the confidence interval is only a weak function of k. The presentation in figure 8 is better suited for making quantative estimates.

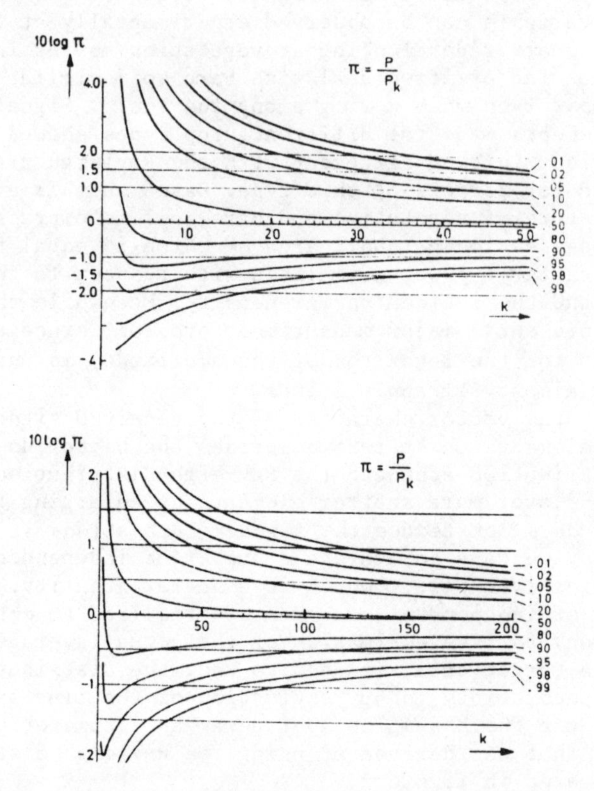

Fig. 8. Constant exceeding-probability curves for the average intensity as estimated from the k-samples average

The figure shows some constant exceeding probability curves as a function of k; the ordinate value is given in decibels. We can find, for example at k=200 the distance between the .95 and .05 curves to be approximately 1 dB, which means that the symmetric 90% confidence interval still is 1 dB when averaging 200 independent samples. This example illustrates that for many applications it will be necessary to average over a considerable number of independent measurements. Independent uncorrelated measurements from a given field can be acquired in several ways. One way is reorientation of scatterelements such as in dynamic targets like the sea surface and vegetated fields under windy conditions.

Another way to achieve decorrelation is by changing the frequency. This is effective because the relative phases of the reflection vectors of individual scatterelements are determined by the electrical distance of the scatterelements to the radar (as long as the frequency dependence of the individual scatteres can be neglected) and this distance is the ratio of geometrical distance and wavelength. Increased accuracy can also be obtained by spatial averaging over one field. Fortunately in this respect a radar having a linear motion such as a SLAR, collects several independent measurements of one single footprint by changing aspect angle. Detailed analysis shows that every time the radar moves one-half antennalength decorrelation is obtained from the previous measurement. The signal fluctuations discussed so far and illustrated with scatterometer records, also show up on radar images. (Figure 9). Since the human eye has the habit of averaging over a homogeneous field the images look better by inspection than they show up in automatic analysis on a pixel-to-pixel basis.

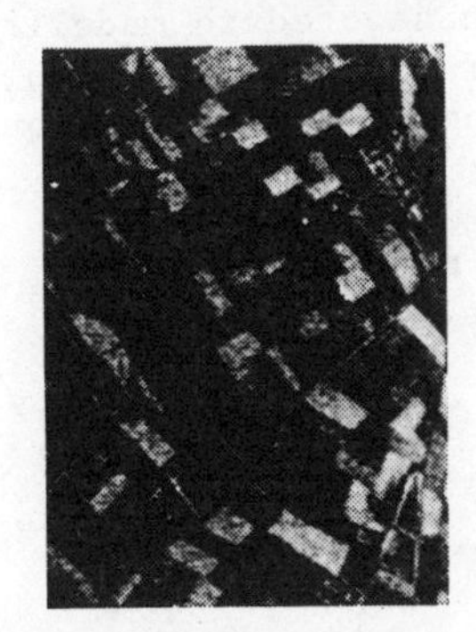

Fig. 9: High resolution radar imagery showing speckle reduction by increasing the number of independent measurements per pixel (From left to right: N = 1,5, 40)

To give an idea of the required accuracy and dynamic resolution for applications over various types of terrain figure 10 presents typical values for σ^o as a function of grazing angle.

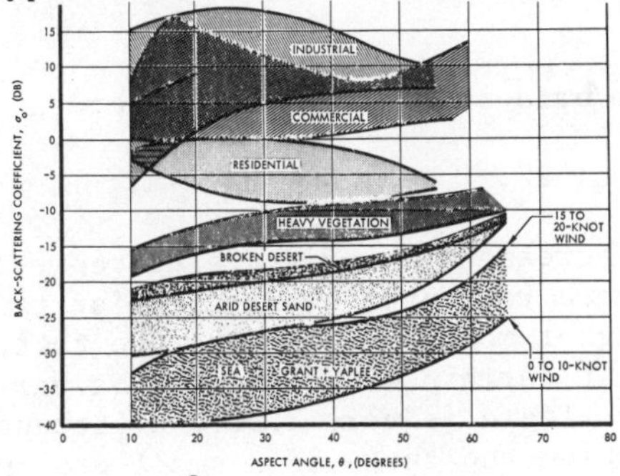

Fig. 10. Typical σ^o ranges for various applications

Evidently the dynamic range for a particular application, such as the monitoring of agricultural crops can be very limited. All the information about the objects is contained in a narrow dynamic window. In this figure the radar cross section per unit area σ^o is expressed as a logarithic value in decibels. The zero dB reference surface is a surface that scatters all the incident energy isotropically. Instead of σ^o also the scattering parameter γ is in use. Gamma differs from σ^o by only a factor equal to the cosine of the incidence angle. There is a slight preference for γ since it is for a certain class of targets a weaker function of the incidence angle.

6. RADAR SIGNATURE DATA

In this section we will review some experimental data in a very qualitative way. More general conclusion will be presented in a subsequent section devoted to modelling.

Starting with the radar cross section of bare soil figure 11 shows the angular response of γ for various values of soil moisture.

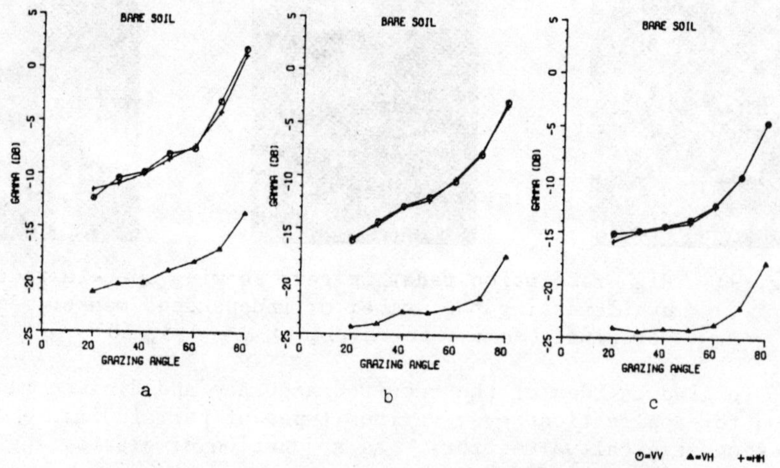

Fig. 11. Radar echo from bare soil at different percentages soil moisture.
a) 13%; b) 7%; c) 2%.

For all angles does σ^o increase with moisture content. Apparently the near vertical incidence region of quasi-specular reflection is more pronounced at higher moisture content. The testfield in this experiment was a flat terrain with rms height variations less than one centimeter. There is very little difference between HH and VV polarization and there is a small crosspolar component at HV polarization.

In vegetation signature data we can roughly recognize two different types of crop. To the first category of crops belong sugarbeets and potatoes. These crops have a more or less closed foliage. Some simple analysis indicates that the transmission of a leaf at 10 GHz is such that only the toplayer contributes significantly to the reflection. The influence of underlying layers, among which the soil,is attenuated strongly.

Crops of the second kind such as cereals and grassland have a more open structure and the incident radiation can penetrate to a significant distance, depending on structure, density and plantheight. For this type of crop we found a more pronounced influence of the underlying soil.

Figure 12 shows some angular responses for sugarbeets.

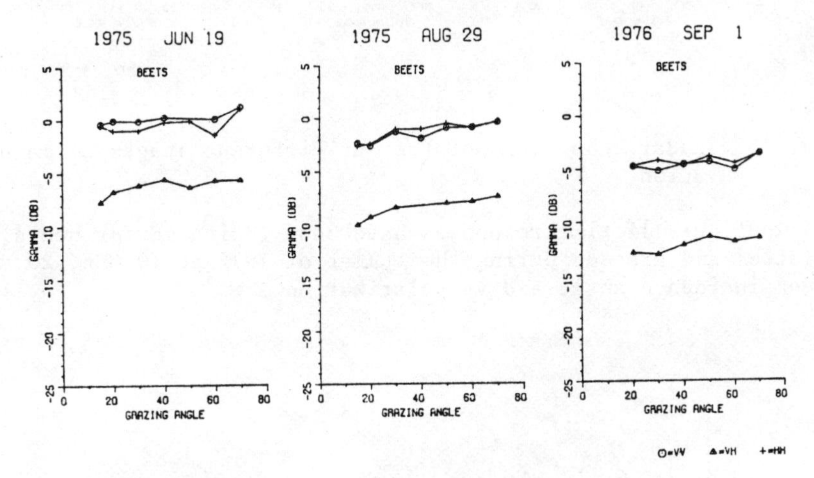

Fig. 12. Radar echo from mature sugarbeets

As can be seen there is little angular effect. It is interesting to note the differences between the 1975 and 1976 results. The summer of 1976 was very dry and the crops did not reach full maturation and complete groundcoverage. This can, apparently, easily be detected by radar at this frequency. Figure 13 shows similar results for potatoes. The left graph represents the angular response of the bare potato field recorded early in the season of 1975. The measurements were carried out in the direction perpendicular to the typical undulations found in potato fields according to common farming practise and the results show clearly the effect compared to the soil response of the flat test plot shown earlier.

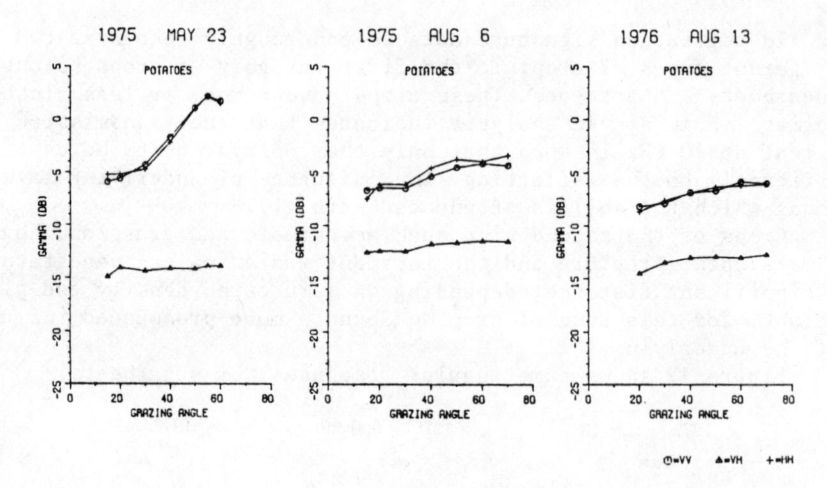

Fig. 13. Radar echo from potatoes at different stages of maturation

In figure 14 time responses have been indicated for beets, potatoes and grasses during the summer of 1975 at 10 GHz, 20 degrees incidence angle and VV polarization.

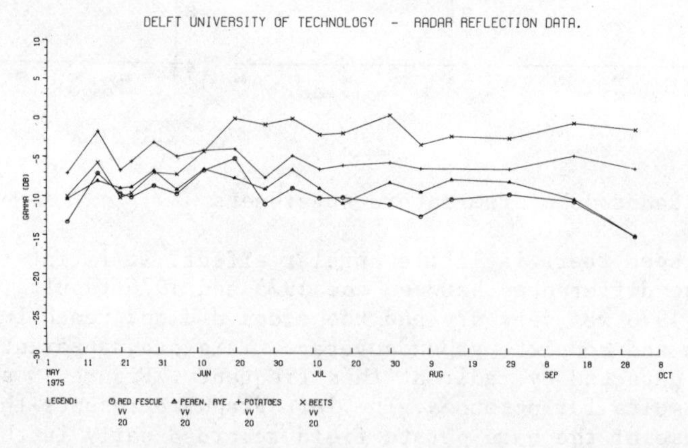

Fig. 14. Radar echo from different crops as a function of time

Starting early in the season the bare fields have nearly the same response except for the potato field discussed earlier. After the crops have grown in June sugarbeets and potatoes show a rather stable time response. Fortunately these crops stabilize at different levels well above the grassland response, facilitating reliable crop discrimination at this angle-polarization combination.

For a discussion of the second kind of crop with the more transparent structure we will have a look at time responses acquired in the season of 1977. In figure 15 some groundtruth taken during the measurements have been indicated. Indicated are plant-height for barley and wheat and soil moisture as a function of time.

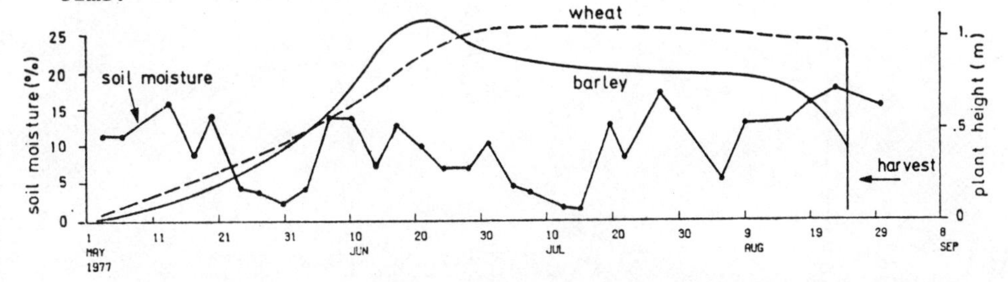

Fig. 15. Groundtruth data taken during radar reflection measurements

Figure 16 shows time responses at 15 degrees VV polarization. In the experiments were included two wheat fields treated in the same way such as to be able to investigate the natural variability of one crop growing under similar environmental conditions.

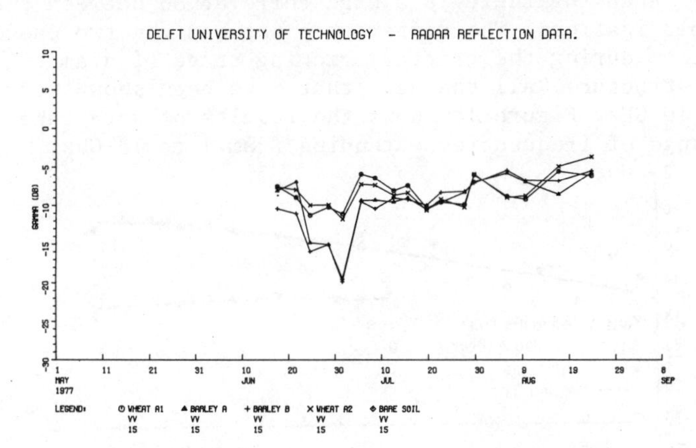

Fig. 16. Radar echo from different crops as a function of time

Two fields of barley were planted with different row spacings in order to investigate its effect on the radar signature.

From the data obtained it appears that there is a certain period ideally suited for discrimination between wheat and barley. Figure 17 shows the time response of the same fields at 80 degrees VV polarization compared to the bare soil echo. Especially late in the season all the fields respond very similar and there is a high correlation with soil moisture, indicating that the vegetation layer contributes little to the reflection, but only acts as a variable attenuating layer for the soil response.

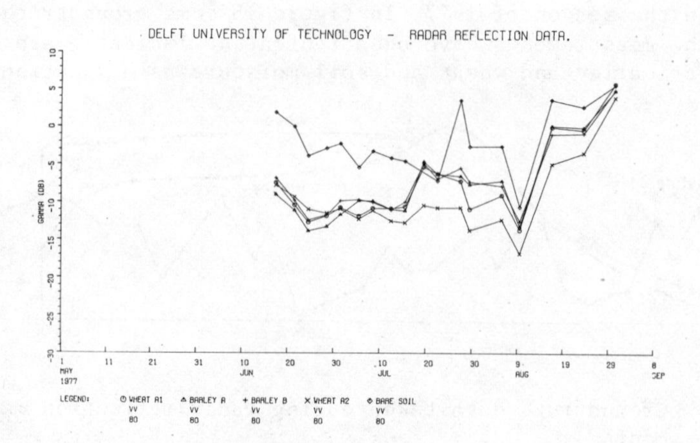

Fig. 17. Radar echo from different crops and bare soil as a function of time

Basically the same responses were obtained at HH polarization. Generally speaking there is a high correlation between the two linear polarizations. The differences between the two cases were at a maximum during the critical growing stage of dramatic change of plant structure. All the data that have been shown sofar were taken at 10 GHz. Figure 18 shows the results of data taken over a broadrange of frequencies extending from 1 to 18 GHz.

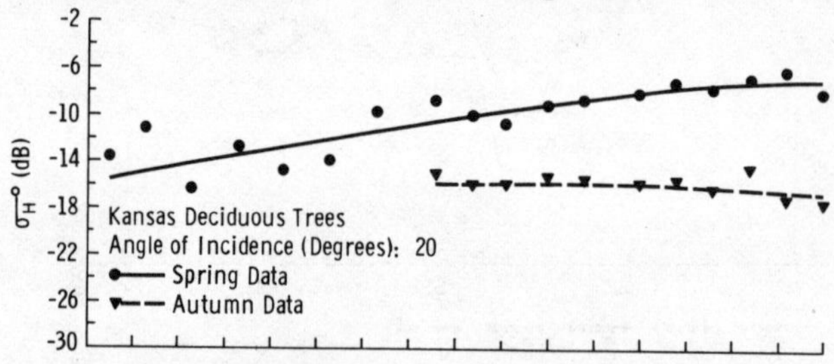

Fig. 18. Radar echo from deciduous trees as a function of frequency (From Bush e.a. [12])

These responses of trees measured in spring and autumn show the kind of frequency dependence typical for the targets we are concerned with. Sofar no evidence has been presented of strong narrow band resonance effects. The results obtained rather feature gradual trends within the frequency range of interest. For foliage we found a decreasing penetration towards higher frequencies. The data confirm the theoretical expectation that smooth surfaces at lower frequencies appear more and more rough to the radar at higher frequencies. This indicates that target structure always has to be considered relative to the wavelength.

7. MODELLING THE RADAR CROSS SECTION

The two objectives of modelling the radar cross section per unit area σ^{o} are: the prediction of σ^{o} for a given set of geometrical and dielectrical parameters on the one hand and the development of schemes for the solution of the inverse scattering problem e.g. the estimation of target parameters from measured values of σ^{o} on the other hand.

Several prediction models have been proposed in the literature ranging from very crude ones to very refined theories involving detailed target properties. In order to satisfy the second objective, which is the most important from an applications point of view, most of the effort is devoted to the development of prediction models that only involve a few predominant target properties. These models will hopefully facilitate the solution of the inverse scattering or remote sensing problem with respect to these predominant target properties.

Prediction models for σ^{o} can be divided into two major categories. The first one is the category of empirical models. These models are not necessarily based on any physical hypothesis for the scattering mechanism involved. They are given in the form of a functional relationship for σ^{o} as a function of frequency, polarization, incidence angle and target parameters. The proposed functions usually involve a number of empirical constants that need to be determined by regression analysis of radar signature data.

In view of the application to remote sensing it is again preferable to limit the number of these constants. An example of a very simple empirical model, surprisingly adequate for some cases, is the so called Lommel-Seeliger law. Figure 19 illustrates this model for the signature of potatoes and sugarbeets.

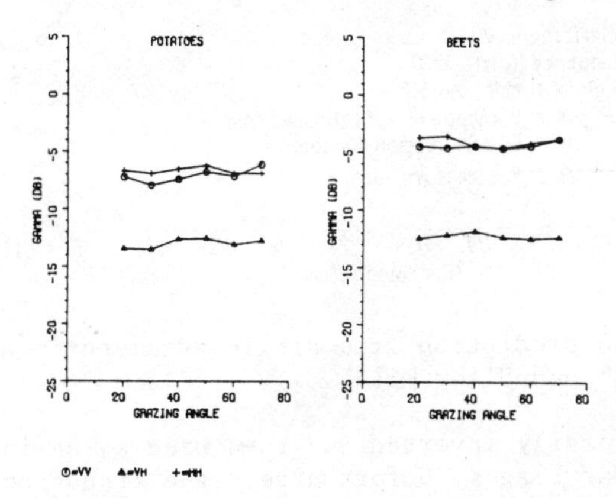

Fig. 19. Radar echo from potatoes and beets at 10 GHz

It has the simple form: γ is a constant specific for a particular crop. Another rather simple and attractive result is presented in figure 20, where σ^o is given as a linear function of a single plant parameter.

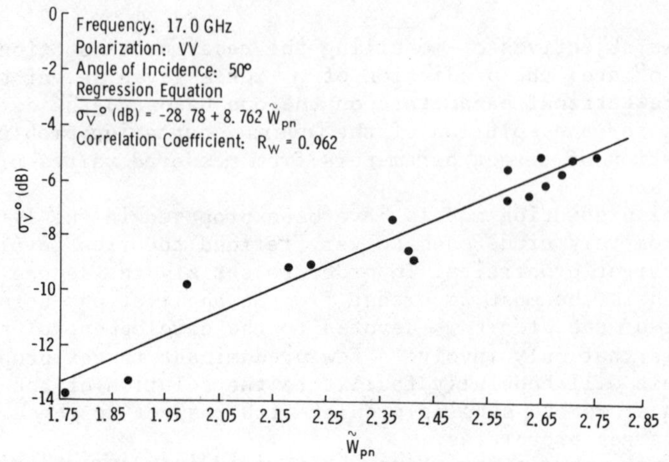

Fig. 20. Single parameter radar signature model (From Bush and Ulaby [6])

For this angle, frequency and polarization a high degree of correlation has been obtained. In figure 21 the measured time response is given in comparison with the predicted response.

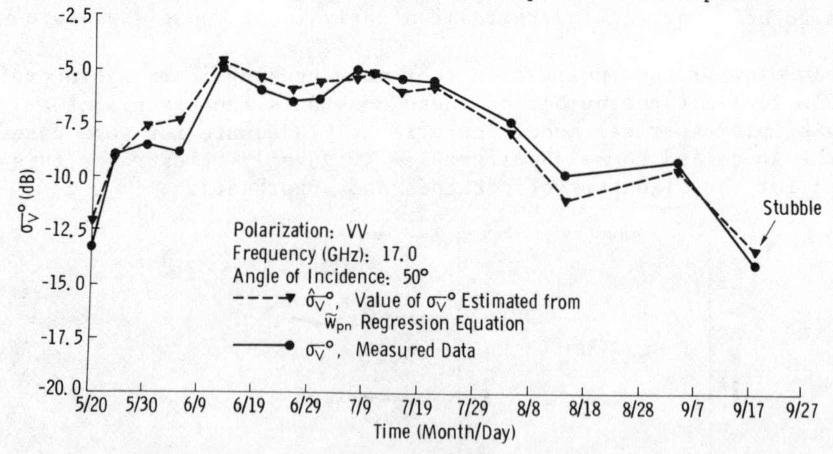

Fig. 21. Radar echo prediction from single parameter model (From Bush and Ulaby [6])

Both models can be easily inverted and then used as an interpretation tool for radar images. Unfortunately the situation in general is far more complex and the prediction of σ^o requires multiple linear regression equations involving many parameters. Still these models can be useful for interpretation purposes. However, some care should be taken in the general application of these models, since they have been derived at for a relative limited number of observations and the statistical significance of the model parameters is not always clear nor is the physical basis of the functional relationships.

In order to provide the theoretical basis for empirical models and to gain insight into the scattering process involved in the radar response of the target, it would be very desirable to have adequate theoretical models available. The development of such models, however, is very much hampered by the difficulties involved in defining the target mathematically. Therefore, the real targets are usually approximated by another physical structure for which the electromagnetic scattering problem can be solved. One of the earlier models was developped by Peake, who represented vegetation by dielectric cylinders occupying a half space [7]. More recently Fung and Ulaby [8] published a scattermodel in which vegetation has been represented by a dielectric half space containing random irregularities. Both models neglect the effect of the soil underlying the vegetation and are therefore limited to very dense vegetation.

Another approach was chosen by Ulaby and Attema [9], while analysing measurements of different vegetation types carried out by the University of Kansas scatterometer. As shown in figure 22 a model has been developped in which vegetation is represented by a cloud of scatterers, randomly distributed throughout the vegetation layer covering the soil.

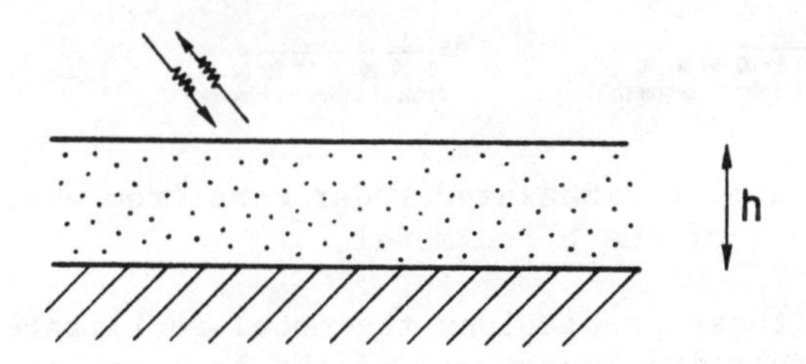

Fig. 22. The geometry of a cloud model for the radar echo from vegetation

The soil reflection is assumed to be proportional to soil moisture (m_s) and the particle density is chosen to be proportional to the volumetric water content (w) of the vegetation layer. Further input parameter is plantheight h. From these the scattering parameter is predicted in the form:

$$\gamma = C_0\{1-\exp(-C_1Wh/\cos\theta)\}+C_2m_s\exp(-C_1Wh/\cos\theta)$$

The model involves the parameters C_0, C_1 and C_2 that were determined empirically using non-linear regression analysis of measured signature data. The figures 23 and 24 present comparisons between predicted and actually measured responses.

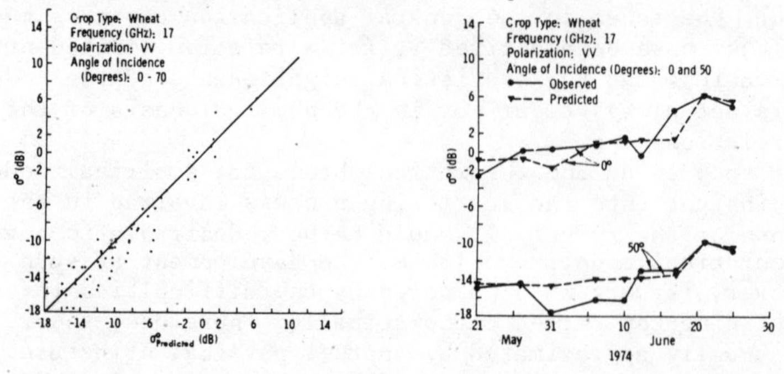

Fig. 23. Comparison of the measured radar echo from wheat and predictions of the cloud model

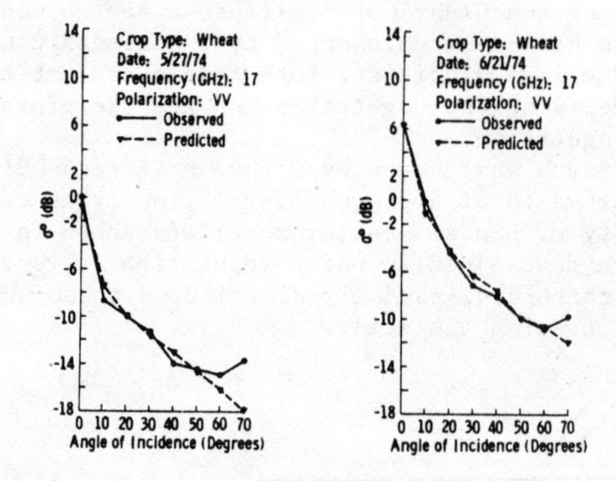

Fig. 24. Comparison of the measured radar echo from wheat and predictions of the cloud model

In addition to these predictions the model facilitates the evaluation of the relative importance of the soil reflection to the total backscattered power. An example of this is given in figure 25.

The model has been successfully applied to measurements made in the 8-18 GHz frequency range including HH and VV polarizations and croptypes alfalfa, corn, milo and wheat.
In the limiting case of very dense vegetation the model reduces to the Lommel-Seeliger law discussed earlier.

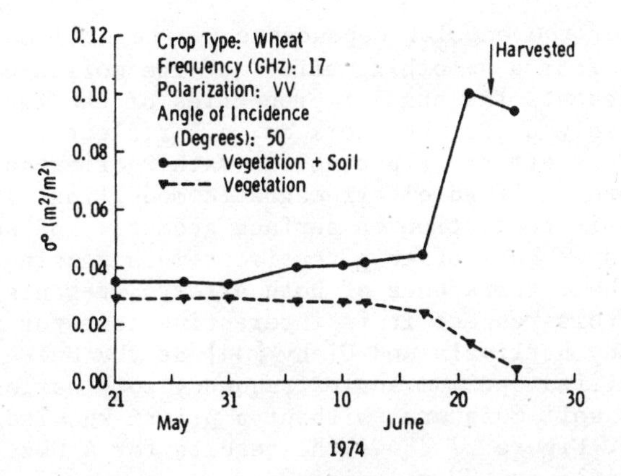

Fig. 25. The soil and the vegetation contribution to the total backscattered power in a wheat field

Radar return from bare soil can be calculated using available theories of scattering by irregular air-dielectric interfaces. This approach requires the assumption that the soil is homogeneous over the penetration depth for the wavelength of interest. The radar cross section γ has the general form $\gamma = G.F(\varepsilon)$. G is a geometrical function of incidence angle and surface properties. F is a proportionality factor that depends mainly on the complex dielectric constant of the soil and hence on soil moisture and salinity. The expression provides the physical basis for the regression equation $[\gamma]dB = a + bm_s$ where m_s is soil moisture. Figure 26 shows the angular dependence of the sensitivity to soil moisture and the angular behaviour of the zero intercept point. This curve shows that there exists a quasi specular component near nadir and a wide angle scattering component. This is typical for a composite surface with fine grain irregularities riding on gentle undulations [10].

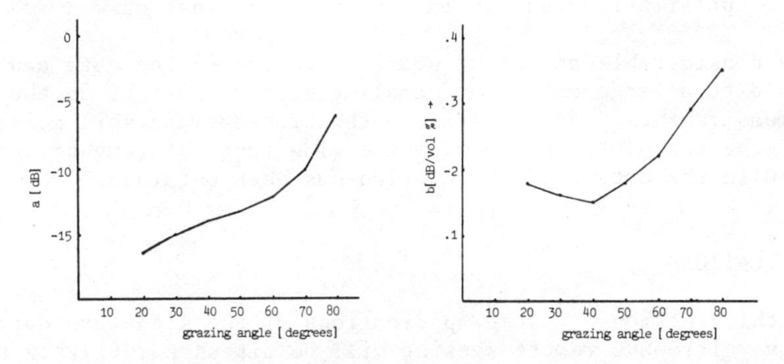

Fig. 26. Regression analysis of soil reflection measurements at 10 GHz based on $\gamma_{dB} = a + bm_s$

The problem whether the angular dependence of the soil moisture sensitivity stems from a smoothing effect of the moisture in the top layer or represents the angular properties of the factor F remains as yet unresolved. Further work on soil reflection modelling is in progress. This effort at present is rather oriented to soil physics than towards refined electromagnetic modelling. The dependence of the soil reflection on surface geometry and soil moisture in the top layer both offer potential remote sensing applications. However, the interference of both effects presents a measurement problem. In this respect it is interesting to refer to the results obtained by Batlivala and Ulaby [11] at the University of Kansas, who identified optimum angle-frequency combinations for the estimation of soil moisture, without a priori knowledge about surface roughness. Figure 27 shows the results for 4 GHz, 20 degrees incidence angle and HH polarization.

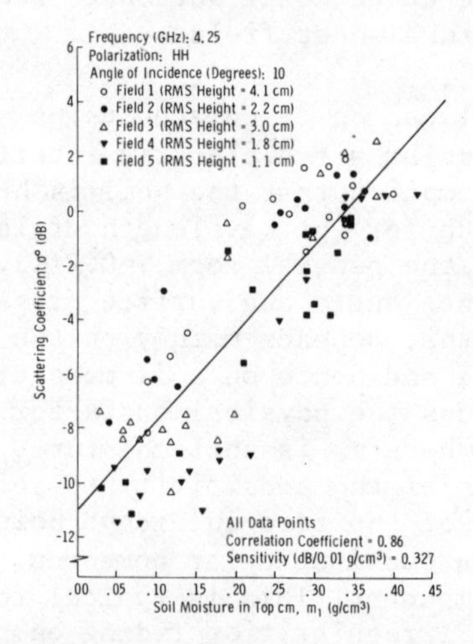

Fig. 27. The soil moisture response of σ^o at an optimum frequency, polarization and incidence angle combination

The considerable amount of scatter present in the data can be explained to a large extend by sampling errors inherent in the point measurements made to estimate the true average soil moisture of the testplot. Considering the wide range of roughnesses included in the data good correlation has been obtained.

8. APPLICATIONS

In this section some applications of the signature data in active microwave remote sensing will be discussed. Firstly the radar signature of various terrains determines the capability to detect and locate geometrical features of a scene. Application areas are mapping, geology, mineral exploration, land use inventory and water resource survey. For those applications for which large coverage in a short time is important and for those that involve dynamic phenomena the all-weather capability and the time-of-day independence is an essential feature.

The potential applications of radar that are based on grey-tone in the image representing the intensity of the backscattering by the object are still of an experimental nature although some have been verified in airborne missions. The major applications in this category to date are soil moisture mapping and crop monitoring.

As has been discussed earlier the knowledge of radar signature has been advanced to the point where large area soil moisture mapping becomes feasible provided that the proper selection of sensor parameters is made.

The analysis by Batlivala and Ulaby indicates that a spaceborne soil moisture mapping radar operating at C-band in the range of incidence angels between 7 and 17 degrees would have a 83% probability of correctly classifying soil moisture content of bare soil into four moisture levels, the associated sensitivity being .3 dB/volume percent moisture [11]. In spite of this evidence the operational introduction of the technique is not expected before the results of large scale experiments with Seasat and the planned radar sensors on board of Spacelab are available.

In regard to the important resources as agricultural crops, grassland and natural vegetation the capabilities of classifying crop type and estimating certain parameters on the basis of echo strength have been mentioned. Application areas are yield forcasting and the detection of stress and decease. Figure 28 shows a SLAR-image verification of the results obtained in groundbased measurements. As expected the contrasts depend on the season. The image at the left obtained in July shows the strongly reflecting white fields with sugarbeets, the dark areas with grasses and cereals and as intermediate the fields planted with potatoes. The image at the right obtained in November shows much less contrasts between the bare fields. Because there are many factors affecting the radar cross section it is very unlikely that a single image will be sufficient to provide all the information about the crop that we are interested in. The discriminating power can, however, be enhanced by applying multi-frequency, multi-polarization and multi-date data in the classification process. An example of the information added by another polarization is shown in figure 29 where scatterplots have been indicated for several polarization combinations.

Spaceborne crop classification has been investigated recently at the University of Kansas, based on radar signature data acquired in the 8 to 18 GHz range at 50 degrees incidence angle [6]. The results indicate that multidate data are essential in this application.

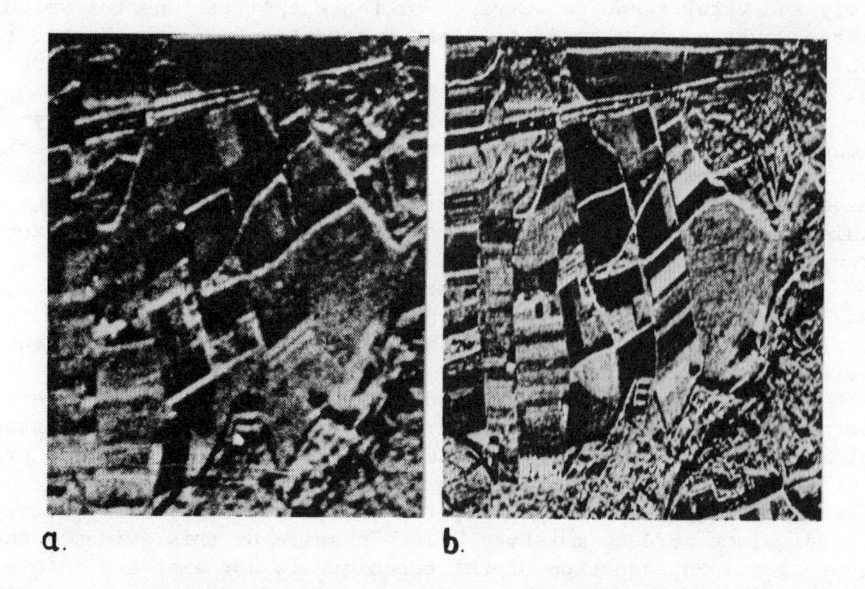

Fig. 28. SLAR imagery of an agricultural area in November (a) and in July (b)

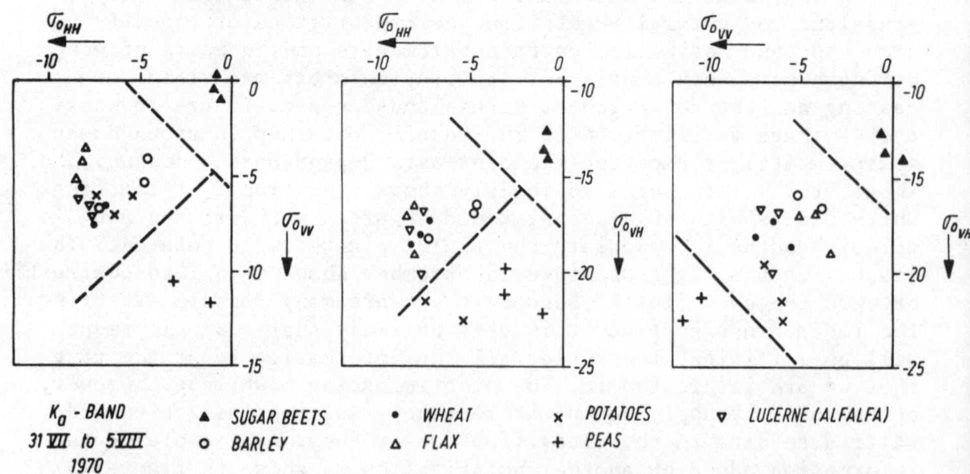

Fig. 29. Vegetation classification by cluster analysis of radar echo using polarization differences

The best choice of sensor parameters for a single frequency instrument turned out to be 14 GHz. The performance of such a sensor has been indicated in figure 30. A remarkable high percentage of correct classification could be reached within a short period for revisit intervals of several days.

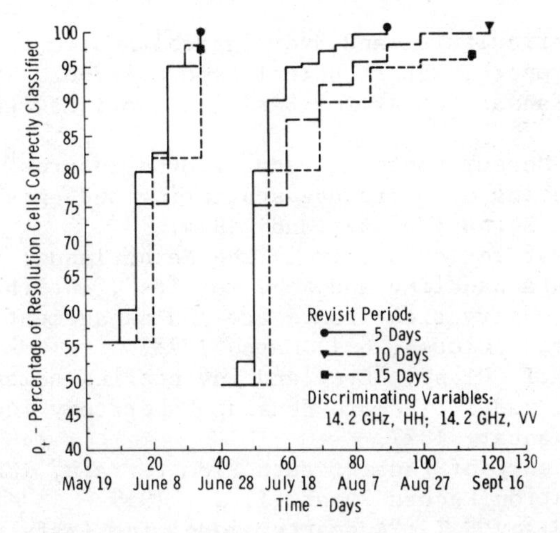

Fig. 30. Simulated crop classification performance of a single frequency dual polarization space borne imaging radar

In this respect it should be noted that the all weather capability makes the proposed revisit time practical, since each overpass of the satellite the area can be covered again.

9. CONCLUSION

The applications of active microwave remote sensing based on radio location are most self-evident and have become operational. Radar signature studies provide the basis for the application of active microwave remote sensing using the intensity of the radar echo and facilitate the selection of optimum sensor parameters for a particular application. Radar signature studies using groundbased scatterometers provide accurate results obtained under well controlled experimental conditions. However, the wide variety of targets of interest and environmental conditions encountered in nature does make the acquisition of data very time consuming.

Therefore, the next step towards the development of operational applications will be an enlargement of the scope of the experiments with continued groundbased basic research in conjunction with airborne and spaceborne tests.

REFERENCES

[1] G.R. Rumney: "Climatology and the World's Climates", Mac Millan, New Yrok 1968.
[2] T.L. Oliver and W.H. Peake: "Radar backscatter from agricultural surfaces", Techn. Report 1903-9, Electro Science Laboratory, Ohio State University, 13 February 1969.

[3] F.T. Ulaby: "Agricultural and Hydrological Applications of Radar", Final report, RSL Technical report 177-62, 1976, University of Kansas Center for Research Inc., Lawrence Kansas.

[4] G.P. de Loor: "Measurements of radar ground returns", URSI specialists meeting on Microwave scattering and emission from the earth, Berne (Switzerland) 1974.

[5] M.K. Smit: "Radar reflectometry in the Netherlands; measurement system, data handling and some results", International Conf. on Earth observation from space and management of planetry resources, Toulouse, 6-11 March 1978.

[6] T.F. Bush and F.T. Ulaby: "Cropland inventories using an orbital imaging radar", Remote Sensing Laboratory Techn. Report 330-4, January 1977.

[7] W.H. Peake, "Theory of radar return from terrain", IEEE National Convention Record 7 part I, 27, 1959.

[8] A.K. Fung and Ulaby F.T.:"A scatter model for leafy vegetation", IEEE transactions on Geoscience Electronics, October 1977.

[9] E.P.W. Attema and F.T. Ulaby: "A radar backscatter model for vegetation targets", Proc. Open Symposium URSI commission F, La Baule (France) 28 April-6 May 1978.

[10] P. Beckmann: "Scattering by composite rough surfaces", Proc. IEEE, August 1965.

[11] P.P. Batlivali and F.T. Ulaby: "Feasibility of monitoring soil moisture using active microwave remote sensing", RSL Technical Report 264-12, January 1977, the Univ. of Kansas Center for Research, Inc. Lawrence, Kansas.

[12] T.F. Bush, F.T. Ulaby, T. Metzler and H. Stiles: "Seasonal variations of the microwave scattering properties of the deciduous trees as measured in the 1-18 GHz spectral range", RSL Technical Report 177-60 June 1976, Univ. of Kansas Center for Research, Inc. Lawrence, Kansas.

MICROWAVE RADIOMETRY APPLICATIONS TO REMOTE SENSING

Erwin Schanda

Institute of Applied Physics, University of Berne,
Berne, Switzerland

ABSTRACT

Physical fundamentals and constraints of passive microwave remote sensing are discussed. Application areas are indicated, the rationale for multi-wavelength operation of passive microwave sensors is reviewed and the state of the art in radiometer systems is summarized. Examples of applications and of basic research in meteorology and hydrology are given and a future utilization in atmospheric science is indicated.

1. PHYSICAL FUNDAMENTALS

Passive microwave sensing makes use of the weak electromagnetic radiation emitted by any medium at temperatures different from absolute zero (thermal radiation). The wavelength region within which radiometers are constructed for the purpose of remote sensing is ranging roughly between one meter and one millimeter. Figure 1 presents the electromagnetic spectrum which is in use for remote sensing and it locates the region where microwave techniques are applied within the wide range of other remote sensing methods.
According to the radiation law by Planck the maximum of the spectral intensity occurs in the infra-red at about 10 microns wavelength for media at usual environmental temperatures (approximately 300 K)

T. Lund (ed.), Surveillance of Environmental Pollution and Resources by Electromagnetic Waves, 253–273.

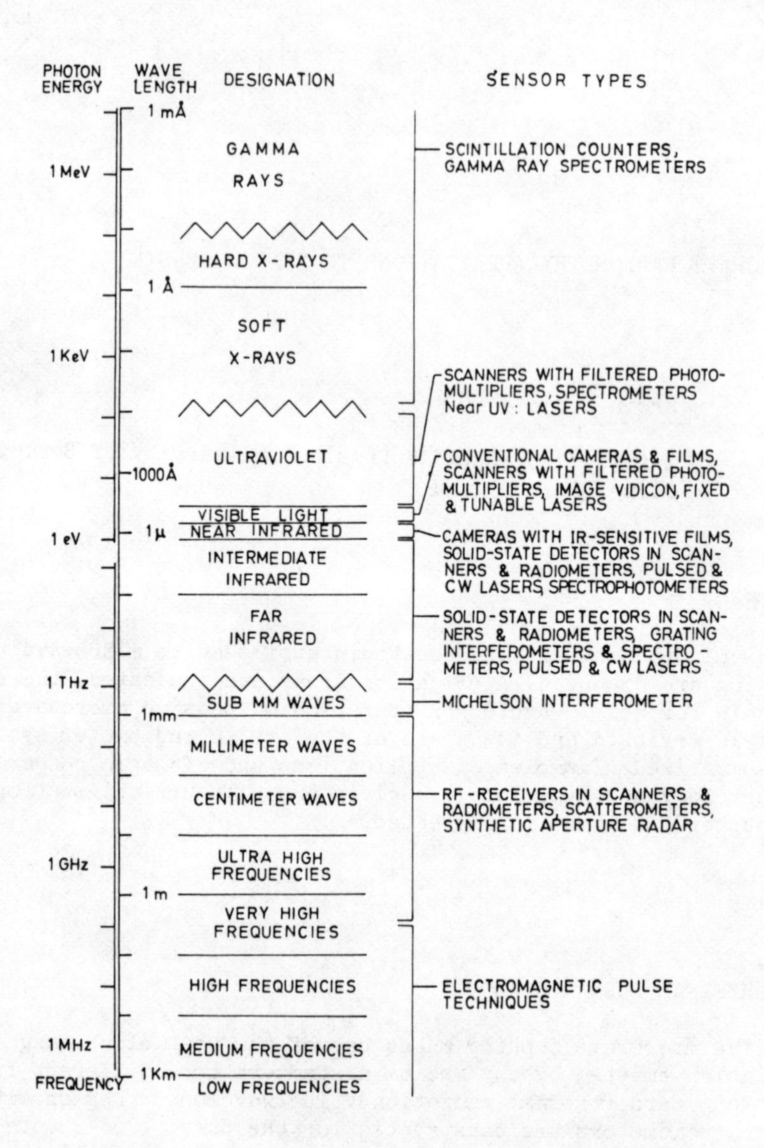

Figure 1. The electromagnetic spectrum in appropriate units and their usual designations. The microwave range is roughly located between the very high frequency and the submillimeter range.

Figure 2 shows that the spectral radiance (intensity per Hertz bandwidth and per unit solid angle) depends on the square of the frequency and is linearly dependent on the temperature at microwaves (below 10^{12} Hz) and at temperatures above 100 K.

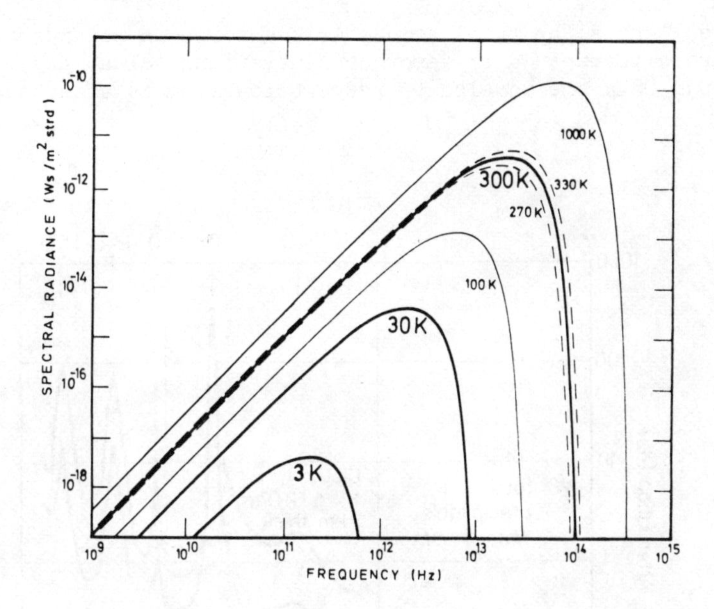

Figure 2. The "black body" radiation intensity per unit bandwidth and unit solid angle between the ultra high frequency (wavelength 30 cm) and the near ultraviolet (wavelength 0,3 mm) ranges of the spectrum for various temperatures of the radiation body.

This linear dependence of the radiated power on temperature allows to introduce the concept of brightness temperature of a surface instead of radiated intensity.
Due to the reciprocity between a radiating surface and a receiving antenna, the so-called antenna temperature can be derived, being the convolution of the brightness temperature distribution of an observed surface and the antenna power pattern.
However the received radiation is in general not originating from the uppermost surface of a medium but the layers at some depth are also contributing to the emitted power. This situation can be treated analytically by the equation of radiative transfer. For a strongly simplified assumption of a homogeneous layer at a temperature T_c constant throughout the layer and absorbing the radiation

along its path through the layer by exp $(-\tau)$ with the opacity τ, a radiating source at brightness temperature T_s behind this layer will cause a resulting brightness temperature

$$T = T_s \exp(-\tau) + T_c [1 - \exp(-\tau)]$$

A situation where this formula applies (or a more sophisticated version) occurs at sensing through clouds or precipitation.

Figure 3 shows the total atmospheric absorption towards zenith for two different water vapor concentrations (clear sky) and with a cloud. Only the wavelength range below 1 cm is affected, but

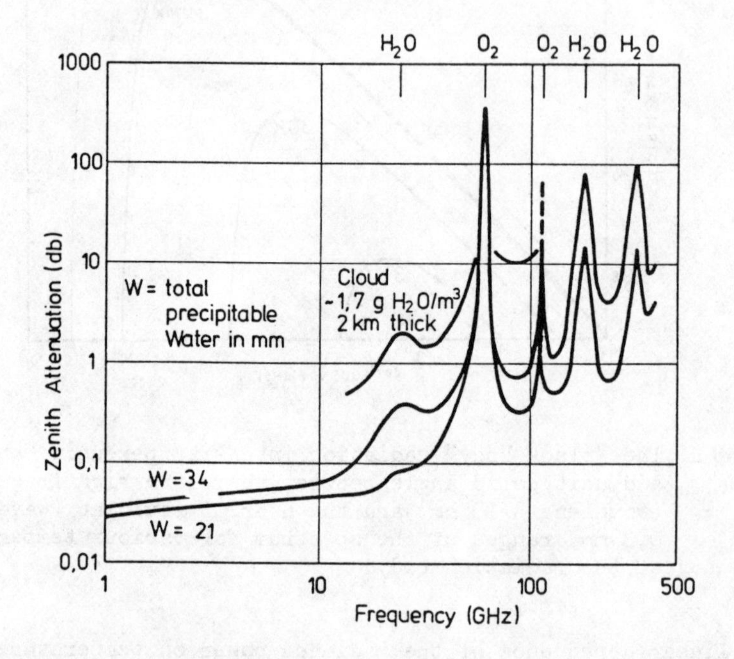

Figure 3. Total attenuation of the atmosphere toward zenith for for two values of water vapor contents and one cloud model. The centerfrequencies of the absorption lines of H_2O and O_2 are marked.

even between the spectral absorption lines of the most abundant atmospheric constituents there are "window regions" of sufficiently high transmission for useful remote sensing applications. On the other hand the existence of molecular absorption lines in this part of the spectrum allows for the remote determination of atmospheric parameters as the vertical and horizontal distribu-

tion of constituents and of temperature.
Figure 4 shows the horizontal attenuation for various rain rates due to a mean model assumption of the drop size distribution.

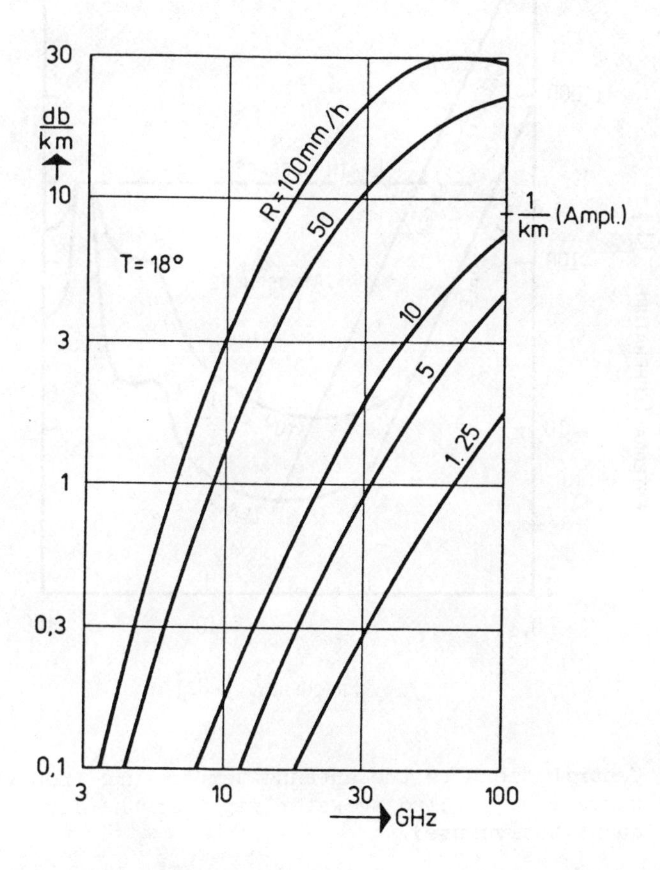

Figure 4. Horizontal path attenuations for various rain rates.

Another source of radiation, limiting the remote sensing applications at the lower frequency part is the emission by the galaxy. Figure 5 gives a crude indication of the resulting antenna temperature in a clear weather condition viewing different parts of the sky.
Important for the application of microwave radiometry for remote sensing of the surface of the earth is the fact that different media exhibit different abilities to emit radiation (emissivity). Therefore two surfaces at the same physical temperature can be discriminated by their different apparent temperatures (brightness temperatures) which are determined by the product of physical temperature and the respective emissivities.

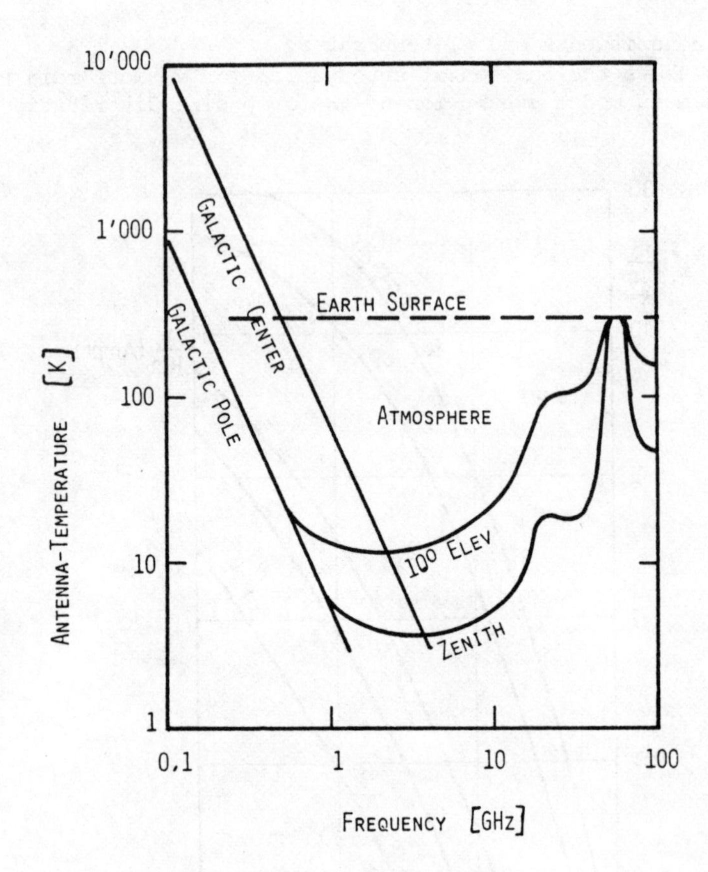

Figure 5. Contributions to the antenna temperature from various directions, including water vapor in the air (not for quantitative use).

The emissivity of a surface on its turn is dependent on the dielectric properties, density and heterogeneity of the underlying medium and on the surface roughness. For the comprehension of the emissivity one has to extend the concept of scattering: First, any radiation impinging onto the surface (eg. the thermal radiation of the atmosphere): part of this radiation is entering the medium (for simplicity we assume it will be dissipated) the remaining part will be scattered on the surface. A smooth surface reflects specularly, on a rough surface (roughness scale comparable or larger than wavelength) the scattered radiation is dispersed, in the case of very rough surfaces nearly uniformly over the hemisphere.
Second, the medium below the surface emits radiation through the surface, not uniformly but continuously distributed over the

halfspace. The principle of energy conservation demands that (for given frequency) the emission coefficient and the reflection coefficient (albedo) are adding up to unity. For the simplest case of smooth surface this is reduced to $e_i(\Theta,\varphi,\nu) = 1 - r_i(\Theta,\varphi,\nu)$ where Θ and φ are polar coordinates, ν the frequency and subcript i marks the polarisation. If T_g is the physical temperature of the ground and T_a (Θ,φ,ν) is the brightness temperature distribution of the atmosphere the resulting brightness temperature becomes $T_{ri}(\Theta,\varphi,\nu) = T_g \cdot e_i(\Theta,\varphi,\nu) + T_a(\Theta,\varphi,\nu) \cdot r_i(\Theta,\varphi,\nu)$

In general the emission coefficient is smaller than unity. This fact is taken into account by the term "grey body" as long as the frequency dependence is negligible within the observed spectral range. The relation between absorption coefficient a and emission coefficient e is given by the law of Kirchhoff $e/_a = e_b$ with e_b the emissivity of the black body which is usually equalled to unity.

The relation between emissivity (emission coefficient perpendicular to the surface) and prenetration depth in numbers of wavelenghts (representative thickness of radiating layer) on the

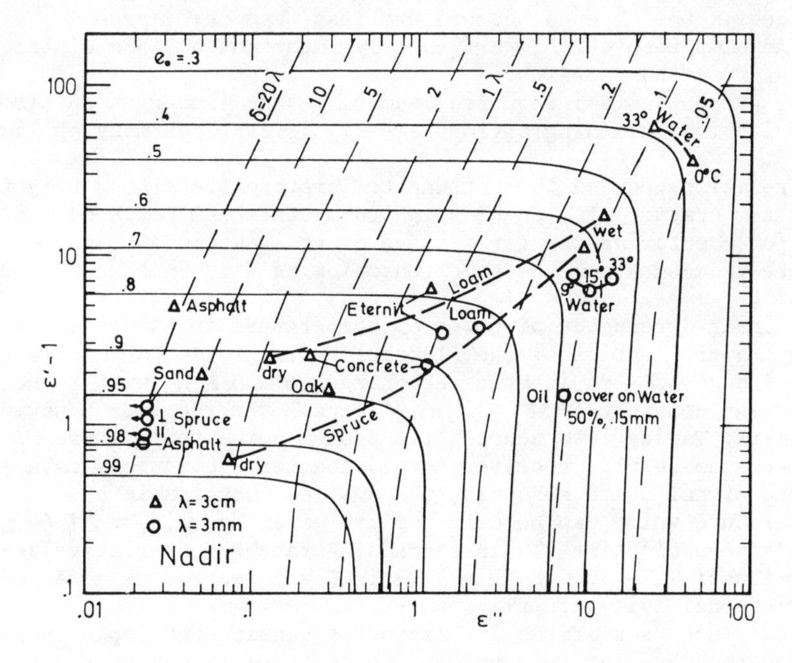

Figure 6. Emissivity and penetration depth as a function of the complex permittivity of smooth surfaces. Measured values of various natural and manmade media are marked.

one hand and the complex dielectric constant $\varepsilon' - i\varepsilon''$ on the other is presented in figure 6 together with the properties of various materials with smooth surfaces.
The differences between horizontal and vertical polarizations of the emitted radiation depends on the viewing angle and is for smooth surfaces (according to the Fresnel formula) very pronounced at angles between 50° and 70° off nadir. This effect decreases however fast for increasing roughness; very rough surfaces display almost no dependence of the emission coefficient on the viewing angle and have unmeasurably small differences between different polarizations.

2. MICROWAVE RADIOMETER SYSTEMS

The most important specifications of radiometer systems are

- frequency; in the first place because of the frequency dependence of the physical properties to be sensed.
- Radiometric sensitivity; it is necessary to detect differences in the antenna temperature less than one degree.
- Angular resolution; particularly important at long distances as from space crafts.

Other properties as: required measuring time (integration time), polarization, scanning options are significant for many applications.
The relation between the antenna temperature (resulting brightness temperature of a scene weighted by the beam pattern) and the power entering the first stage of the radiometer receiver can be deduced along the same reasoning as used in the previous section:
The linear dependence of power on temperature in the long-wavelength approximation of the Planck formula yields for a temperature change ΔT an input power change $\Delta P = k B \Delta T$ with k the Boltzmann constant ($1.38\ 10^{-23}$ Ws/K) and B the receiver bandwidth in Hertz. Taking into account the system noise temperature T_n (antenna noise plus receiver noise) and the statistical nature of the signal to be measured, the smallest changes in antenna temperature which can be detected are given by $\Delta T_{min} \approx 2 T_n (B\tau)^{-1/2}$ with integrationtime τ in seconds. A rather conservative assumption of T_n = 1000 K, B = 4 MHz, τ = 1 sec yields a radiometric sensitivity of $\Delta T_{min} \approx 1°K$.
The main drawback of passive microwave sensing from spacecrafts is the poor angular resolution. Due to the diffraction the resolution is given by the ratio of wavelength and diameter of the antenna. Figure 7 presents the angular resolution of microwave antennas according to the approximate relation $W/R \approx \lambda/L$ with range R, width W of the beam at R, wavelength λ and linear

antenna dimension L.

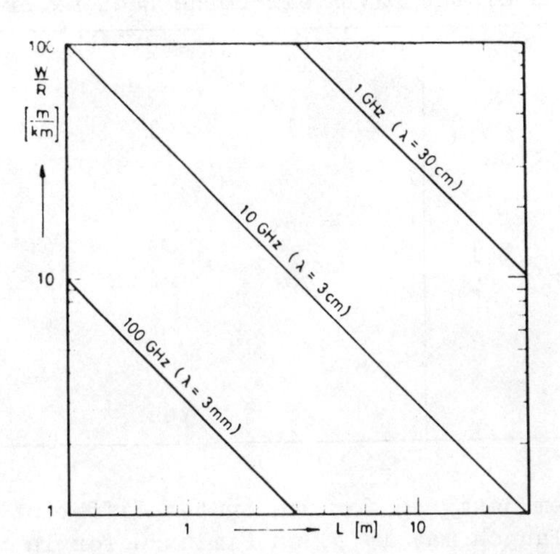

Figure 7. Diffraction-limited resolution, presented as width W (in meters) of the antenna beam at range R (in kilometers) as a function of the linear dimension L of the antenna cross-section, for three wavelengths.

Flight levels of automatic satellites (eg. Nimbus series: approximately 1000 km) or of geostationary satellites (approximately 36000 km) result in ground resolutions useful for large-scale observations, eg. in meteorology and oceanography, but are far too coarse for any fine-scale studies, eg. in agriculture, from space. Passiv microwave sensing is based on natural emission of radiation which is incoherent like noise. Therefore no phase relations exist between signals from different locations in the scene which would enable an improvement of the resolution as in the synthetic aperture technique.
A mission can be optimized by the proper choice of the receiver frequencies (eg. multi frequency receivers 5 to 35 GHz for snow determination and between 53 and 60 GHz for temperature profiles of the atmosphere). The sensitivity can be improved with the square root of the measuring time, but for a satellite-borne experiment (speed over ground approximetely 7 km/sec) the measuring time is limited to below 0.1 sec if for the purpose of mapping some scanning (perpendicular to the ground track) is applied. The resulting radiometric sensivity is between several tenths and one degree Kelvin.
Table 1 presents the main specifications of an airborne radiometer system in use for thematic mapping by the Deutsche Forschungs- und Versuchsanstalt für Luft- und Raumfahrt (DFVLR, personal communication K. Grüner, 1978)

Table 1

Main specifications of the DFVLR air-borne Radiometer System

Frequency [GHz]	11	32	90
Sensitivity [°K] (at 10 ms integration time)	1	1	2.1
Groundresolution [ft] (3db width for 1500 ft flight altitude)	72	54	54/24 exchangeable antennas
Scanning option	no	yes	yes

A radiometer system designed for use on two different satellites (on Seasat-A for launch may 1978, on Nimbus-G for launch fall 1978) is the Scanning Multifrequency Microwave Radiometer (SMMR) a five channel instrument for applications in oceanography, meteorology, climatology and related studies of land surfaces (Table 2, [1]). The absolute accuracy of the sensed brightness temperature is claimed to be 2°K. A conical scan will be used with constant earth incidence of 49°, covering a swath about 577 km wide below the spacecraft in a 4 seconds period. An off-set parabolic reflector of 79 cm projected antenna aperture will be used for dual linear polarisation (Figure 8).

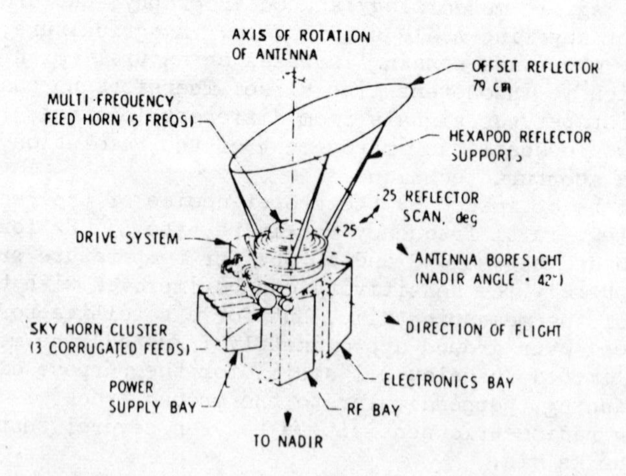

Figure 8. The scanning multichannel microwave radiometer instrument configuration on Nimbus-G and Seasat-A, [1]

Table 2

Main specifications of the scanning multichannel microwave radiometer for Seasat-A and Nimbus-G satellites [1]

Frequency [GHz]	6.6	10.69	18	21	37
Sensitivity [°K] (1σ, 77 K target)	0.34	0.48	0.67	0.72	1.09
Integration time [msec]	126	62	62	62	30
Footprint size $(\frac{\text{major axis}}{\text{minor axis}})$ [km] for satellite altitude 794 km	$\frac{121}{79}$	$\frac{74}{49}$	$\frac{44}{29}$	$\frac{38}{25}$	$\frac{21}{14}$

The data rate of passive microwave sensors is usually very low (eg.: 2kbps total for the SMMR instruments). This allows storage on board and transmission to only a few ground stations.
A typical meteorological radiometer package for sounding the atmospheric temperature profiles on the operational weather satellite Tiros N (for launch in 1978) has frequencies centered at 50.3, 53.74, 54.96 and 57.95 GHz [2] . A channel bandwidth of 220 MHz and integration time of 1.8 seconds yield 0.3°K sensitivity; the absolute accuracy on long term is 2°K. A cross track swath over approximately 2320 km with a nadir resolution of 102 km (circular) is appropriate to the large scale meteorological objectives.

3. APPLICATIONS OF MICROWAVE RADIOMETRY

Many successful applications of passive sounding from air- and space-borne platforms have demonstrated the usefulness of this method in meteorology, ocean studies, snow cover determination and others. Several new projects for satellite-borne use - even at operational systems as Tiros N - are planned for the near future or are about to be realized. From what has been said on the angular resolution it is evident that only large scale features (10 to 100 km or more) can be observed as long as the antenna dimensions are limited to one or a few meters.
As an example of microwave imagery Figure 9 shows the effect of a water surface as a highly reflecting medium [3,4]. The low brightness temperature of the lake is due to the low emission

Figure 9. a) Thermal image at λ = 3 mm of a scene with a lake, distance to the moutain peak 9,5 km [3],
b) optical identification of a),
c) the same mountain scanned from a place close to the shore with rain clouds [4].

coefficient (about 0.4) for horizontal polarization and almost grazing angle at 3 mm wavelength and due to the low sky brightness temperature (about 150 K) even at this low elevation angle. Figure 9c shows a cloud layer appearing at about the altitude of the mountain peak.

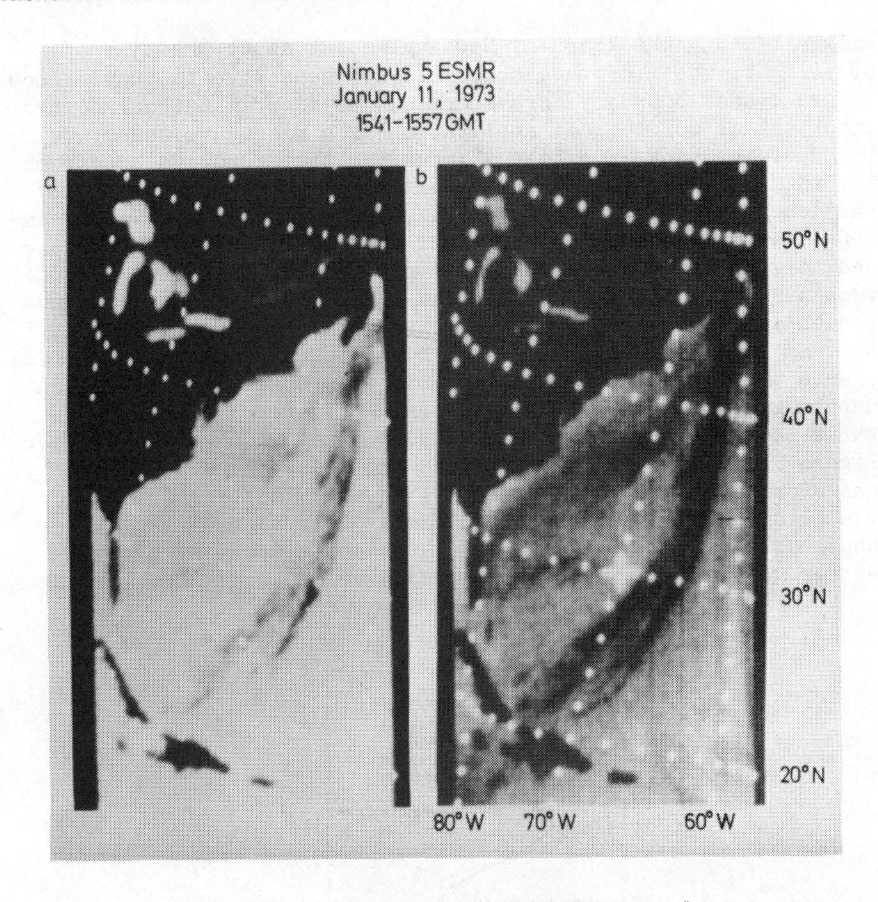

Figure 10. Maps of the east coast of the US, obtained by the Nimbus 5 scanning radiometer
a) brightness temperature range 190-250 K
b) range 130-200 K
The rows of dots indicate lines of geographic longitudes and latitudes [5].

Figure 10 is an excellent example of a meteorological application [5]. The electronically scanned microwave radiometer (ESMR) on board of Nimbus 5 was able to detect various rain intensities within a front over the North Atlantic area east of North America which reaches from Nova Scotia, where precipitation is heaviest, to Puerto Rico. The presentation of the more sensitive measuring range b) depicts the gradations of lighter rain while a) presents the heaviest precipitation yielding the highest brightness temperature and the areas of light rain do not show up in this image. From these images there does not appear any significant rain at the coast but surface observations reported snowfall along the coast of New England at about the time of overflight. The microwave sensor is not sensitive to snow, hence the rain/snow boundary in a storm can be discriminated. Another important meteorological application of a microwave sensor on board of the same Satellite (Nimbus 5), the microwave spectrometer (NEMS) has been demonstrated. Among other applications as: the determination of the total water vapor and liquid water content over the oceans [6], three of the five channels were designed to yield vertical temperature profiles of the atmosphere between surface and 25 km height. The three receiver channels are placed at frequencies along the oxygen line spectrum at about 5 to 6 mm wavelength and according to different absorption coefficients they are sensing the radiation temperatures at different depths in the atmosphere. An inversion algorithm transfers the three measurements at three altitudes into a continuous profile. Figure 11 presents the temperature profiles [7] as obtained by the microwave spectrometer (full line) over a place in arctic winter and a tropical place and compares these profiles with those obtained by the Radiosonde at times Δt and at places ΔS different from the respective microwave measurements. Very sharp

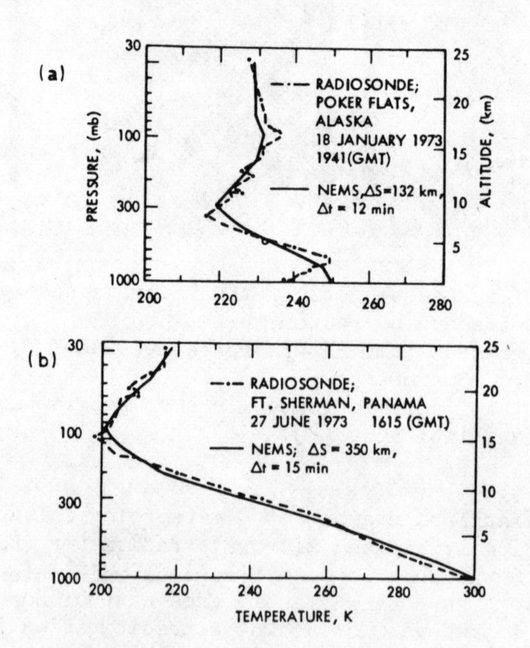

Figure 11. Atmospheric temperature profiles as obtained by the inversion of the Nimbus-5 microwave spectrometer data (Oxygen channels) over
a) a polar
b) a tropical latitude [7].

changes of the profile due to inversions and other small scale features are not sufficiently represented in the microwave result because of an insufficient number of channels and a coarse height resolution (about 8 to 10 km).
Airborne and satellite-borne mapping of land and water surfaces would be useless if not systematic investigations of the relationship between radiation behavior and conventional parameters of the surface features were performed on controlled experimental fields and by theoretical modelling.
A large scale surface feature of considerable impact on climate and on hydrology is the snowcover in the high latitude regions of the world. Important parameters are: the surface extent, the quantity of stored water, the starting time of melting.

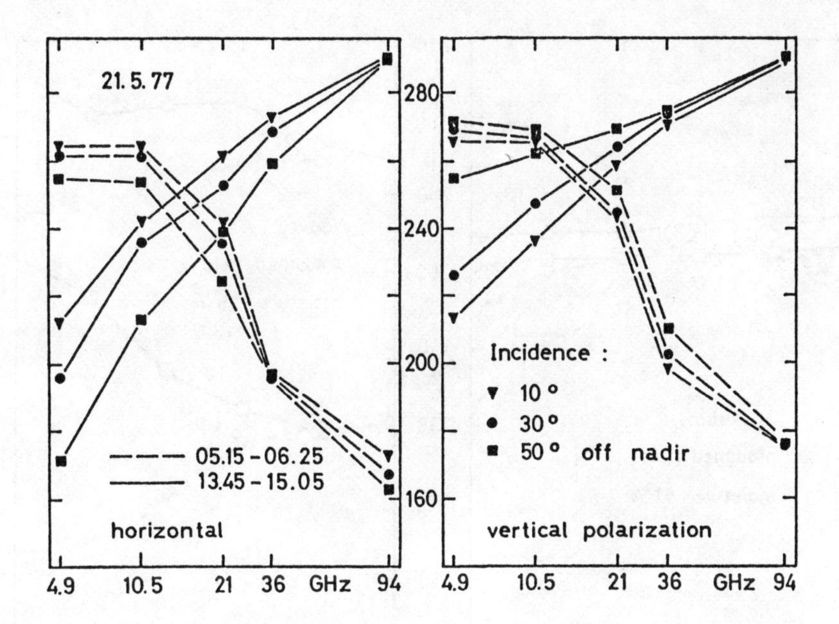

Figure 12. Microwave Spectrum of the brightness temperature of the snow cover (about 2 meters thick) on May 21st 1977 at Weissfluhjoch Davos, 2540 m altitude [8].

Microwaves are penetrating several meters of dry snow and are very sensitive to changes of the moisture content (small amounts of free water between the ice cristals) of the surface layers. Figure 12 shows the measured microwave spectra of a typical spring situation short before the melting process of the whole snow cover (about 2 meters thick) sets in [8] . The afternoon spectrum represents a moist surface layer of about ten centimeters due to sun shine, the early morning spectrum is due to a

refrozen surface. This drastic change of the microwave emission indicates the potential for surveying the state of the snow cover. Models of the absorption and scatter behavior of snow [9] enable the qualitative interpretation of the microwave brightness temperature maps of the polar regions [10].
Agricultural areas are small scale features which cannot be resolved by microwave antennas from space, but low flying airplanes can execute important surveying of the soil moisture during the phase of growth by passive microwave sensing. The effects of moisture and of surface roughness on unvegetated fields [11] are demonstrated in Figure 13. The brightness temperature as a func-

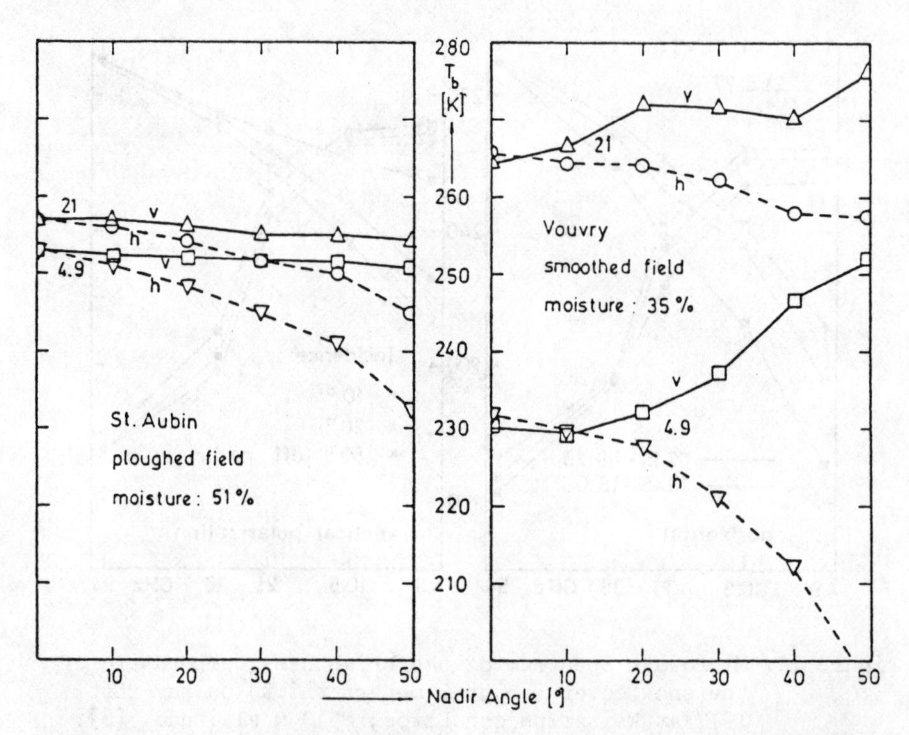

Figure 13. Microwave brightness temperatures of two bare fields at 4.9 GHz and 21 GHz for vertical (v) and horizontal (h) polarizations as a function of the viewing angle. Moisture contents per weight of dry soils respectively [11].

tion of the viewing angle of the smoothed field (r.m.s. roughness about 3 cm) shows a Fresnel-like behavior with clear splitting of the horizontal and the vertical polarizations (more pronounced for the wavelength longer than the roughness scale) and with distinct sensitivity on the moisture contents versus frequency: closer to black body behavior at 21 GHz, closer to the properties of water at 4.9 GHz as compared to the ploughed field. The ploughed field (r.m.s. roughness 10 to 15 cm) exhibits only very little dependence on the viewing angle and no significant difference for the two orthogonal polarizations because of multiple scattering on the heterogeneous surface structure. This is also the reason that the brightness temperature cannot attain the low value which would be expected according to the high degree of moisture content (per weight of the dry soil). Figure 14 shows the temporal development of the moisture content and the corresponding emissivities (ratio of brightness temperature and physical

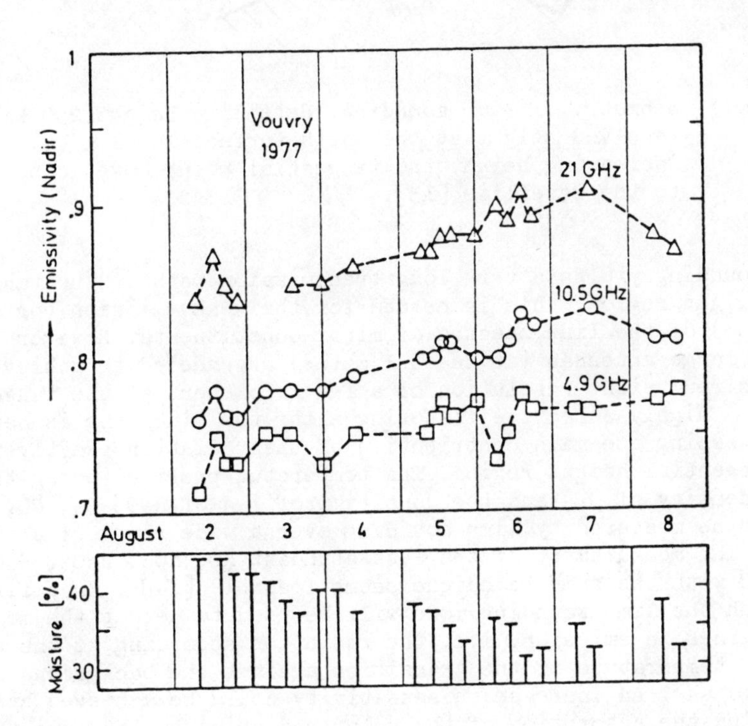

Figure 14. Microwave emissivities and moisture content during one week of observation [11].

temperature) measured by the three channel radiometer in nadir looking position during one week. Decreasing moisture corresponds with increasing emissivity.
The last example of applications of microwave radiometry is taken from a proposal [12] to fly a multichannel microwave spectrometer (together with infrared sensors) on Spacelab for investiga-

tions on stratospheric and mesospheric parameters in the configuration of limb sounding (Figure 15). The main objectives of the planned mission - as far as the microwave part of the experiment is concerned - are presented in Table 3.

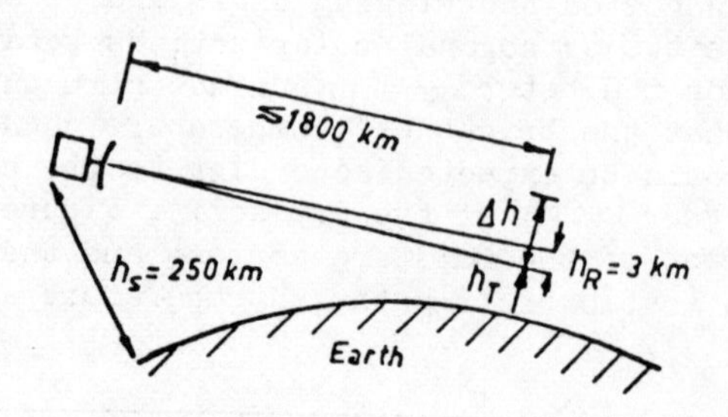

Figure 15. Geometry of limb-sounding. Satellite height 250 km, resolved height at the earth surface level 3 km. For increasing height the tangential point moves closer to the satellite [13].

Limb sounding yields a very long transmission path in the tangential height region, this is needed for the weak emission (or absorption) of the line spectra of minor constituents. However rather large antennas (at least 1 meter) are needed to achieve the desired height resolution of a few kilometers at the tangent point. A highly selective filtering along the lineshape is needed for obtaining the main contribution of the radiation exactly from the tangential height region. The temperature profile (using the known density of O_2) and the density profiles of H_2O, O_3, CO, N_2O can be measured by limb sounding over a wide range of altitudes. The measurement of the less abundant NO, NO_2, HNO_3, H_2O_2 and ClO with the same technique seems feasible [14] . In a first approach the limb sounding mode will be used to detect the molecular lines in emission, i.e. the radiation according to the effective temperature of the transition against the background of the cold sky. An increase in sensitivity could be achieved by measuring the absorption of the solar radiation by the atmospheric constituents [14]. However the short measuring time available (due to the needed relative positions of satellite, earth and sun) and the fact that only the situations of sunrise and sunset can be sensed are essential disadvantages for atmospheric investigations. The introduction of beacon transmitters on subsatellites as radiation sources instead of the sun could be an essential step towards improved sensitivity.

TABLE 3 Scientific and Technical Main Objectives of the Microwave Limb Sounder [12]

Radiometer Frequency [GHz]	Quantity to be measured	Height Range [km]	Height Resolution at tangent point [km]	Radiometer Noise temperature [°K]	Radiometric Sensitivity expressed as rms noise [°K] : max (at line center)/min (at the wing) 2)
63	O_2 (pressure reference)	35- 70	9.4	1200	1.7/0.2
118	O_2 (kinetic temperature) O_2 density O_2 winds 1)	20-100 90-120 70-100	5.0	1000	1.0/0.2
167	ClO	30- 45	3.5	1700	0.25
184	H_2O, O_3	20- 90	3.2	3200	3.2/0.5
230	CO	20-110	2.6	4000	4.0/0.6 3)

1) will be attempted if adequate pointing data is available
2) The bandwidths of the filter channels are 2 MHz at line center and 50 MHz at the far wing channels 150 MHz off line

centers, except: for 167 GHz: all channels are 15 GHz wide. The integration time is assumed 2 seconds, except for 63 GHz: 0.25 sec.

3) At the CO line center a much narrower bandwidth (the order of 100 kHz) will be used. The resulting degradation of the sensitivity has to be compensated by a corresponding increase of the integration time.

REFERENCES

[1] P. Gloersen, Scanning Multichannel Microwave Radiometer, Jet Propulsion Laboratory, Sensor Capability Handbook and Data Sheets, Vol. II.Performance Data Sheets, PM 78-1, May 1977

[2] A. McCulloch, Microwave Sounder Unit, Jet Propulsion Laboratory, Sensor Capability Handbook and Data Sheets, Vol. II. Performance Data Sheets, PM 78-2, May 1977

[3] G. Schaerer, Passive Sensing Experiments and Mapping at 3.3 mm Wavelength, Remote Sensing of Environment, 3, 117-131, (1974)

[4] G. Schaerer, E. Schanda, Deteriorating effects on 3 mm wave passive imagery, Proc. 9th Internat. Symp. Rem. Sensing of Environment, Ann Arbor, Mich., 1593-1602, (1974)

[5] T. Wilheit, J. Theon, W. Shenk, L. Allison, Meteorological interpretations of the images from NIMBUS 5 electrically scanned microwave radiometer, Preprint X-651-73-189, NASA GSFC, Greenbelt, Maryland, (1973)

[6] D.H. Staelin, K.F. Kuenzi, R.L. Pettyjohn, R.K.L. Poon, R.W. Wilcox, J.W. Waters, Remote Sensing of Atmospheric Water Vapor and Liquid Water with the NIMBUS 5 Microwave Spectrometer, J. Appl. Meteorology, 15, 1204-1214, (1976)

[7] J.W. Waters, R.L. Pettyjohn, R.K.L. Poon, K.F. Kuenzi, D.H. Staelin, Remote Sensing of Atmospheric temperature profiles with the NIMBUS 5 Microwave Spectrometer, J. Atmospheric Sciences, 32, 1953-1969, (1975)

[8] E. Schanda, K.F. Kuenzi, R. Hofer, Microwave Experiments for Spacelab and NIMBUS G, Proc. Coll. GDTA l'utilisation des satellites en teledetection, St. Mandé, France, Sept. 1977

[9] H.J. Zwally, Microwave emissivity and accumulation rate of polar firn, J. of Glaciology, 18, 195-215, (1977)

[10] H.J. Zwally, P. Gloersen, Passive microwave images of the polar regions and research applications, Polar Record, 18, 431-450, (1977)

[11] E. Schanda, R. Hofer, D. Wyssen, A. Musy, D. Meylan, C. Morzier, W. Good, Soil moisture determination and snow classification with microwave radiometry, Proc. 12th internat. Symp. Remote Sensing of Environment, Manila, Philippines, April 1978

[12] J. W. Waters, Microwave Infrared Mesospheric Stratospheric Experiment, Proposal for the first Spacelab mission, submitted to NASA, 11 June 1976

[13] E. Schanda, A millimeter-wave sensor for Spacelab astrophysical and atmospheric investigations, Proc. 16. Convegno sullo spazio, 23. Rassegna internat. elettr. nucl. ed aerospaziale, Rome, Italy, March 1976

[14] E. Schanda, J. Fulde, K.F. Kuenzi, Microwave limb sounding of strato- and mesophere, in: Atmosph. Physics from Spacelab J.J. Burger etal edit., D. Reidel Publishing Company, Dordrecht, Holland, 135-146, (1976).

CORRECTION OF AIRBORNE IR-SCANNER DATA*

G. A. Becker

Deutsches Hydrographisches Institut
Hamburg, Federal Republic of Germany

ABSTRACT. The radiant emission of water surface recorded by IR-scanners is afflicted with various errors. Sources of error are described and suggestions for their elimination are given. In one example a radiometrically determined temperature distribution is compared with one measured conventionally.

The radiant emission of a water surface recorded by an IR-Scanner is afflicted with various errors. In many application orientated cases, it is sufficient to extricate the horizontal temperature gradients from the IR images so that only the scan angle dependent limb darkening effect has to be empirically corrected. On the other hand, if one is interested in quantitative temperature values, there are two courses that can be adopted. In order to avoid extensive numerical procedures, quite frequently only an empirical correction is applied, which is statistically determined from a number of sea-truth observations. A second, more exact, method is the determination or estimation of the different physical effects which can cause errors in the IR images.
Some examples of the latter method are given in the following.

1. ERROR SOURCES AND CORRECTIONS

In general, four sources of error can be quoted which play a role in the acquirement of temperature distributions with the aid of IR-scanners.

* This work was carried out as part of the Erdwissenschaftliches Flugzeugmeßprogramm, sponsored by the Federal Ministry for Research and Technology in Bonn, FRG.

T. Lund (ed.), Surveillance of Environmental Pollution and Resources by Electromagnetic Waves, 275–282.

1.1 Errors caused by the instrument

Principally, faulty calibration of the sensor resp. the calibration black bodies can be specified here. During the conversion of the radiation intensities, received by the sensor, into temperature values, the calibration black bodies are used according to the following formula [1]

$$T_A = T_u + \frac{(I_A - I_u)(T_o - T_u)}{(I_o - I_u)},$$

linearity presupposed. With

T_A = actual temperature of an image element (pixel)

T_u = temperature black body 1

T_o = " " " 2

I_u = intensity " " 1

I_o = " " " 2

I_A = " of the actual image element (pixel).

Owing to the necessary intensification of the intensity signal, this is superimposed by instrument noise. In unfavourable cases, there follows a low signal to noise ratio. Fig. 1 shows, as an example, the intensity values of a scan line from an M^2S scanner (Bendix).

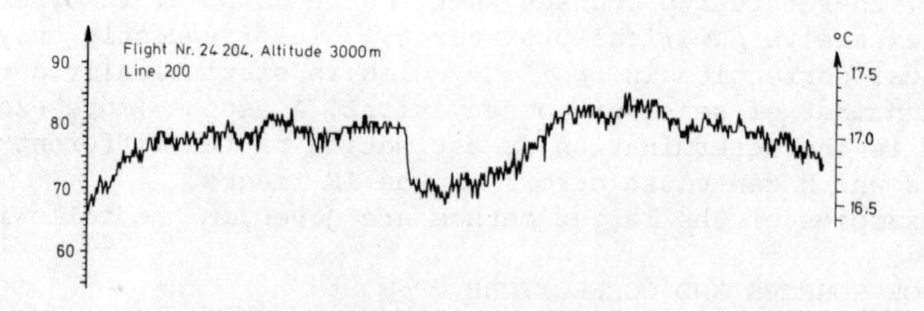

Figure 1. Pixel intensity of a line. M^2S-Scanner; 29.8.76; German Bight; 3,000 m.

A mean amplitude spectrum of the intensity variations [2] is shown in Fig. 2a. One can clearly see the peak caused by noise at wavelengths of 6 relative wavelengths (30 to 50 m). This noise can be eliminated by digital filters without difficulty. Fig. 2b shows the spectrum of such a filter.

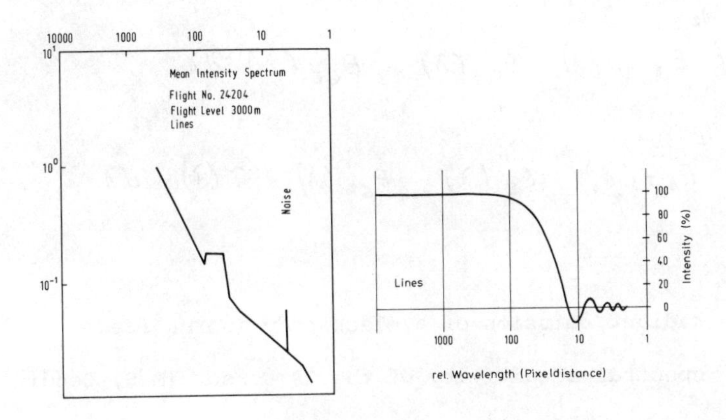

Fig. 2a Mean intensity spectrum, Lines

Fig. 2b Spectrum of a filter

1.2 Atmospheric attenuation

Errors up to several degrees Kelvin happened, due to the effect of the atmospheres absorbing and emitting radiation within the infra-red atmospheric window (8 to 14 µm). Various models are available for the calculation of the transmittance of the atmosphere. A simple model of Davis and Viezee [3], which is valid for wavelengths of 8.3 to 12.5 µm describes the transmittance as

$$\tau = \exp\left\{-(k_\nu W P)^{\alpha_\nu}\right\}$$

with k_ν, α_ν empirical coefficients

W water vapour

P effective pressure

An extensive numerical atmospheric model is put forward by Selby and McClatchey [4]. This complicated model determines the transmittance in a broad spectral band with high resolution. Different standard atmospheres and also radiosonde values can be used.

However, the model demands a relatively high quantity of numerical processing.
The atmospheric correction factor [5] is determined by

$$D = \frac{I_1}{I_2} \quad \text{with}$$

$$I_1 = \int_{\lambda_1}^{\lambda_2} E_{\lambda,T}(\lambda) \cdot E_S(\lambda) \cdot E_{GE}(\lambda)\, d\lambda$$

$$I_2 = \int_{\lambda_1}^{\lambda_2} E_{\lambda,T}(\lambda) \cdot E_S(\lambda) \cdot E_{GE}(\lambda) \cdot \tau(\lambda)\, d\lambda$$

whereby

$E_{\lambda,T}$ radiant emission of a black body (normalized)

E_S spectral sensitivity of the IR sensor (M^2S, Bendix)

E_{GE} spectral sensitivity of the germanium filter (M^2S, Bendix)

τ transmittance of the atmosphere.

In Fig. 3, I_1, I_2 are shown as a function of the wavelength for a M^2S-scanner and a particular meteorological situation.

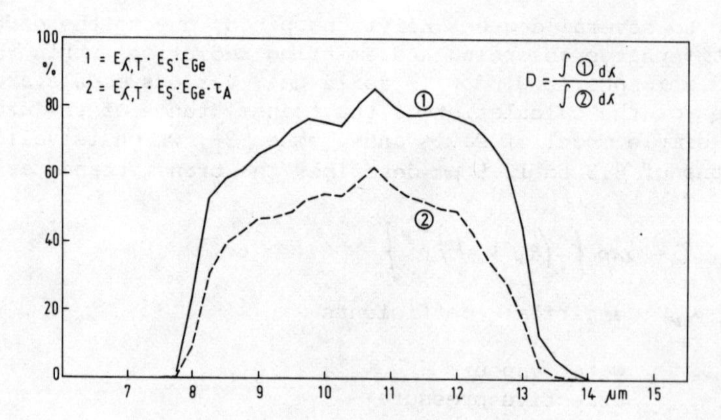

Fig. 3, I_1, I_2 as a function of the wavelength

A further source of error is the sky radiation, which has a relative minimum [5] between 8 to 14 μm. The reflectivity of a water surface also has a minimum at 11 μm. Fundamentally, however, the sky radiation cannot be neglected; especially in the case of a cloudy sky, because clouds practically radiate as black bodies. Suitable Tables [7] are available for estimation of the reflectivity.

1.3 Surface effects

In the determination of the atmospheric correction factors (1.2) the emissivity of the water was equal to that of a black body. However, the emissivity of water depends not only upon the wavelength but also upon the angle of view. The emissivity at 10 μm for an angle of view of $\varphi = 0^o$ is 0.990084 [7]. The emissivity at $\varphi = 50^o$ reduces to 0.98113.
Investigations about the mean emissivity of a defined water surface (pixel dimensions) as a function of different wave spectra are not known. In order to compare remote sensed temperatures with those conventionally measured in situ, one is forced to take into account the temperature difference between the skin temperature and the subsurface temperature. In general, the skin temperature exhibits a higher temporal and spatial variability, which are guided by the mostly upwards directed heat flux.

The temperature difference can be estimated with a formula given by Hasse [8]

$$T_o - T_w = C_1 \frac{H}{u} + C_2 \frac{Q}{u} \quad \text{with}$$

C_1, C_2 empirical coefficients

H heat flow at the surface

Q short wave radiation

U wind speed

It is to be taken into account that T_w is dependent upon the depth, even in the uppermost metres, so that comparison values or standard values for radiometrically determined temperatures must be measured very carefully. An example of a near-surface temperature profile is shown in Fig. 4.

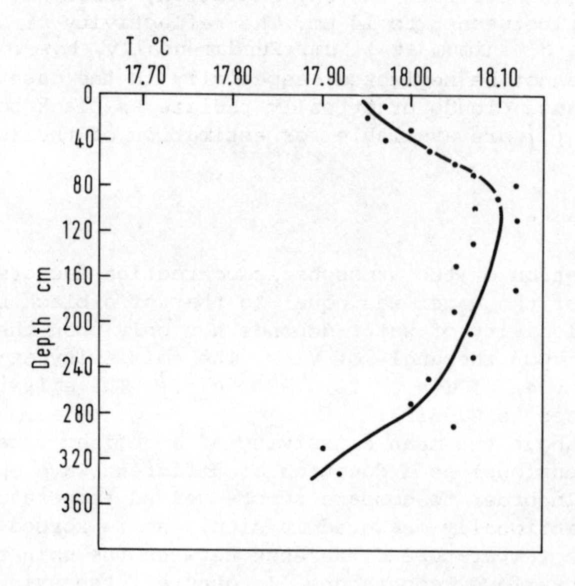

Fig. 4 Near-surface temperature profile
Surface Rider thermistor chain 29.8.1976, 15.12 GMT.

1.4 Geometrical corrections

The panorama distortion correction is an essential, relatively simple, correction. This corrects the effects of the scan angle upon the pixel dimensions. The correction can take place by variable enlargement along the scan line. The enlargement factor can be calculated according to the following formula [1]

$$v_f = \frac{1}{\cos^2 \alpha}$$

2. COMPARISON OF RADIOMETRICAL AND CONVENTIONALLY MEASURED TEMPERATURES

In a test area in the German Bight an investigation was made into in how far radiometrically determined surface temperature distributions are comparable with those measured conventionally. The IR data were taken by a DFVLR aircraft, which was equipped with a multispectral scanner (M^2S, Bendix). For correction (radiosonde) and comparison of the IR data, extensive sea-truth material was also obtained. The processing of the IR data was carried out using a Digital Image Analysis System at the DFVLR. After carrying out all the corrections specified in 1.1 to 1.4, the skin temperature distribution shown in Fig. 5b was obtained. Fig. 5a, as comparison shows the T_w distribution measured conventionally with a thermistor from a moving ship.

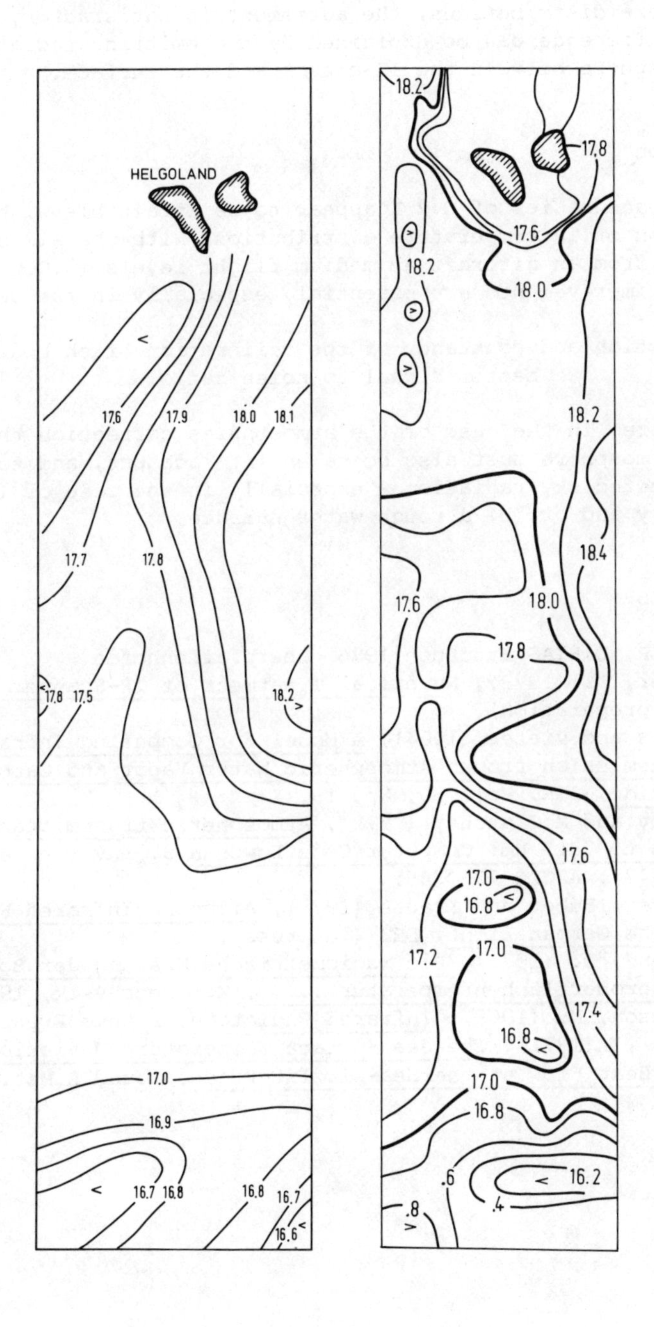

Fig. 5a Fig. 5b

The mean difference was about 0.4 K. Taking into account the tidal movement of the water-masses and the time difference between both temperature distributions, the agreement is satisfactory. A part of the difference can be explained by the emitting radiation of the atmosphere between the IR-sensor and the surface.

Conclusion

Absolute accuracies of 0.2 K appear to be attainable with the extraction of IR temperature distributions with the aid of scanners operated from an aircraft at medium flight levels (3,000 m). However, improvements are essential, especially in the sensors

calibration and constancy of the calibration black bodies;
better signal to noise ratio.

Furthermore, in the case of the atmospheres correction the emission of the atmosphere must also be taken into account, and to consider the reflected sky radiation - especially in the case of (a) a cloudy sky and/or (b) a rough water surface.

REFERENCES

1. DFVLR, DIBIAS Handbuch 1976, Oberpfaffenhofen.
2. Huber, K. (1978), Numerical Treatment of IR-Scanner Images (in preparation).
3. Davis and Viezee (1964), A Model for Computing Infrared Transmission trough Atmospheric Water Vapor and Carbon Dioxide, JGR, Vol. 69, No. 18.
4. Selby and McClatchey (1972), Atmospheric transmittance from 0.25 to 28.5 μm: Computer Code Lowtran 3, Environ Res. Paper No. 427, AFCLR-72-0745.
5. Becker, Huber and Krause (1978), Airborne Infrared Radiometry in the German Bight, DHZ (in press).
6. Lorenz, D. (1973), Die radiometrische Messung der Boden- und Wasseroberflächentemperatur ....., Z.f.Geophysik, 1973, Bd. 39.
7. Bramson, M. (1968), Infrared Radiation, Plenum Press, New York.
8. Hasse, L.(1971), The Sea Surface Temperature Deviation and the Heat Flow at the Sea-Air Interface, Bound.L.Met. 1 (1971).

LASER - INDUCED FLUORESCENCE TECHNIQUES FOR SOUNDING OF THE HYDROSPHERE

F. Günneberg

Bundesanstalt für Gewässerkunde, Koblenz.

ABSTRACT. Feasibility studies and design considerations for an airborne laser - fluorometer are discussed. The principle of operation allows the detection and in many cases the identification of oil - spills, chemical deposits and algae, and perhaps the measurement of important indicators, e.g. $SO_4^=$. The discrimination of different types of fluorescence is discussed. The application of multichannel spectroscopy is recommended. Sensors and an operational system are described.

1. PRINCIPLE OF OPERATION

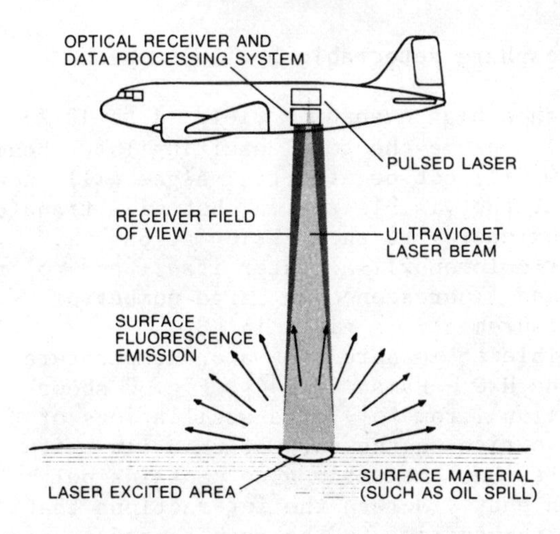

Fig. 1. Airborne laser fluorosensing system. (Barringer, [1])

T. Lund (ed.), Surveillance of Environmental Pollution and Resources by Electromagnetic Waves, 283–290.

From any type of platform, an aeroplane or a ship, laser - light pulses of a well-defined wavelength are sent to the water surface. The backscattered light is admitted through the aperture of the receiving apparatus.

1.1 Observed kinds of scattering

Specular reflectance from the water surface will not change wavelength or polarisation. A part of the laser-light energy will be scattered elastically by the turbidity of the water, without a change in wavelength. Another part of the light energy will be converted to a wavelength-spectrum according to the characteristic energy level scheme of the molecules, e.g. of oil or of chlorophyll in algae. A fourth part of the incident light will interfere with the vibrational spectrum of the molecules encountered, and the frequency-shifted light will be re-emitted, (Raman-effect). So the components of light seen by the receiver will be those

a) reflected from the surface, wavelength and polarisation retained
b) scattered by turbidity, wavelength retained
c) transformed to specific spectra by fluorescence
d) Raman - scattered with a specific wave number shift.

A survey of the effects implied is given by Collis and Russel in [2]. The elastic scattering and the specular reflectance and the Raman-scattering will result in the re-emission of light synchronously with the exciting laser light. The inelastic fluorescence (c) will exhibit an accumulation of exciting light and an exponential decay of the emission of transformed light, with time constants of the order of magnitude of picoseconds to microseconds.

1.2 Substances in the hydrosphere detectable by fluorescence

Oil will fluoresce with such a high quantum - yield (1 to 10 %) that even a thin layer will consume the total exciting laser beam energy, and the water below will not be affected. Algae will show some fluorescence throughout the visible region, but will transfer a great part of the exciting energy to an emission at 685 nm, which is characteristic for chlorophyll-a. Water itself and polar ions like $SO_4^{=}$ will show Raman fluorescence. A large number of interesting examples of measurements is given in [3].
Moreover it should be possible to measure the water temperature from the line profile of the H_2O - Raman line [4]. Fig. 2 shows that line with high resolution (from [5]). The oscillations of monomer water molecules give rise to the part of this line with the higher wave number shift. But the water also contains polymer $(H_2O)_n$ molecules, which partly screen the interactions that determine the oscillation frequencies. So the resulting frequen-

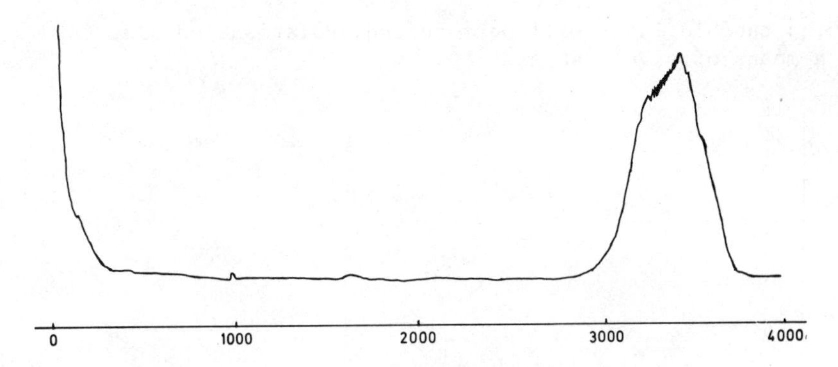

Fig. 2. Raman spectrum of sea water (from [5])

cies are lower and the wave number shift is smaller. The proportion of monomer to polymer water is highly temperature-dependent; the proportion of the Raman line will change correspondingly. In clear Pacific water a depth sounding of the water temperature profile by gating the receiver with nanosecond resolution seems feasible.
On looking closer at Fig. 2, another small line with a shift of 982/cm can be found; it is due to the $SO_4^=$ ion in sea water. If the detection and measurement of this weak line could be made feasible under rough field conditions and with somewhat turbid water, the salinity of estuaries could be mapped. Unfortunately the Cl^- ion, which is present at much higher concentrations, cannot oscillate in any form, and thus cannot give rise to any fluorescence or Raman shift.
Generally speaking, the remote sensing of water quality is not as advanced by far as that of air pollution. This is in part due to the fact that the spectra of substances dissolved in water cannot be as sharp as those of gaseous molecules, because of the interaction of the molecules in the liquid phase.

1.3 Separating out the signal of interest

Fluorescences of different decay times can be separated by time-gating the reciever or by fast waveform recording. As the Raman-fluorescence is emitted synchronously with the exciting light, separation from inelastic scattering can be done by time gating, if the decay time of that fluorescence is long compared with the pulse length of the laser. To discriminate between very fast normal fluorescences and Raman-fluorescences, the exciting wavelength can be shifted. Then the normal fluorescences will stay where they are, while the Raman-lines will be shifted by the same wavenumber as the exciting beam. The procedure is schematically shown in Fig. 3. To achieve the shifting of the exciting wavelength, a tunable laser will be required. Polarisation also may offer a means of signal separation.

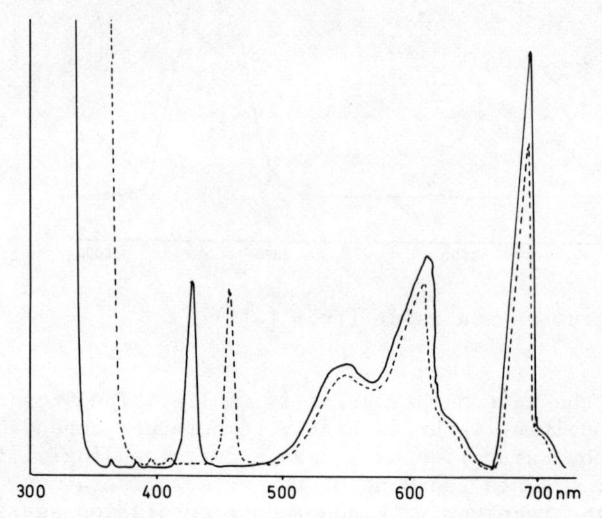

Fig. 3. Shifting the Raman-lines by shifting the wavelength of the exciting laser

1.4 Calibration and handling of data

The Raman line of H_2O, which has a shift of about 3400/cm, i.e. it fluoresces at 381 nm when excited by a nitrogen laser at 337 nm,

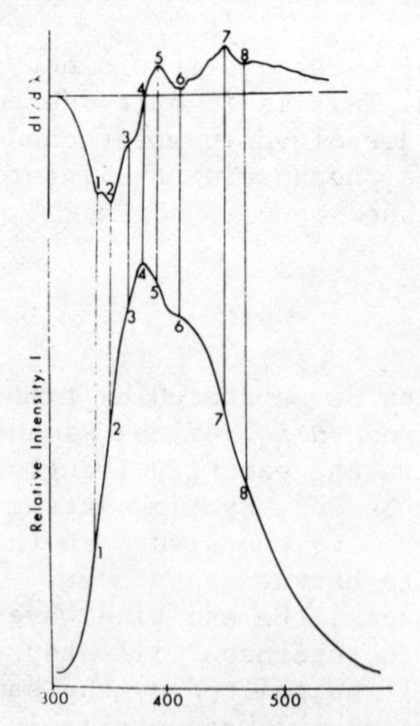

Fig. 4. Fluorescence spectrum of a crude oil and first derivative

is of great value in standardising all other received signals. It gives information about the volume of water really affected by the exciting laser beam, subtracting the losses of fluorescence light by scattering. Moreover ground truth measurements will be essential for any quantitative interpretation, as fluorescences can be quenched by other chemicals and the fluorescence of algae depends on the oxygen content of the water and other factors.
In some cases it may be of value to form the derivatives of a spectrum. For example, Fig. 4 shows the fluorescence spectrum of a crude oil and its first derivative [6].It fills almost the entire visible region, and has very little detail. The derivative does not in principle contain any new information, but the percentual differences are much more pronounced than those of the primary spectrum. Organic analysts sometimes make use of these derivatives for the quantitative analysis of mixtures of substances.

2. THE MEASURING TECHNIQUES

2.1. The single pulse technique

The interaction between a laser-beam and the sea-surface is schematically shown in Fig. 5. The reflection loss will depend on the angle of incidence, and this in turn will depend on the wave slope that we happen to meet. Moreover the excited volume is of some extent, and considering some small angle scattering, the return light will pass through a certain patch of the water surface, which has a certain extent. This patch will be curved by the wavyness of the water surface and by the wind ripple. Thus it can have a focussing or defocussing power and can correspondingly

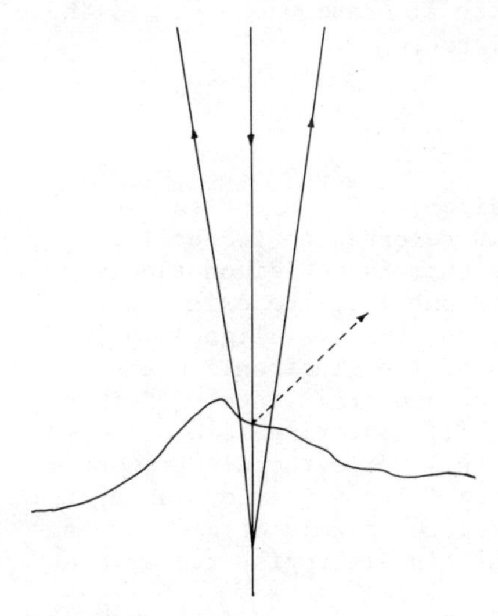

Fig.5. Laser beam interacting with sea surface

modify the intensity reaching the receiver aperture. So even if the pulse-to-pulse repeatability of the laser is very good (Manufacturers brochures quote ± 3 to 10%) and the fluorescence of the water body is stable, there still will be a pulse-to-pulse fluctuation of the received signal. To overcome this, you can try a statistical approach. For example if you fire ten pulses for every wavelength setting, and you need 500 settings to scan the spectrum, that's a total of 5000 pulses. If you have a powerful and fast laser, you can fire 100 pulses per second. then you will need 50 seconds for your spectrum, and your aeroplane, cruising at least about 150 km/h will move 2,5 km during that time. So you would miss the plume of a wrecked tanker. If as an alternative you take a helicopter as your platform and stay over the same point, the algae will fade out, so that their response to the exciting flashes alters, thus spoiling your spectrum. Thus there are strong arguments for obtaining the total spectrum with a single laser pulse, using a multichannel analyser.

2.2. Multichannel analysers.

Such instruments for example are the Optical Multichannel Analyser (OMA) of Princeton Applied Research Corporation or the OSA of BM-Spektronik GmbH (8031 Puchheim) or the Barringer Research (Toronto) fluorosensor.
These measuring systems take the signals from their detector arrays, digitalise them, store them in one of at least 4 memories, handle them in a microcomputer and feed them to a CCT recorder. Thus you can correct for the spectral response of the complete receiving apparatus, or you can subtract the background radiation or you can form the derivatives of the fluorescence spectra. There is also a video-display for surveillance.

2.3. Multichannel detectors

Different types of detectors are available. For an extensive state of the art report the reader is referred to two brilliant articles by Talmi [7] .The most common form is a Vidicon tube, as used for television cameras, which is put into the exit focal plane of the monochromator. This Vidicon tube is aligned in such a direction, that the scanning lines of the electronic beam on the silicon target are parallel to the spectral lines. (If this alignment is not perfectly done, it will give rise to a Moireé-structure and will decrease resolution.) While the electron-beam scans one line, it averages over detectivity variations along that strip of the sensitive surface or over the pixels of a discrete diode array. The average value of one line is registered as one

digital value; here you will not need a video-recorder band-width. An advanced form of the Vidicon tube is the S.I.T. (Silicon Intensified Television) tube, which has an image intensifier ahead of the silicon target, that is scanned by the electron beam. This tube has a very high dynamic range of 1 to 10^4 and has a 300 fold SNR-capability compared with the ordinary Vidicon.
These OMAs with SITs were originally designed for low-level photon couting, and the producers are still proud to report integration times of hours when the SITs are cooled with liquid nitrogen. For our purposes this integration capability even for targets at room temperature is disadvantageous. There is a lag, an image retention on the secondary target, of 50 % after 0,15 sec. To get a correct measurement, the photo cathode can be flashed with a small xenon lamp or the target erased by several scans with the electron beam. A full scan of all lines will take 30 ms, so that the apparatus needs a long recovery time. The image intensifier of the SIT can be gated with nanosecond resolution, so that fluorescence decay times can be measured. SITs are sometimes said to suffer from microphonics.
There are other types of photosensors, the silicon-photodiode-arrays. [7, 8] They range from linear arrays of e.g. 1000 elements to two-dimensional area arrays of e.g. 320x512 elements. The charge coupled devices (CCD)s are integrated circuits, that include shift registers, by which the charge packets generated in the photo-diode-pixels can be read out. The charge injection device (CID) utilises two arrays of charge storing capacitors, that allow x-y random readout. Clock rates up to 10 MHz are reported. Developments in this field are very rapid.

2.4 The Barringer Laser Fluorosensor

This system is adapted to large scale routine measurements. The number of spectral lines is reduced to 16, with 20 nm linewidth each, except for a special line of 6,5 nm for the water Raman line.

One channel is centered at 685 mm and is 10 nm in width, to coincide with the chlorophyll a-fluorescence band. In the output-focal plane of the monochromator the light is imagesliced into the channels mentioned and the energy of each channel guided by fibre optics to a microchannel plate (MCP) image intensifier. The intensified light again is conducted to a discrete diode for each channel by fibre optics. These 16 diodes each have a separate sample and hold and background subtraction circuit. The signals are processed by a microcomputer, that feeds them to a digital tape recorder and to a CRT display. A fraction of the collected light signal is split off to measure the fluorescence decay times in two channels defined by interference filters, the resolution being in nanoseconds. A 1 MW pulsed nitrogen laser, 337.1 nm,

4 nsec pulsewidth, 100 pps repetition rate, is utilised. Flight hight is 1000 feet in full daylight.

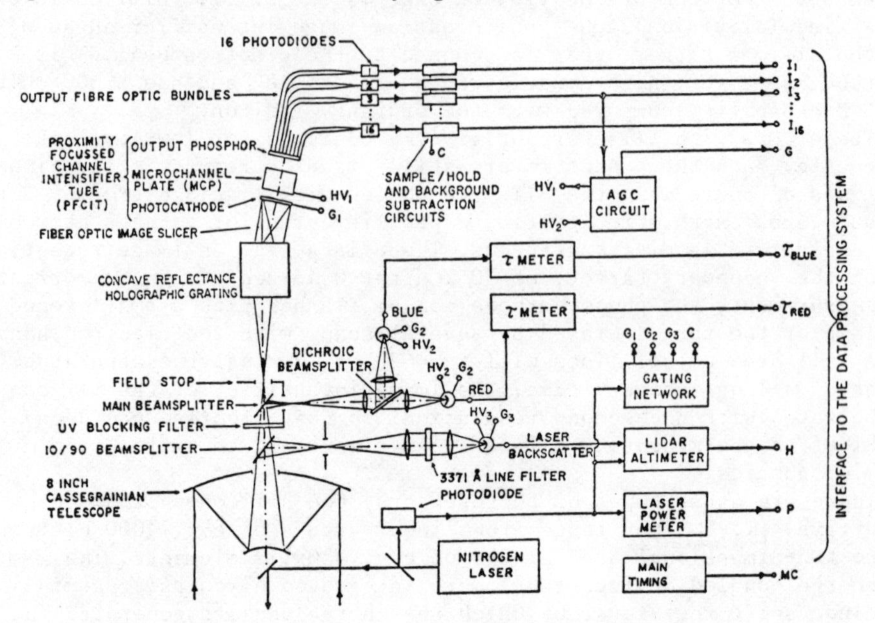

Fig. 6. The electrooptical part of the Barringer system

REFERENCES

1. Barringer, manufacturers brochures, Toronto 1976
2. E. Schanda, Remote Sensing for Environmental Sciences Ecological Studies 18, Springer Verlag 1976
3. Celander, Fredriksson, Galle and Svanberg, Investigation of Laser-Induced Fluorescence ... Göteborg Institute of Physics Reports GIPR-149, February 1978
4. Chang, CC, Young, L.A.: Remote measurement of water temperature by Raman scattering. Proc. Eight Symp.Rem.Sen.Env. II, 1049-1068 Ann Arbor 2.-6. October 1972
5. A. Davies, M. Bristow and J. Koningstein: Raman Spectroscopy as a Water Quality Indicator. American Water Resources Association, Proc. No. 17, June 1973
6. G.L. Green and T.C. O'Haver: Derivative Luminescence Spectrometry Analyt. Chemistry, Vol. 46, No. 14, 2191-2196 December 1974
7. Y. Talmi, Applicability of TV-type multichannel detectors to spectroscopy. Analytical Chemistry, Vol. 47, No. 7, 1975 659A - 670A and 697A - 709A
8. Keyes, Optical and Infrared Detectors. Vol. 19 of Topics in Applied Physics. Springer Verlag 1977.

MICROWAVE MEASUREMENTS OVER SEA IN THE NETHERLANDS

Ir. E.P.W. Attema * and Ir. P. Hoogeboom **

* Microwave Laboratory, Delft University of Technology
** Physics Laboratory TNO. P.O.Box 96864
2509 JG the Hague, the Netherlands

Abstract. In collaboration with the NLR (National Aerospace Laboratory) the Physics Laboratory TNO carries out microwave measurements at sea for Rijkswaterstaat (Department of Water Control). The purpose of the measurements is mainly the development of an operational system for oil detection, ships traffic control and estimation of seawave characteristics. Up to now groundbased measurements took place in a wavetank and on a platform near the coast of Noordwijk. Also many X-band SLAR flights are available. Some results will be discussed.

1. INTRODUCTION

In the Netherlands Rijkswaterstaat is responsible for the coastal zone and the Dutch continental shelf. This task includes for instance the detection (and clearing away) of oil spills of tankers and the inspection of the use of shipstraffic lanes. Especially under bad weather conditions and by night a lot of breaches occur. Therefore a surveillance system is needed, which is capable of monitoring under these conditions (ref.1).

A SLAR can be a useful instrument for this application. However, a lot of research still has to be done on the interaction between seawaves and microwaves. This research is done in collaboration with the Physics Laboratory TNO.

For a better understanding of the phenomena, groundbased radar backscatter measurements are carried out on the North Sea. Simultaneously groundtruth (seatruth) is gathered with different kinds of sensors. These measurements were taken during October and November 1977 on the platform Noordwijk, 5 miles off the coast of Noordwijk. During the measuring period SLAR flights have also been

T. Lund (ed.), Surveillance of Environmental Pollution and Resources by Electromagnetic Waves, 291–298.

made.

In 1974 measurements have been performed in a wavetank. The measurements consisted of radar backscatter measurements on several types of waves and on oilslicks (ref.3).

2. DESCRIPTION OF THE MEASURING EQUIPMENT

Several instruments are used for the radar backscatter measurements at sea. At short range measurements are made with an X-band reflectometer, based on the FM-CW principle. It uses a pencil shaped beam. Maximum range is about 100 m, depending on the reflectivity. Its characteristics are given in table I.

Table I: characteristics of the short range X-band scatterometer.

RF frequency: 8-12 GHz
Output power: 2 W (with external TWT)
Modulation: triangular, 25 Hz
Frequency deviation: 400 MHz
Antenna beamwidth: 5°
Sensitivity: -30 dB relative to 1 m^2 at 40 m distance
Polarization: VV, HH and cross

At the platform Noordwijk the radar is mounted at 13 m above sea level. Given the range of the system, the grazing angle may vary down to 10°. The construction of the platform limits the higher grazing angles to about 55°.

A Q-band rotating beam shipsradar is used to provide a general overview of the wave pattern around the platform. The range is limited to about 1000 m. Display is on a PPI. Its properties are given in table II.

Table II: characteristics of the Q-band rotating beam radar.

RF frequency: 33 GHz
Pulse peak power: 20 kW
Pulse length: 40 nsec
PRF: 2500 Hz
Antenna beamwidth: 0.6^o
Polarization: HH; adapted to VV and cross

The PPI photographs can be used for finding directional spectra of the waves. In the future the system will be modified to measure wave spectra with this system with the antenna in a fixed position.

A long range reflectometer has been used at X-band frequencies. This system uses a pulse radar (ref.2). The minimum detection range is ± 600 m.

Its characteristics are given in table III.

Table III: characteristics of the long range scatterometer.

RF frequency: 9.375 GHz
Pulse peak power: 50 kW
Pulse length: 0.12, 0.25, 0.5 or 1 μsec
PRF: 2000 Hz - 1000 Hz (depending on the pulselength
Antenna beamwidth: 2.1° (pencil beam)
Polarization: HH, VV

At the platform Noordwijk this radar was mounted at 19 m above sea level. The grazing angle varies from 2.1° to almost zero.

SLAR images are made with an EMI X-band SLAR. The characteristics of this system are given in table IV (ref.1).

Table IV: characteristics of the X-band SLAR.

RF frequency: 9.6 GHz
Pulse peak power: 80 kW
Pulse length: 0.2 μsec
PRF: 1260 Hz
Resolution across track: 30 m
along track: 16 mrad
Output:on film and videotape

3. SOME RESULTS

3.1 Measurements in the wavetank

In 1974 measurements were carried out in a wavetank at the Delft Hydraulics Laboratory (ref.3). Wavespectrum and the windspeed can be varied. Oil slicks were made to find out the contrast produced in radar backscatter. A contrast of 6 dB was measured at a windspeed of 10 m/s. At lower windspeeds (5 m/s) the contrast was even higher than 6 dB. The influence of the thickness of the oillayers on the radar backscatter could not be established by these measurements. Fig.1 shows a recording of the short range scatterometer. When the oil enters the radar beam, the signal decreases. The thickness of the oillayer is only 10 microns.

SLAR flights later showed that the contrast was high enough for applications at the North Sea. Fig.2 shows two examples. The distance between the range markers is 1 n.m. The left hand image shows an oil slick observed under moderate wind conditions (4-8 m/s). The right hand image shows an oil slick under slightly lower wind conditions (2-5 m/s). The clearly visible wave pattern is caused by sanddunes on the bottom of the sea (waterdepth is

20-30 m). The strong tidal current makes these dunes visible in the image by modulation of capillary waves on the surface (ref.1; see also fig.5). Under both conditions the oil slick remains clearly visible.

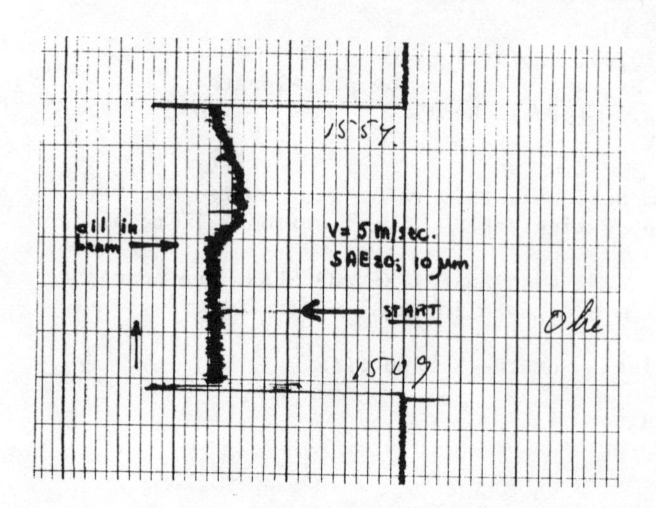

fig.1: recording of short range scatterometer: reflection of oil-slick.

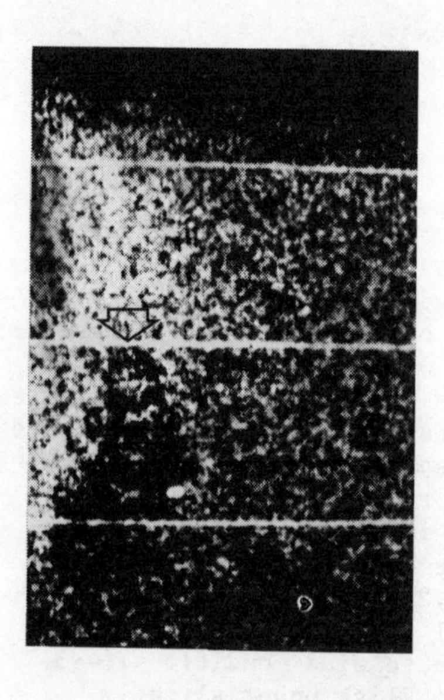

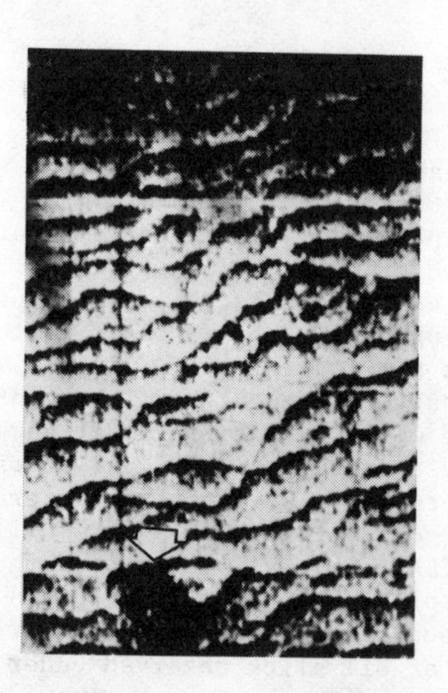

fig.2: SLAR images of oil slicks at sea. Arrows indicate oilslicks.

Measurements in the wavetank on waves under varying conditions showed good agreement with values known from literature.

3.2 The project "Noordwijk 77"

The platform Noordwijk was used for radar backscatter measurements from September-November 1977. The platform itself has different kinds of sensors attached to it.

Windspeed and winddirection are measured at a height of 25 m. A wavestaff, DATAWELL waveriders and current meters give general information on the seawaves.

The short range and long range scatterometers were used to measure radar backscatter coefficient for different antenna polarizations and as a function of grazing angle and at different angles with respect to winddirection and wavedirection.

Long recordings were made to produce wavespectra. These can be correlated with spectra of the wavestaff signal, which was recorded simultaneously. The measurements are now being worked out. An example is given in fig.3 (on the next page). Here spectra are shown of the short wave scatterometer and the wavestaff. The data were taken during 1 hour immediately after a heavy storm. The windspeed is still very high (18 m/s) and the significant waveheight is 3.5 m. These spectra show good agreement, especially for the principal component of the waves.
Fig.4 shows an image (PPI photograph) of the rotating beam Q-band radar. The windspeed is here 16 m/s. The wavepattern is clearly visible. The distance between the rangemarkers is 300 m (total range: 1200 m); antenna polarization is HH. Perpendicular to the main wavepattern a second pattern with longer wavelength can be noticed.

A start is made with the investigation of the relation between the radar backscatter and the position on the wave. This work will be continued.

During the measurements SLAR flights were made. In the past few years also many SLAR flights have been made and a lot of experimental work is done at SLAR images (ref.1). A very nice example is given in fig.5, which shows an image of the coast near Hoek van Holland. Different kinds of phenomena are to be seen here simultaneously (see explanation in subscript of fig.5).

4. FUTURE PROGRAMMES

Groundbased measurements at the platform Noordwijk will continue this year and next year (1979). Measurement periods will be from September-November. In 1979 the measurements come in the framework of the SURGE/MARSEN project.

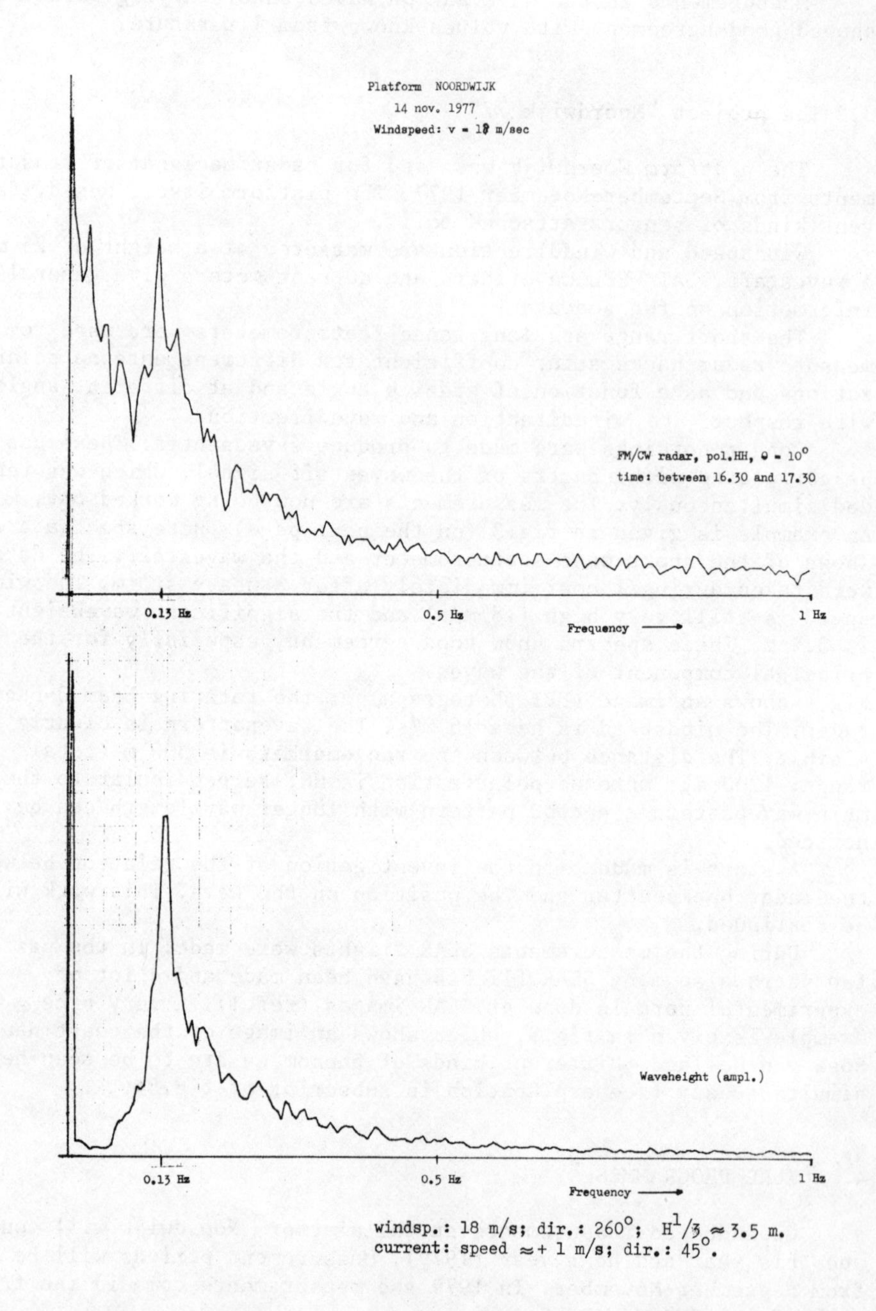

fig.3: simultaneously taken spectra of short range scatterometer and wavestaff.

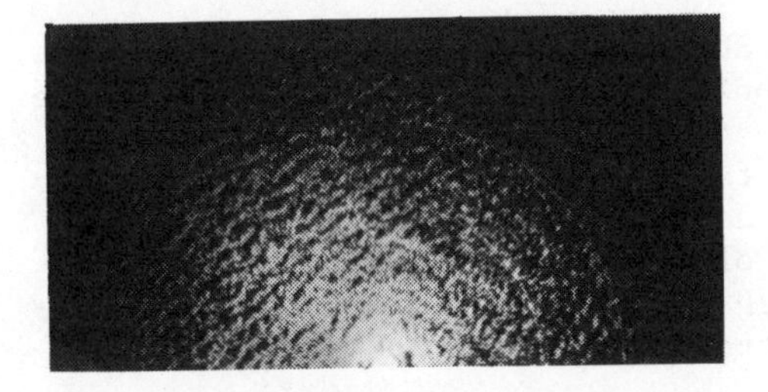

fig.4: image of rotating beam Q-band radar.

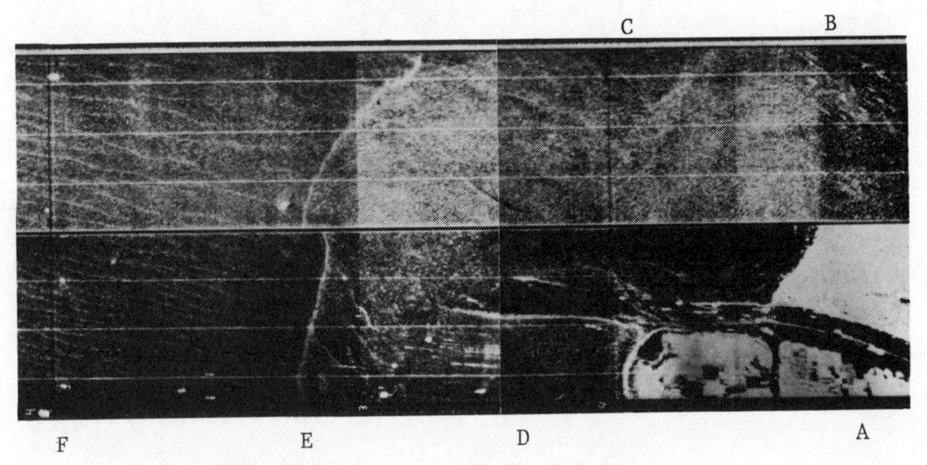

fig.5: SLAR image of the coast near Rotterdam.
The centre of the film is the aircraft track. The gap in the centre is 2 n.m. The range is 2x6 km.

A: the gain setting of the system is increased here for observation over sea. The land echoes now saturate.

B: wavepattern; waves travelling to the coast.

C: see details in the water. Waves, ships tracks, etc.

D: edge of current: Rhine flowing into the sea.

E: edge between the fresher water of the Rhine and the salter water of the sea.

F: megaripple (sanddunes on bottom; see 3.1) becomes visible through its effect on the surface through the tidal current.

After more general investigations of the interaction between seawaves and microwaves we now hope to go into more specific problems concerning this subject. Also new experiments with oilslicks will be carried out.

A new digital calibrated SLAR system is being considered, to make quantitative comparisons between groundbased and airborne measurements possible. This system will have a digital output, which is recorded on magnetic tape. The properties of the objects to be seen were considered in designing this SLAR, in particular the Rayleigh fading character of land- and sea echoes. This is done in order to achieve a high accuracy (± 0.5 dB) in the determination of the radar backscatter coefficient and its modulation due to waves and oil slicks.
The system will be in operation in 1979.

References

1. De Loor, G.P. and Brunsveld van Hulten, H.W.: 1978, "Microwave Measurements over the North Sea", in: R.H. Stewart (ed.) "Radio Oceanography", special issue of Boundary Layer Meteorology, 13, pp 69-81.

2. Sittrop, H.: 1974, "X-band and K_u-band Radar Backscatter Characteristics of Sea Clutter", Proc. URSI, Specialists Meeting on Microwave Scattering and Emission from the Earth, Berne, Sept. 23-26, pp 25-37.

3. Van Kuilenburg, J.: 1975, "Radar Observation of Controlled Oil Spills", Proc. 10th International Symposium on Remote Sensing of Environment, Ann Arbor, Oct. 6-10, pp 243-250.

APPLICATIONS OF REMOTE SENSING BY CONVENTIONAL RADARS

P.D.L. Williams

Decca Radar Limited, Development Laboratory,
Chessington, Surrey, England.

Abstract

There is a widespread availability of surveillance radars used primarily for navigation and collision avoidance, apart from larger equipments used for air traffic surveillance and defence purposes. Examples of the use of these radars for sensing air-sea interactions are given with comments on interpretation of the displays.

Typical Sea-Going Radars

Civil marine radars are fitted to various craft ranging from pleasure boats to the largest ships. The smaller sets cost less than a modest car and are capable of detecting targets as small as 10 m^2 Radar Cross Section (RCS) out to 2 nautical miles (n.mi). These usually operate on a wavelength of 3.2 cms though some operate on 10 cm. Large ships fit two or even three radars each specialising in particular features such as long range landfall, better performance in precipitation or better spatial resolution than that offered by the basic set capable of meeting the U.K. M.o.T. Marine Radar Performance Standard or other national basic specifications[1].

The larger sets may have 20 or 30 dB improved performance, thus able to detect sea birds at several miles and in general have a performance which is horizon limited. There are some 100,000 such radars at sea and this paper invites their more widespread use for remote sensing purposes. These radars are nearly all pulse sets with rotating aerials having a fan beam some 20° in elevation with the nose set to the horizon and azimuth

T. Lund (ed.), Surveillanceof Environmental Pollution and Resources by Electromagnetic Waves, 299-308.

resolutions from three-quarters to 2 or perhaps 3 degrees for the really small sets. Their range resolution is switchable from 0.05 μs (7.5 m) to about 1 μs (150 m) and they scan the sea surface between 20 to 30 times per minute, giving an immediate CRT presentation in PPI form with range scales variable from about 0.25 n.mi to maximum scales from 24 to 60 n.mi. Data recording using either a polaroid camera or ciné film taking one picture per aerial revolution is convenient and when the ciné film is projected at faster rates the time lapse record is quickly scanned for interesting phenomena, particularly if the equipment has been left unattended for a few weeks, taking perhaps one photograph per hour, as carried out by Hardcastle of the Coastal Sedementation Unit at Taunton, U.K.

The choice of wavelength for marine radars is complex, depending very much on (i) targets (whose wavelength dependency varied from the ratio of $(\lambda_1/\lambda_2)^2$ through zero for spheres large compared to either wavelength used, where the RCS equals the optical cross section, to $(\lambda_1/\lambda_2)^2$) to (ii) rain returns where the RCS varies as the fourth power of the wavelength change, i.e. the small spheres lie in the Rayleigh region. (iii) For surface to surface radar operation not only has the radar horizon a dominating effect but the Lloyd's Mirror effect caused by an aerial and its image beneath the sea is subject to the sea surface roughness, these various conditions are tabulated for a range of typical situations by Williams[2] 1975.

As an illustration of the use of two radars similar in all respects except their operating wavelength, Fig.1 shows a radar view of the English Channel viewed from Dover. It will be noted that while the two PPI scenes are not widely dissimilar the sea (clutter) return is greater on the shorter wavelength, as theory suggests[3].

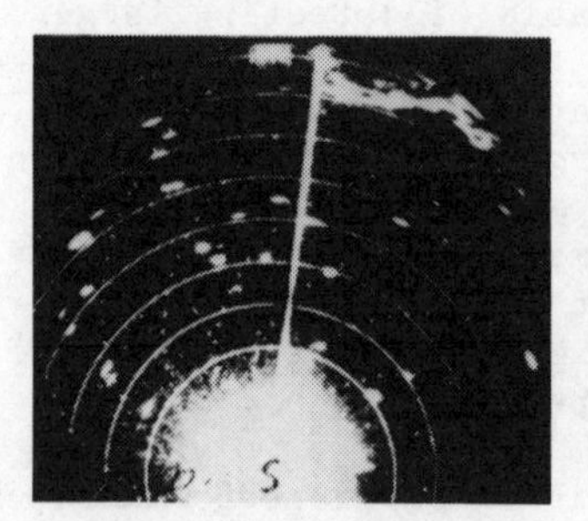

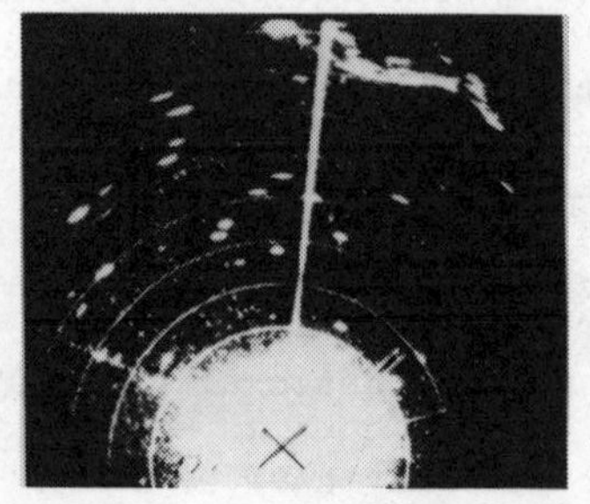

Fig.1 Dual radar presentation of the English Channel from twin radars at Dover operating on 3.2 and 10 cm.

Most marine radars use a common horizontally polarized aerial for both transmit and receive. This choice is based on the lower amplitude return experienced in low sea states by the use of HP at low grazing angles. However, as soon as the sea surface is sufficiently disturbed the difference tends to zero, this topic is reviewed by Skolnik[3] and illustrated by 2 pairs of figures showing a marked difference of sea return in calm weather followed by similar returns from either polarization as Figs. 2 to 5. Here the sea was viewed from an experimental marine radar on the South Coast of England looking out over the Channel. The sea return in calm weather was so slight that a long pulse of 1 μs had to be used to show any clutter at all.

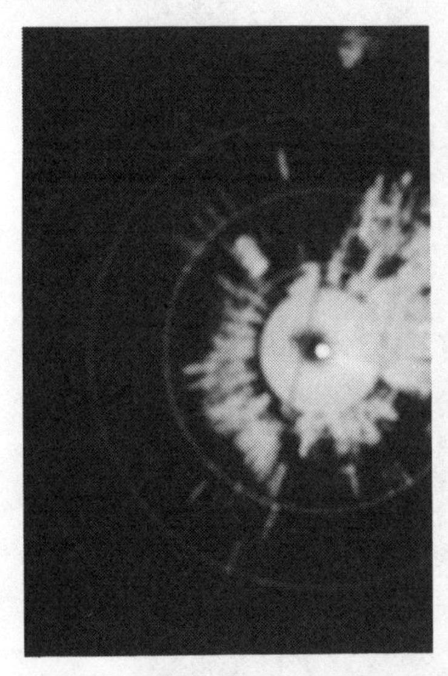

Fig.2 Light sea clutter using HP-HP (1μs) to show returns in a calm sea. ¼ n.mi cal. rings.

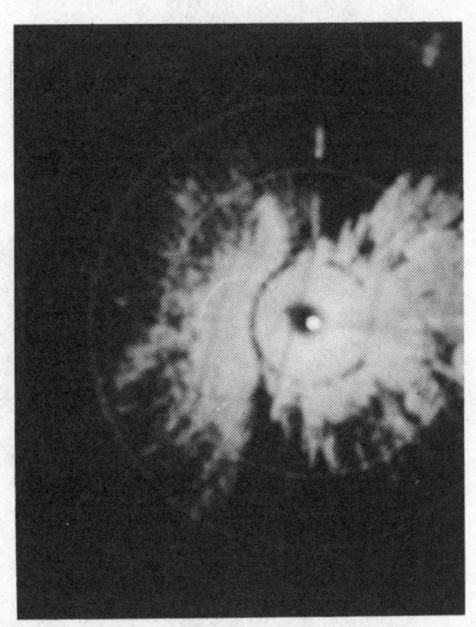

Fig.3 The same clutter using VP-VP. ¼ n.mi cal. rings.

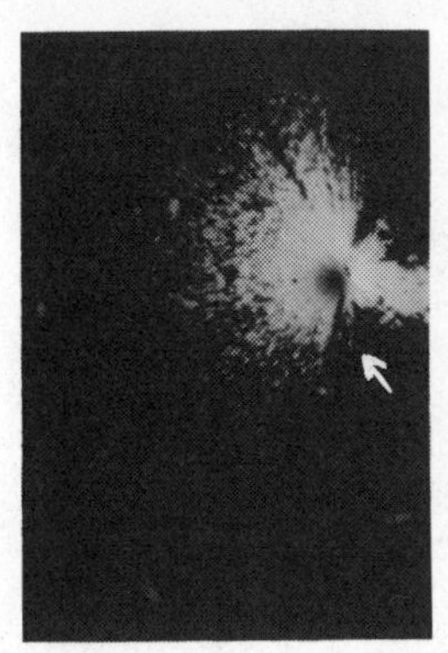

Fig.4

Sea clutter using HP-HP on a longer range than Figs.2 & 3 in a rougher sea.

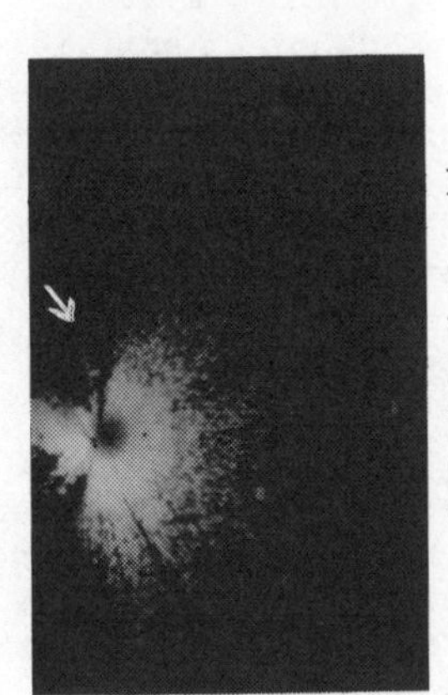

Fig.5

Sea clutter using VP-VP looking at the same sea shortly after Fig.4.

The sea surface return may be modified by various features, the ones of current interest are oil and ice. The detection of small pieces of ice at sea is important to high speed craft. Problems of ice being detected are discussed by Budinger[4] 1959 and Williams[5] 1973. However, in severe winter conditions when large areas of sea are ice-bound, radar provides a useful tool for finding ice leads, particularly under a covering of dry snow which masks the visual image. Fig.6 shows a PPI photograph taken aboard an ice-breaker in the Baltic during April 1978 without any special equipment. Other examples of sea ice photographs are to be found in the Sea Ice '75 reports[6].

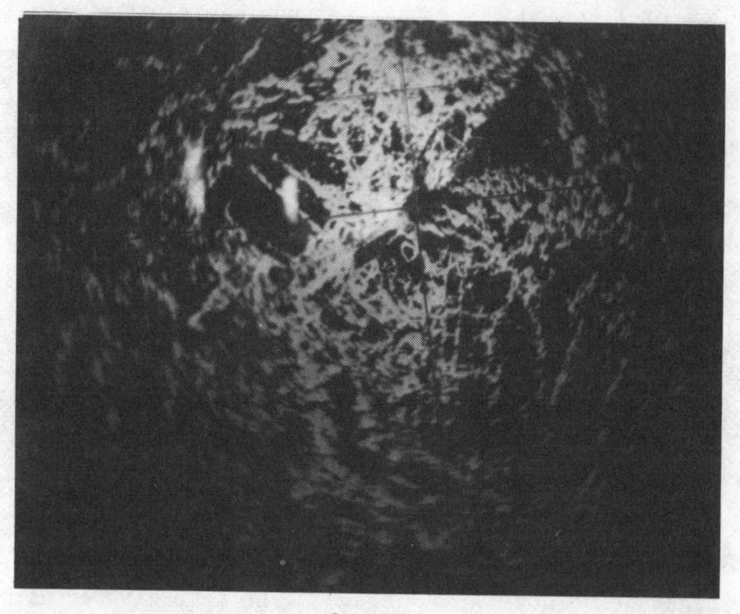

← 8 n.mi →

Fig. 6 View of the Baltic ice from a marine radar aboard an ice-breaker. April 1978.

A Marine Radar Used in an Airborne Role

To illustrate the flexibility of these small sets Figs. 7 and 8 show one mounted in a Llama Alouette III helicopter. This type of aircraft is used by the Swedish Administration of Shipping and Navigation on a daily basis during the Winter season to provide their ice-breakers with good short term forecasts of the often rapidly changing ice-field in the Baltic. In clear weather the visual contact is perfectly adequate, but in poor visibility a radar in the helicopter enables the observer to relate the display to the visual scene a hundred metres below and then follow ice features out to 5 or 10 n.mi to examine possible leads to be used by the ice-breaker over the next twentyfour hours. Vibration in the helicopter prohibited good quality PPI photographs being taken during a short trip in April this year, but good airborne photographs may be found in the Sea Ice '75 report[6].

Fig. 7

A small marine radar aerial and RF head mounted beneath an Alouette III.

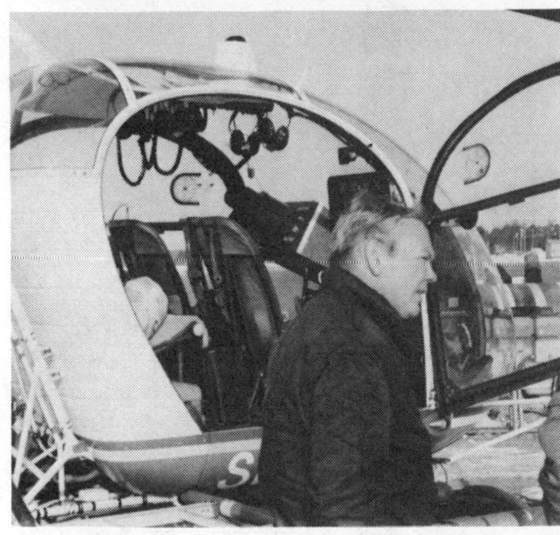

Fig. 8

The 110 display with light visor to the left of the pilot's position.

Size limitation prohibits the use of real aperture antennas to about 1 metre when scanning, but results from an interesting real 3 metre aperture SLAR have been presented by Ekengren[7]. Here the large, static, 3 m, X-band, array was carried slung under a light aircraft fuselage providing angular resolution of half a degree.

Coastal Radars

Free from aircraft and ships limitations regarding aerial size, apertures up to 8 m are currently in use. At X-band this yields an azimuth resolution of 0. 3°. An example of a swell wave pattern is shown in Fig.9. However, at low grazing angles the phenomenon of sea surface reflection is only recently being studied and Fig.10 shows the disappointing result obtained by the same high definition radar viewing the same patch of sea on a different day with what appears by eye to be a similar sea state.

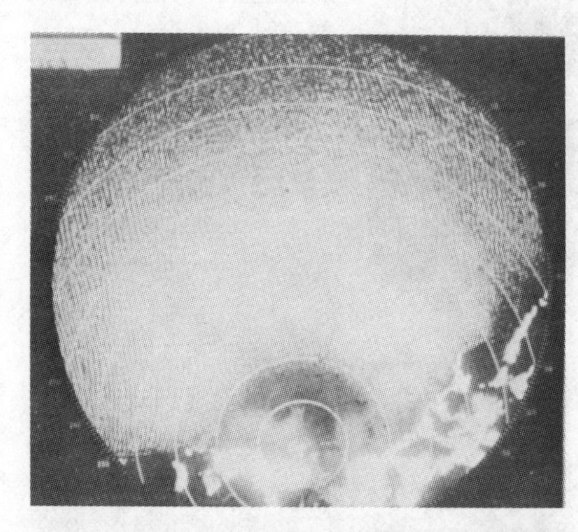

Fig. 9

Swell wave pattern shown on a coastal radar whose details are as for Fig.10.

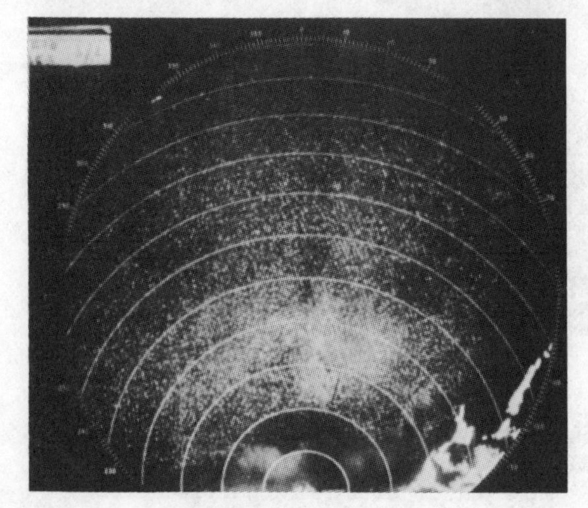

Fig.10

The same sea state as viewed in Fig.9 by the same radar but with poor swell pattern. For Figs.9 and 10 the radar had an azimuth resolution of 0.3° and a range resolution of 8 m, situated 150 m above sea level.

In rougher weather (sea state 4), the PPI photograph takes on a more noise-like and incoherent appearance as shown in Fig.11.

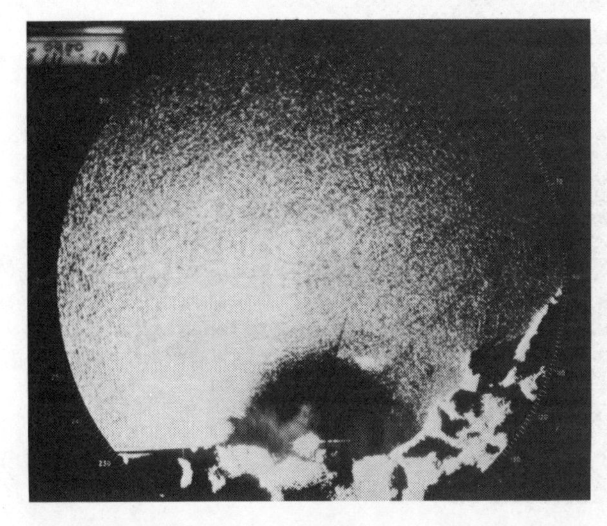

Fig. 11

Incoherent sea clutter return in sea state 4. Radar used as in Figs. 9 and 10.

Normally, the sea clutter return falls off as R^{-3} within its transition range, increasing to an R^{-7} dependency beyond. However, this is very much dependent on the evaporative surface duct. If, therefore, a patch of sea return is observed clear of the normal clutter region this often provides a useful indication of local disturbance such as tide rips. Another example of discrete areas of remote sea clutter is shown in Fig.12 where the broken water above the barely submerged Goodwin Sands East of Dover in the English Channel shows a distinctive pattern.

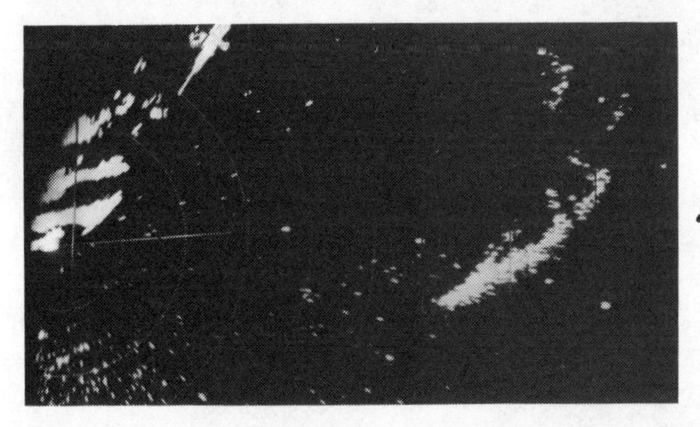

Fig. 12 High gain.

The radar display of the broken water over the Goodwin Sands to the East of the radar at Dover.

Fig.13 is taken from the same high resolution radar at Dover on a longer range to show the heavy shipping in the Straits and a large extent of the North French Coast. On rare occasions the wind is sufficiently strong to produce sufficient sea clutter return to stretch to Calais at maximum system gain, under which conditions oil slick detection over a very large area would be possible.

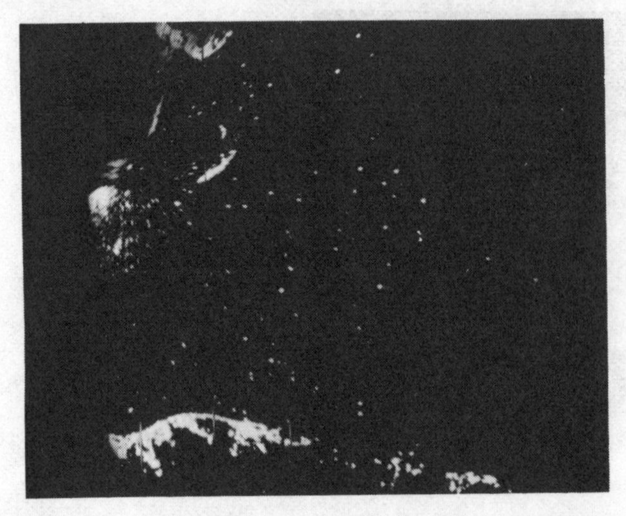

Fig. 13

A radar view of the Straits of Dover and French Coast from Dover taken with a high definition radar.

At 3.2 cms the sea return is heavily dependent on the local wind and Figs.14 and 15 show the relationship of the extent of sea clutter and its direction of maximum range as measured over one year from a marine radar situated at Dungeness also on the South Coast of England[8].

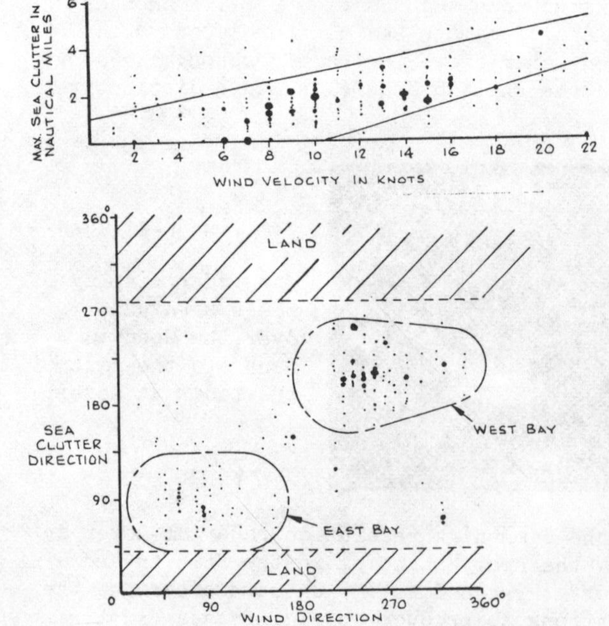

Fig. 14

Wind velocity versus maximum range of sea clutter.

Fig. 15

Wind direction versus direction of maximum sea clutter in coastal waters.

However, interpretation of radar displays in coastal waters is far more likely to ambiguity than in the open sea, particularly at low grazing angles.

Problems due to Using Radar at Low Grazing Angles

Most of the work reported over the last thirty years has been concerned with the sea return viewed from elevated platforms. At low grazing angles the effects of wave shadowing, surface ducts and difference between the radar and optical horizon all contribute to the variation in apparent radar performance over the sea during what appear to be similar environmental conditions.

Two of the papers by Wetzel[9-10] 1977 discuss these effects as well as the appearance of patchy phenomena having periods of the order of several minutes. Fig.16 shows a radar sea clutter picture with a striated appearance, the spokes do originate at the radar and the possibility of the effect being an instrumental one should, in this case, not be discounted. An example of false images being caused by sets of hyperbolae is illustrated in Fig.17.

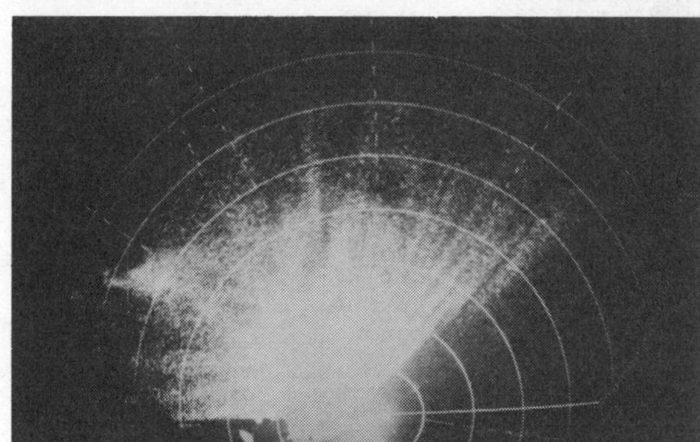

Fig. 16

Striations in sea clutter viewed from a coastal site.

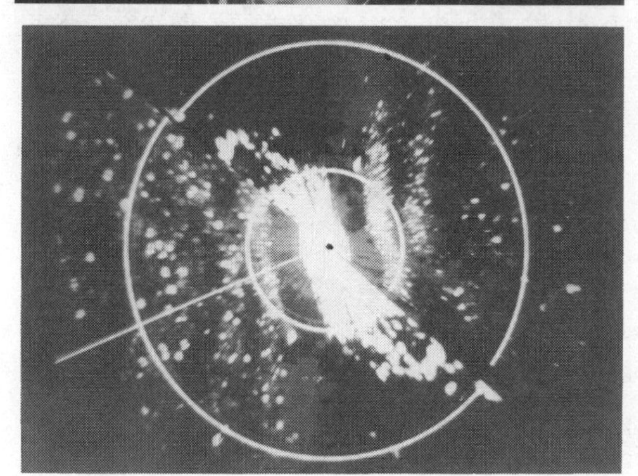

Fig. 17

False hyperbola pattern shown by integrating sea clutter return over several revolutions of the aerial. Two sets of pattern are shown on this double or split display[2].

This particular phenomena has been shown to be sensitive to the radar wavelength used by Taylor[11] 1976 and Croney et al[12] 1975.

When using surface radars at low grazing angles it is important to deploy several calibration targets in the areas of interest at the same time, as large variations occur over tidal cycles and also over longer periods. An example of the variability of performance against a reputedly non-fading Luneberg lens target at a constant range is shown as Fig.18[8]. Whilst the long term performance changes, heavily smoothed to remove tide height variations, is shown as Fig.19 from the same source[8]. The use of the scattering matrix[13] to help identify particular targets at sea is corrupted by variations caused by the multipath from the ever-changing sea surface, illustrations of which are given by Williams[8] et al.

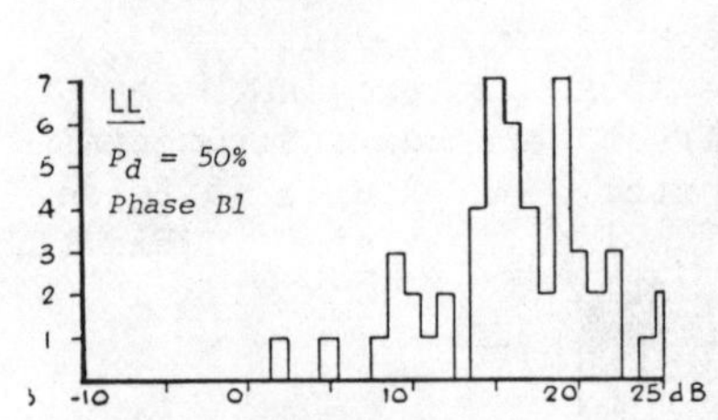

Fig.18 Luneberg Lens performance variations.

Fig.19 Performance of control targets by Phase.

References

1. Croney,J. Civil Marine Radar. Ch.31,Radar Handbook Ed.Skolnik.
2. Williams,P.D.L. Limitations of Radar Techniques for the Detection of Small Surface Targets in Clutter. Radio & Electronic Engineer, 45,8,1975.
3. Skolnik,M.I. Sea Echo. Ch.26 of Ref.1.
4. Budinger,T.F. Ice Detection by Radar. Coastguard Bull.45,1959.
5. Williams,P.D.L. Detection of Sea Ice Growlers by Radar. IEE Conf. Publication 105, 1973.
6. Blomquist,A. Sea Ice '75. Swedish/Finnish Report 16-9.
7. Ekengren,B. Oil Slick Detection with Real Aperture Side-Looking Radar. Ibid.
8. Williams,P.D.L. The Passive Enhancement of Navigation Buoys at Sea --. Radar '77. IEE Conf. Pub. 155.
9. Wetzel,L.B. A Mode for Sea Backscatter Intermittency at Extreme Grazing Angles. Radio Science,12-5, 749-756, 1977.
10. Wetzel,L.B. On the Origin of Long Period Features in Low-Angle Sea Backscatter. Radio Science 13-2, 313-320, 1978.
11. Taylor,R.G. Stationary Patterns in Radar Sea Clutter. Radio & Electronic Engineer, 45-3, 1976.
12. Croney,J. Further Observations on the Detection of Small Targets in Sea Clutter. Radio & Electronic Engineer, 45-3, 1976.
13. Vogel,M. Scatterometry. Ibid.

ERICSSON SLAR

Birger Ekengren

L.M. Ericsson, Mi Division,
Fack, S-431 20 Mölndal, Sweden

ABSTRACT. L.M. Ericsson has developed and delivered an inexpensive real aperture Side Looking Airborne Radar (SLAR) to the Swedish Space Corporation. The radar is intended for ocean surveillance with special emphasis on oil slick mapping.

Figure 1 Ericsson SLAR mounted on Cessna 337.

T. Lund (ed.), Surveillance of Environmental Pollution and Resources by Electromagnetic Waves, 309–318.

This paper describes the SLAR under the following topics: 1. General Background. 2. Design Philosophy. 3. System studies. 4. Experience from Flight Testing. 5. Conclusions about the usefulness of Ericsson SLAR for remote sensing.

1. GENERAL BACKGROUND

The Swedish Board for Space Activities is responsible for promoting remote sensing in Sweden. The Board, in its policy, has given priority to the development of operational remote sensing systems for

- oil spill surveillance
- sea ice mapping
- vegetation mapping
- analysis of effluents from industrial smoke-stacks.

Various techniques have been investigated for the accomplishment of these tasks. The sidelooking airborne radar has been identified as a valuable part of any system aimed at the first two applications.

By contract from the Swedish Space Corporation L.M. Ericsson has developed an inexpensive SLAR that is now installed in an aircraft belonging to the Swedish Coast Guard. (See Figure 1). The main purpose behind this prototype equipment is to test its ability to detect oil slicks on the ocean surface. Other applications are sea ice mapping and monitoring coastal economic and fishery zones. Flight tests during the spring of 1978 will give operational experience and the final design of operational equipment will then be frozen in December 1978.

2. DESIGN PHILOSOPHY

Remote sensing research has often been carried out using military equipment. It is, however, obvious that very few remote sensing end users can afford to buy and operate these elaborate systems. Successful testing of such equipment then does not automatically lead to operational use of the investigated method.

The basic philosophy behind the Ericsson SLAR is to design a system that is sophisticated enough to accomplish the given task, yet not too expensive for operational use. To find the best compromise between the "technically possible" and the "economically feasible" the consequences of varying radar parameters had to

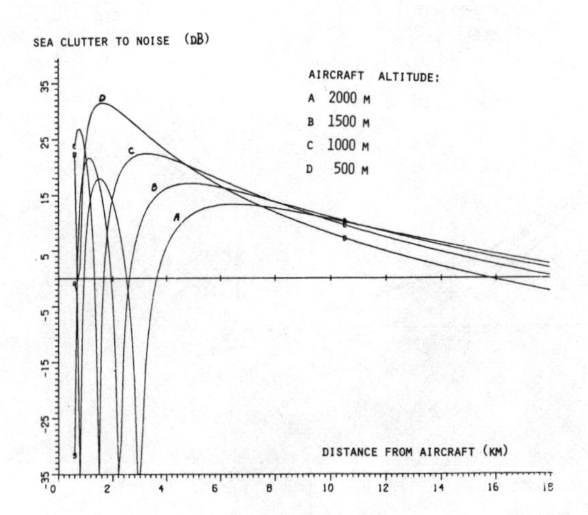

Figure 2 Average sea clutter power at different aircraft altitudes.

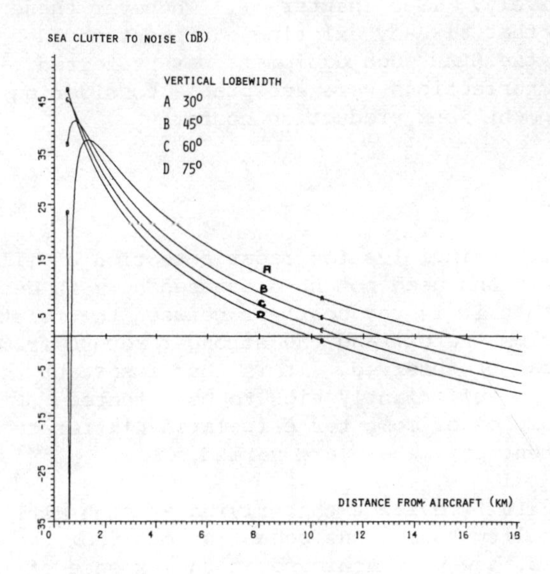

Figure 3 Average sea clutter as a function of vertical antenna lobewidth.

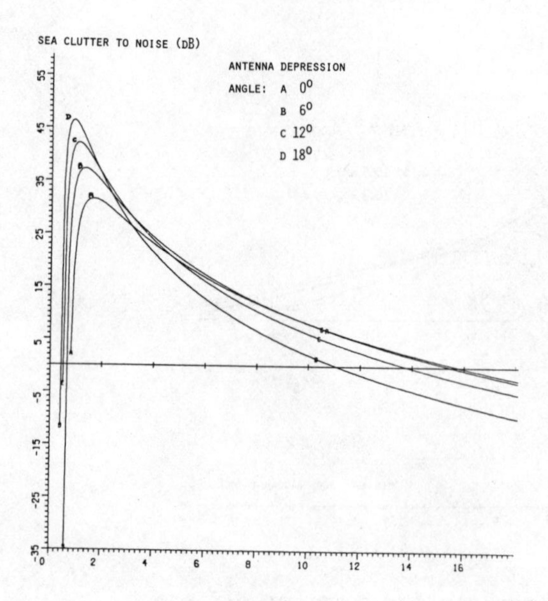

Figure 4 Dependence of average sea clutter return on antenna depression (Aircraft Roll Angle).

be studied extensively. (See chapter 3). Whenever these calculations indicated that already existing equipment could be used as a subsystem in the SLAR such equipment was preferred. The resulting minor imperfections were acceptable considering the savings in development- and production costs.

3. SYSTEM STUDIES

The fundamental principles for radar detection of oil on water are well known and need not be discussed here in detail. Let us just note that it is the contrast between the relatively weak backscatter from the oil and the stronger return from the sea surface that can be observed. It is thus important that the sea clutter level is sufficiently high to be detected. Figures 2, 3 and 4 are examples of computer calculated clutter to noise ratios when different parameters are varied.

From Figure 2 it is evident that flying at high altitudes gives increased ability to map the ocean far out from the aircraft. The gain is, however, achieved at the expense of loosing mapping capability close to the aircraft where the radar antenna gives the best resolution. For a prototype where resolution is more important than area coverage, 500 m is a good choice of

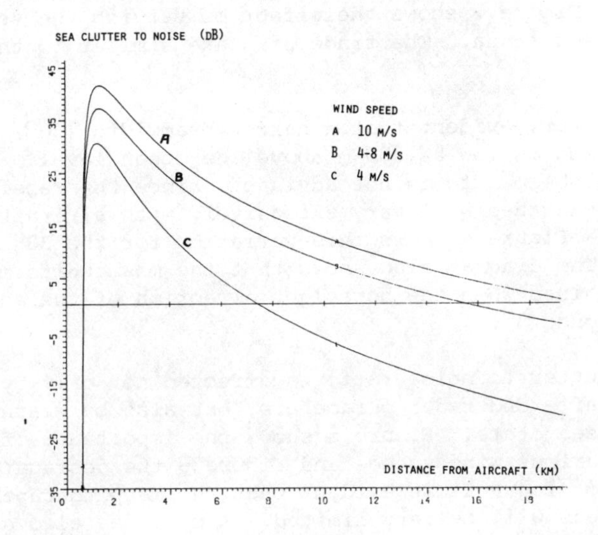

Figure 5 Average sea clutter return at different wind speeds.

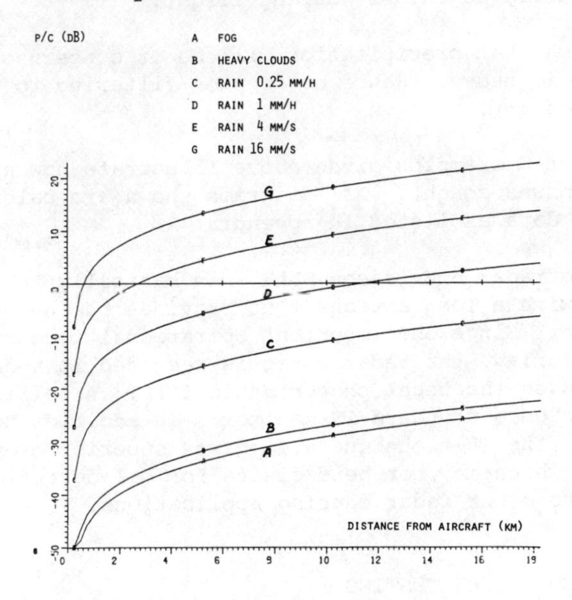

Figure 6 Back scatter from precipitation (P) Compared to average sea clutter (C)

flying height. Figure 3 shows the effect of varying the vertical beamwidth of the antenna. The trade offs are similar to the previous case.

It is, however, evident that a narrow beamwidth (30°) can be chosen when flying as low as 500 m above the ocean level. A smaller vertical beamwidth is not advisable since the received sea clutter signal then will vary extensively with aircraft rolling motion. Figure 4 shows this variation for the 30° beamwidth chosen. The diagram also shows that the most uniform illumination is obtained when the boresight direction of the antenna is depressed around 5°.

The sea clutter to noise ratio is affected not only by controllable aircraft- and radar parameters, but also by weather conditions and sea state. Figure 5 shows the importance of the sea state for various wind speeds and Figure 6 the degradation caused by rain or fog. It is evident that the mapping capability at low wind speeds will be very limited. A calm sea also gives a radar return that diminishes more rapidly with range than the radar return from heavy sea. These variations with range can be compensated in the radar receiver to form a radar map with uniform intensity. This type of receiver gain control has to be included and should be manually adjustable during flight.

Figure 6 shows that precipitation will be of concern only when the rainfall is heavy. Hence no doppler filtering to remove rain echos is necessary.

The few selected examples given above illustrate how an optimum radar design was sought. All diagrams shown are calculated using final radar data as listed in appendix.

The resulting radar - Ericsson SLAR - is basically an ordinary pulse radar with a long antenna that provides the necessary angular resolution. Since one important operational requirement was a real time display, the radar video is recorded in a digital memory rather than on the usual photographic film. A rolling radar map is shown on a standard TV-monitor. In addition to real time presentation, the TV-technique also gives superior grey scale resolution, which may not be decisive for oil detection, but is essential to other radar mapping applications.

4. EXPERIENCE FROM FLIGHT TESTING

No paper study, however careful, can give all the answers to the design of a new radar system. In this particular case the following questions were left for flight testing:

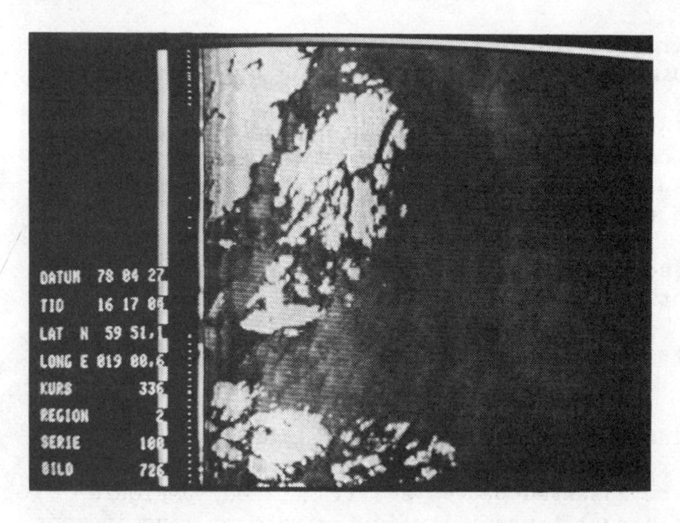

Figure 7 Radar map of coastline area. Coverage 20×20 km^2.

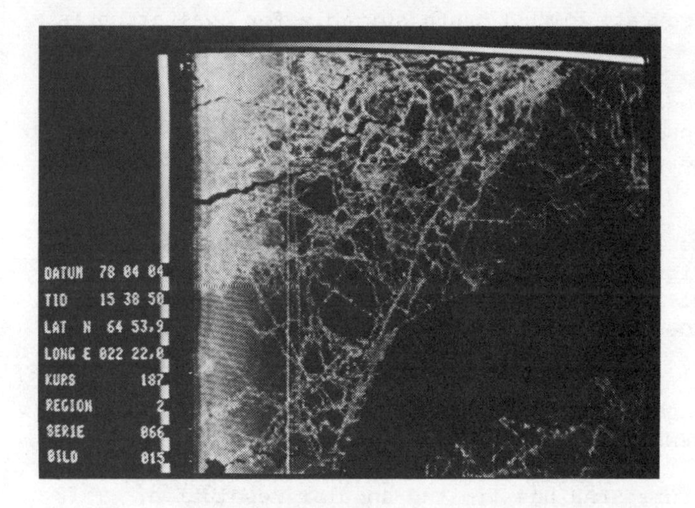

Figure 8 Radar map of ice in Gulf of Bothnia. The darkest part of the image represents open water while white lines are ice ridges. Area coverage 20×20 km^2.

1. Is a small aircraft stable enough in the air to allow a fixed SLAR antenna with no aircraft motion compensation?

2. Is it possible to match receiver gain characteristics and antenna vertical gain pattern to the range dependence of the sea clutter in order to get a radar map with even intensity regardless of range?

3. Is the selected radar resolution (75×100 m^2) sufficient to detect all interesting oil slicks?

4. Which possible uses other than oil slick mapping does this system have?

The flight testing of Ericsson SLAR is scheduled for spring 1978. The first results have been very encouraging. Questions 1 and 2 can both be answered by yes as illustrated by Figure 7. The worst case of aircraft stability is flying over land in poor weather. Under these conditions radar resolution is deteriorated, but the resulting radar maps are still useful.

No organized test against oil has been conducted as of April 1978. Accidental spills have, however, been detected on several occasions. This oil has been found and checked by Coast Guard vessels but no accurate ground truth suitable for evaluation is available.

The SLAR uses digital integration of returns from a large number of radar pulses and this has proved to give a very even background of sea clutter. Thus the possibilities to detect oil as well as ships of all sizes should be excellent.

The SLAR has also been flown over sea ice in the Gulf of Bothnia. (See Figure 8). It is relatively easy to distinguish open water, level ice and ice with ridges. Other registrations show that ice floes and drifting ice are equally well detected.

5. CONCLUSIONS ABOUT THE USEFULNESS OF AN INEXPENSIVE SLAR FOR REMOTE SENSING

A real aperture SLAR has limited angular resolution. This type of equipment accordingly is best suited for mounting in low-flying aircraft for applications that do not require extensive area coverage.

Both theoretical studies and practical experience of Ericsson SLAR has shown that it is well suited for ocean surveillance such as oil slick detection and sea ice mapping. The radar can also be used to study large scale variation of sea state and wave patterns but the resolution is not sufficient to detect individual waves. Another promising application is control of sea charts since an archipelago can be mapped with accuracy. The information obtained is limited to features above and just below the water surface since electromagnetic waves at radar frequencies do not penetrate water.

When flying over land it is possible to map large scale features such as urbanization, waterways and railroads. It is also possible to distinguish farming areas, forests and mountaineous terrain. The TV display gives excellent grey scale reproduction which could be useful for seasonal studies of vegetation. For this application it would, however, be desirable to increase the ground resolution an order of magnitude, which is easily done.

At least until pictures from satellite synthetic aperture radar with regular overflights become generally available there is no doubt that inexpensive SLAR is a cost effective and powerful remote sensing instrument. For operational use that requires real time display, surveillance at small grazing angles, or extremely high flexibility and availability, the SLAR will remain competitive for the foreseeable future.

APPENDIX

L.M. ERICSSON SLAR

BASIC TECHNICAL DATA

Antenna

Horizontal lobe width	0.52°
Vertical lobe width	33°
Gain	31 dBi
Polarization	vertical

Transmitter

Frequency	9.4 GHz (X-band)
Peak power	10 kW
Pulse repetition frequency	1.0 kHz
Pulse width	0.5 μs

Receiver

Logaritmic amplifier
Automatic or Manual Gain Control
Variable Sensitivity Time Control (STC)

Signal processing

Variable number of pulses integrated
Digital display memory (400 kbit RAM)
Video thresholded in 64 levels (6 bit)

Display

Rolling map, real time display
Standard TV Monitor
Contour mapping and level mapping
Positive or negative picture
Grey scale calibration
Alfanumeric information, mixed into picture

Registration

Standard video tape recorder

WORKSHOP REPORT

WORKING GROUP: OCEAN WAVES

PARTICIPANTS: Bartsch, Kjelaas, Parr, Thomas, Valenzuela, Vogel, Williams

CHAIRMAN: O.H. Shemdin

1. INTRODUCTION

Remote sensing techniques are being recognized as valuable tools for obtaining information on wave height and direction. Significant advantage lies in the capability to obtain information rapidly over large areas of the ocean. Satellite borne sensors can now provide such information on a global scale.

Information on wave climate is of importance to most human activities in offshore areas. In deep water the primary impact is on shipping where significant travel time reductions can be achieved by careful routing of ships around storm areas. In shallower waters waves impact a broader range of activities which include construction and operation of offshore structures, fishing, recreation and nearshore processes. Both extreme and mean (long term) levels of wave heights are needed. This information is not available in many coastal areas of the world which are not populated. The need is nevertheless present for exploring natural resources thought to be present in offshore areas.

Active microwave sensors have been shown to be more useful for measuring waves compared with passive techniques. The HF technique is ground based and can provide direct measurement of the wave height along the direction of look of the radar antenna. The wave length is determined by the frequency of the radar. Second-order HF techniques can potentially provide the entire directional wave height spectrum. The first order technique can also provide information on wave current. Dual-frequency radars at microwave frequencies also provide information on the dominant wave in the ocean via the modulation of short gravity waves.

T. Lund (ed.), Surveillance of Environmental Pollution and Resources by Electromagnetic Waves, 319–326.

Active microwave techniques provide useful wave height information from ground based systems and more extensively from airborne and spaceborne systems. The latter is optimally done by the synthetic apperture radar (SAR) which utilizes both intensity and phase of scattered signals from the ocean surface. The SAR provides a map of the ocean surface undulations which contain directional information on waves. The wave height can be potentially derived from the SAR but has not been conclusively demonstrated yet. The significant wave height can be obtained more easily from a different space sensor, the Altimeter. This sensor operates in the nadir mode. The scattering information from the sea surface can be directly related to the significant wave height (defined as the average of the one third highest waves).

2. MECHANISMS OF EM INTERACTION WITH OCEAN WAVES

a) HF-Bragg

For frequencies less than 100 MHz the sea surface behaves as a true "diffraction grating" for all sea conditions. In first order, of surface height, a single wave of unique wavenumber contribute to the EM scattering in any given direction for a planar transmitted wave (Bragg scattering). At grazing incidence and for backscatter, ocean waves one-half the EM wavelength, travelling parallel to the line of sight contribute, and the first order scattered power is proportional to the energy of the Bragg resonant ocean wave. Away from grazing incidence the backscatter is due to longer ocean waves of wavenumber

$$K_w = 2k_o \sin\theta$$

where k_o is the wavenumber of the EM radiation in free space and θ is the angle of incidence, with respect to the normal, of the mean surface. Accordingly, in first order Bragg scattering, the frequency spectrum of the return is a narrowdoppler line shifted from the transmitted carrier frequency by the frequency of the Bragg resonant ocean wave. The polarization ratio (i.e. $\sigma_{oVV}/\sigma_{oHH}$) for first order Bragg backscatter away from normal incidence is relatively large for water surfaces. At HF frequencies (3 - 30 MHz) the vertically polarized ground wave is used for off-shore monitoring application, while for extended coverages the ionospheric refracted OTH sky mode is used. As the ocean becomes rougher with wind, time and fetch "higher order" contributions to the scatter become significant and these introduce replicas of the low frequency spectrum of the ocean on each side of the first order Bragg doppler lines.

Perturbation EM scattering theory predicts quite well the amplitude and shape of the doppler spectrum at these low EM frequencies and from this a number of oceanographic parameters may be recovered (wave-height, currents, two-dimensional spectrum and phase speed).

b. Microwaves: tilt, straining, etc

At higher radar frequencies toward the decimeter and centimeter regions the first order Bragg resonant ocean wave becomes much shorter than the dominant wave of the ocean and the EM scatter is the result of a two-scale rough surface. "Patches" of Bragg scatterers (short gravity waves) are "lifted" and "tilted" by the presence of the dominant wave in the ocean. Now the backscattered power is an average (over the distribution of slopes of the dominant waves) of the power predicted by first order perturbation theory. A reduction of the polarization ratio is the result.

The doppler spectrum of this return is a "continuum" spectrum caused by the advection of the Bragg scatterers by the fluid velocity of the dominant wave of the ocean. The long gravity waves modulate the short gravity waves in the ocean (straining), and this is crucial in the resulting transfer function of SAR imaging radars and dual-frequency (coherent) radars.

At normal incidence the backscatter is "specular" and away from normal incidence it is diffuse, produced by the Bragg scatter mechanism. Then, as a first approximation, one may add the backscattered power (cross-sections) by geometrical optics and perturbation scattering properly averaged over slopes.

3. TECHNIQUES FOR SEA STATE DETECTION

a) Imaging of Waves

Wave images can be obtained by airborne radars in two modes, the real aperture and synthetic aperture. The real aperture radar detects modulation of the radar frequency in the EM signal return, along the profile of long waves. The modulation is determined by the variation of the ocean surface roughness as well as the local slope of the long wave. The mean ocean roughness is affected by wind speed, atmospheric stability and presence of local currents. These factors make it possible to image current boundaries and to infer a measure of the wind speed.

The synthetic aperture radar (SAR) uses the phase information of the returned EM signal in addition to magnitude. It is therefore possible to achieve additional contrast in wave imaging since the orbital velocity of long waves induce a relatively high modulation in the phase history of the returned signal. Experimental verifications of wave imaging concepts have been obtained in the Marineland and West Coast Experiments conducted in the US. The results indicate that, for a dominant swell condition, the SAR measures the wave length accurately and the wave direction to 1-2 degrees. The focussing process of the signal film yields wave direction and phase speed of the waves. In this respect the SAR potentially provides all of the desired wave properties except the wave height.

b) Two-frequency technique - Nadir and off nadir

Dual frequency scatterometers couple selectively to periodic backscatter intensity variations in range, according to the beat-pattern periodicity of the two electromagnetic waves with the adjustable frequency difference Δf. Dual-frequency scatterometers can, by varying Δf, directly analyse the spatial ocean wave spectrum in a slant angle look direction, if the relation between the fractional modulation of the short wave backscatter by the long wave amplitude is known.

The data rate requirement is low; compared to the wave analysis, by transformation from high resolution SAR images. However, the directional wave spectra have to be measured successively by time consuming scanning in angle and in frequency difference.

For a look to nadir no periodic features of the ocean wave exist in range. Dual frequency scatterometry relates in this case the spread in range caused by the effective wave height with the cross correlation between the two returns versus frequency difference.

c) Radar altimeter

The SEASAT-A Radar Altimeter, for example, will operate at a frequency of 13.5 GHz (K-Band) corresponding to a wavelength of 2.2 cm. A nadir viewing paraboloid of 1 m diameter and 1.5° beamwidth is utilized to provide an effective foot print of 1.6×12.0 km. Altitude resolution is quoted as $\pm$ 10 cm achieved through the use of pulse compression techniques on 3 μs pulses with a PRF of 1500 per sec. The specified objective is the measurement of significant wave height, $H_{1/3}$, (in the range 1-20 m) to an accuracy of $\pm$ 0.5 m or $\pm$ 10% whichever is the greater. It is known from high resolution aircraft radar altimeter measurements that a large part of the observed standard deviation of the measurements ($\sim$ 10%) is attributable to effects arising from the finite area illuminated by the radar. The radar altimeter, apart from measuring significant wave height ($H_{1/3}$ (x,y)), basically provides high precision sea surface topography measurements (departures from geoid) and geoidal heights above a reference ellipsoid.

A number of topics require consideration for ocean surface topography. These include

- spacecraft orbit parameters
- spacecraft altitude variation + measurement accuracy
- geometrical considerations - earth oblateness, earth rotation, radar wavefront curvature
- target motion and target backscatter coefficient.

In order to determine the calibration constant for the altimeter, particularly height bias errors, a good knowledge of the spacecraft orbit is essential. Previously, typically four high precision ($\sim$ 5 - 10 cm) lasers have been deployed to determine orbital parameters in the measurement area with increased accuracy. For the radar altimeter not only increased orbital accuracy but improved allowance for effects such as those due to winds and tides will be important. One approach to improve orbit determination accuracy is to use the altimeter data themselves to reduce the RMS errors.

For some types of studies, accuracies of order ± 1 cm are required ideally - e.g. those concerned with major ocean currents and eddies.

4. IMAGE PROCESSING

Numerous pre-processing transforms must be applied to "raw" radar data (amplitude or intensity, range and phase measurements) to allow for geometrical corrections, before the observations take a form suitable for widespread dissemination to the user community. Such pre-processing transforms are not discussed herein. We confine ourselves to a brief discussion of the various types of subsequent transforms or manipulations of the grey-level data which are performed to facilitate the extraction of the key information in the image in the most economical way both from the overall cost effectiveness point of view and in making feasible the automatic processing of very large amounts of data. The application of such transforms and data manipulation techniques is generally described as imagery processing.

Imagery processing may be applied, for example, to enhance the contrast between adjacent parts of the image. Density slicing, contouring, edge enhancement, detection and enhancement of linear features are examples.

There are three types of information in an image - spectral, contextual and textual. The latter is concerned with those detailed relationships which exist between the grey level values in adjacent or neighbouring resolution cells and which, taken together, give the image its characteristic appearance or texture. The quantification of these relationships is known as texture analysis and it is possible to provide statistical, mathematical or information theory measures of texture suitable for extraction in classifiers, i.e. in software pattern recognition algorithms for assigning data to specific groups, clusters or classes.

It is possible to identify four types of approach to texture analysis

- transform techniques or spatial frequency analysis
- first order grey level statistics
- second order grey level statistics or co-occurrence matrices (p-matrices)
- shape measures or other techniques.

The first of these provides, for example, Fourier measures of texture - Fourier rings and wedges. Simple Fourier transforms of radar imagery of the sea-surface (derived either optically or digitally) provide a ready way of producing information on ocean wavelengths present in the image and have been extensively utilized by the JPL group. The use of Fourier wedge measures might be valuable in determining preferred wave direction in a complicated image. Transforms such as the Hadamard transform may well be useful in reducing the computing effort as compared with Fourier transforms.

In the routine processing of large amounts of sea imagery it may well prove possible to by-pass the production of an image for visual inspection and to utilize the Fourier domain as the basic image.

Complicated sea surface imagery with areas of varying wave travel directions may well be amenable to analysis using second order statistics which are often well suited for determining the preferred textural alignment directions. Here the emphasis is on the probability of occurrence of specified grey level dependencies in some pre-determined pixel geometry relationship for a particular resolution cell. Measures are then defined and calculated from the p-matrix elements. Measures exist for a variety of textural properties including the degree of homogeneity or complexety of an image, directional properties of an image and so on.

As far as is known, texture analysis using 1st order grey level statistics has not yet been applied to sea imagery obtained by radar. It may well be important to apply 1st order techniques in addition to the other methods mentioned earlier to investigate the extent to which sea state parameters such as wave height, wave shape, etc. can be obtained from such imagery.

5. SUMMARY

The state-of-the art as far as measurements of ocean waves using active microwave sensors is summarized in Table 1.

6. RECOMMENDATIONS

a) Further theoretical and experimental work should be pursued on the modulation of short gravity waves by longer waves as a function of wind speed and wave direction.

b) Further work should continue on the derivation of the transfer functions for the ocean response of dual-frequency and SAT imaging radars.

c) Further work should continue with multifrequency (more than two) radars as well as looking into the possibilities of a more complete use of the ocean wave information contained in the time and frequency fine structure of the backscattered echoes.

d) Pursue collaborate efforts with users to determine utility of data.

e) Encourage development of SEASAT-A data received and processing facilities to accelerate/expedite dissemination of data in suitable form to user community.

f) Promote interaction between the remote sensing user community and available majory processing capability.

g) Encourage collaboration by the user community in experiments such as the MARSEN experiment.

OCEAN WAVE INFORMATION FROM ACTIVE MICROWAVE PROBES

Technique	Principle, typical parameters	Information obtained/accuracy			Routinely available to users: When?	Comments. Difficulties foreseen Effort needed. Advantages/disadvantages
		Sea state	Freq. spectrum	Direc.spectrum		
Monostatic ,HF,VHF Ground wave OTH	3-100 MHz		$T \leq 7$ sec	1.order: one component		Large antenna needed Also information on currents Intersection of lines of constant Doppler and range
Bistatic HF	Doppler	2nd Order-Directional Wave-height spectrum	"	2.order: full spectrum		
HF SAR			"	one wavenumber all directions		A quantitative treatment of orbital motion and modulation of Bragg scatterers regulated
Altimeter	13.5 GHz $t = 3\ \mu$sec pulse limited	H_s to 1-20 m resolution ± 10 cm	none	none	1978	Satellite
SAR -SEASAT -Aircraft	$\sim$ 25 m 3 m	Not yet achieved	$L \geq 50$ m $L \geq 6$ m	Resolution within 2° Information for $L > 50$ m accuracy?		Only high resolution imaging system from space
Dual-frequency -CW -pulsed	Differential phase L,X-Band	Directional spectrum	5 - 500 m wavelength	Yes (azimuth and energy)		Determination of transfer function requested
Coastal Surveillance Radar	X-Band	No	No	Yes	Now	

Table 1

PART IV : METHODS OF INFORMATION RETRIEVAL

THE FOURIER TRANSFORM: PROPERTIES AND APPLICATIONS

Anton G. Kjelaas

Royal Norwegian Council for Scientific and Industrial Research,
Remote Sensing Technology Programme,
P.O.Box 25, N-2007 Kjeller, Norway

1. INTRODUCTION

Fourier analysis has been widely used in communication theory and electrical engineering for many, many years and is very well treated in text books covering these fields, but Fourier analysis is also very useful in many other aspects and it is increasingly being used in all branches of the physical science.

To have the full benefit of Fourier analysis a good knowledge about different representations of both Fourier series and Fourier integrals is needed in addition to their basic properties.

In this contribution we will present some useful formulas and discuss properties and application which may be of some use for people not having their educational background in communication theory or electrical engineering.

It is not our intention to present a complete mathematical description, but more to treat the Fourier analysis as a tool helping us to understand the physical processes and easing the retrieval of information burried in noise. We will only discuss application of Fourier analysis on aperiodic and random functions which means that we will limit the discussion to Fourier transforms and not include Fourier series which only can be used in analysis of periodic functions.

T. Lund (ed.), Surveillance of Environmental Pollution and Resources by Electromagnetic Waves, 329–351.

2. TIME AND FREQUENCY RELATIONS - BASIC FORMULAS

2.1 Aperiodic functions

Let g(t) be a aperiodic function with time, i.e. a function with finite energy. The Fourier transform of g(t) is then given by

$$G(\omega) = \int_{-\infty}^{\infty} g(t)\, e^{-j2\pi ft}\, dt \tag{2.1}$$

and the original function may be obtained from

$$g(t) = \int_{-\infty}^{\infty} G(\omega)\, e^{j2\pi ft}\, df = \frac{1}{2\pi} \int_{-\infty}^{\infty} G(\omega)\, e^{j\omega t} d\omega \tag{2.2}$$

where f is the frequency and $\omega = 2\pi f$.

Eqs (2.1) and (2.2) are now called a Fourier transform pair and gives the relation between time and frequency representation of a aperiodic function. $G(\omega)$ is called the Fourier transform of g(t) and g(t) is the inverse transform. G( ), which we will call the amplitude spectrum, is a continuous function.

Often it is very convenient to introduce two operators F and F^{-1} to denote the transform and the inverse transform, i.e. $FF^{-1} = 1$.

Another symbolic form is using

$$g(t) \leftrightarrow G(\omega)$$

In this paper we will use $\leftrightarrow$ to denote transform pairs. Because of the symmetry in Fourier transforms it is independent which function is the transformed and the inverse transformed i.e. $G(t) \leftrightarrow g(\omega)$.

Any aperiodic function g(t) can be written as a sum of an even component $g_e(t)$ and an odd component $g_o(t)$ as

$$g(t) = g_e(t) + g_o(t) \tag{2.3}$$

Eq (2.1) can then be written as

$$G(\omega) = \int_{-\infty}^{\infty} \left(g_e(t) + g_o(t)\right) e^{-j\omega t}\, dt$$

or on trigometric form as

$$G(\omega) = \int_{-\infty}^{\infty} g_e(t) \cos \omega t\, dt - j \int_{-\infty}^{\infty} g_o(t) \sin \omega t\, dt$$

$$= A(\omega) + j\, B(\omega) \tag{2.4}$$

The amplitude spectrum $G(\omega)$ is complex with a real part

$$A(\omega) = \int_{-\infty}^{\infty} g_e(t) \cos \omega t \, dt \qquad (2.5)$$

and an imaginary part

$$B(\omega) = - \int_{-\infty}^{\infty} g_o(t) \sin \omega t \, dt \qquad (2.6)$$

Often it is more convenient to write $G(\omega)$ as

$$G(\omega) = |G(\omega)| e^{j\phi(\omega)} \qquad (2.7)$$

where

$$|G(\omega)| = \sqrt{A(\omega)^2 + B(\omega)^2}$$

and the phase spectrum is

$$\phi(\omega) = \tan^{-1} \frac{B(\omega)}{A(\omega)} \qquad (2.8)$$

Before proceeding further it is of interest to see that the transformations may be simplified if $g(t)$ has special properties

if $g(t)$ is real — $A(\omega)$, $B(\omega)$, $\phi(\omega)$ real

$G(\omega)$ comples

$G(\omega) = G^*(-\omega)$ $\quad A(\omega) = \frac{1}{¶} \text{Re } G(\omega)$

$G^*(\omega) = G(-\omega)$ $\quad B(\omega) = -\frac{1}{¶} \text{Im } G(\omega)$

(* denotes complex conjugate)

if $g(t)$ is purely imaginary — $A(\omega)$, $B(\omega)$ imaginary

$\phi(\omega)$ real

$G(\omega)$ complex

$G(\omega) = -G^*(-\omega)$ $\quad A(\omega) = \frac{i}{¶} \text{Im } G(\omega)$

$G^*(\omega) = -G(-\omega)$ $\quad B(\omega) = \frac{i}{¶} \text{Re } G(\omega)$

if $g(t)$ is even — $A(\omega) = \frac{1}{¶} G(\omega)$ $\quad B(\omega) = 0$

$\phi = 0$, $G(\omega) = G(-\omega)$

if g(t) is odd $\quad A(\omega) = 0 \quad B(\omega) = \frac{i}{\P} G(\omega)$

$$\phi(\omega) = - \P/2$$

$$F(\omega) = - F(-\omega)$$

2.2 Random function

Although we in some mathematical sense up to this point have limited the discussion to aperiodic functions, the definitions are also valid for other functions such as random functions.

To ensure that our function has finite energy we define a new function

$$g_T(t) = \begin{cases} g(t) & g(t) \leq |T| \\ 0 & g(t) > |T| \end{cases} \qquad (2.9)$$

where subscript T denotes trancation and later let $T \rightarrow \infty$, the Fourier transform in this case is then given as

$$G_T(\omega) = \int_{-\infty}^{\infty} g_T(t) e^{-i\omega t} dt = \int_{-T}^{T} g(t) e^{-j\omega t} dt \qquad (2.10)$$

3. SOME BASIC PROPERTIES OF FOURIER TRANSFORM PAIR

To have the full benefit of using Fourier transforms it is of importance to know some basic properties of transform pair. These are listed in Table 1.

Function	Fourier transform
$g(t)$	$G(\omega)$
$g(-t)$	$G(-\omega)$
$g^*(t)$	$G^*(-\omega)$
$g^*(-t)$	$G^*(\omega)$
$g(t/a)$	$a\ G(a\omega)$
$g(at)$	$\frac{1}{\lvert a \rvert} G(\frac{\omega}{a})$
$a\ g(t)$	$a\ G(\omega)$

cont.

cont.

$g_1(t) + g_2(t)$	$G_1(\omega) + G_2(\omega)$
$g(t-t_o)$	$G(\omega) \cdot e^{-j\omega t}$
$g(t)\, e^{-j\omega_o t}$	$G(\omega+\omega_o)$
$\frac{d^n g(t)}{dt^n}$	$(j\omega)^n\, G(\omega)$
$(-jt)^n\, g(t)$	$\frac{d^n G(\omega)}{d\omega^n}$
$\int_{-\infty}^{t} g(\tau)\, d\tau$	$\frac{G(\omega)}{j\omega}$
$\frac{g(t)}{-jt}$	$\int_{-\infty}^{\omega} G(x)\, dx$

Table 1 Basic properties of Fourier pairs.

Notice that the effect of stretching the time function, shows up as a narrowing of the amplitude spectrum.

This is a very important characteristic of Fourier transforms which gives the relationship between bandwidth (width of the spectrum) and time duration of the function (or time series) to be transformed.

In practical life this implies that a very wide bandwidth is needed to generate a very sharp pulse in the time domain or vice versa. This is illustrated in Figure 1.

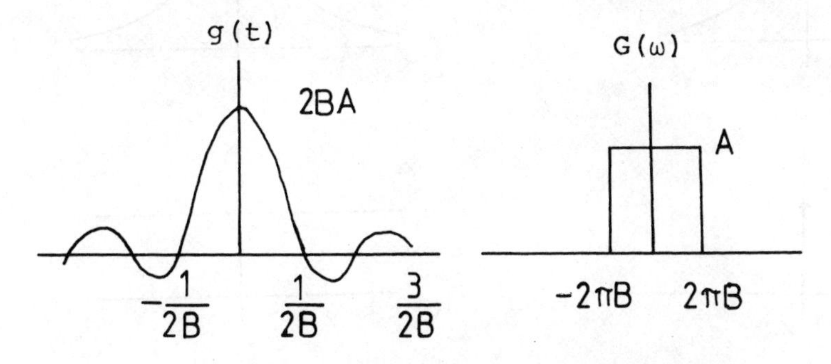

Figure 1 Relationship between pulse duration and spectral width.

3.1. Graphical representation of some often used Fourier pair

Table 2 shows the graphical representations of some commonly used transform pairs.

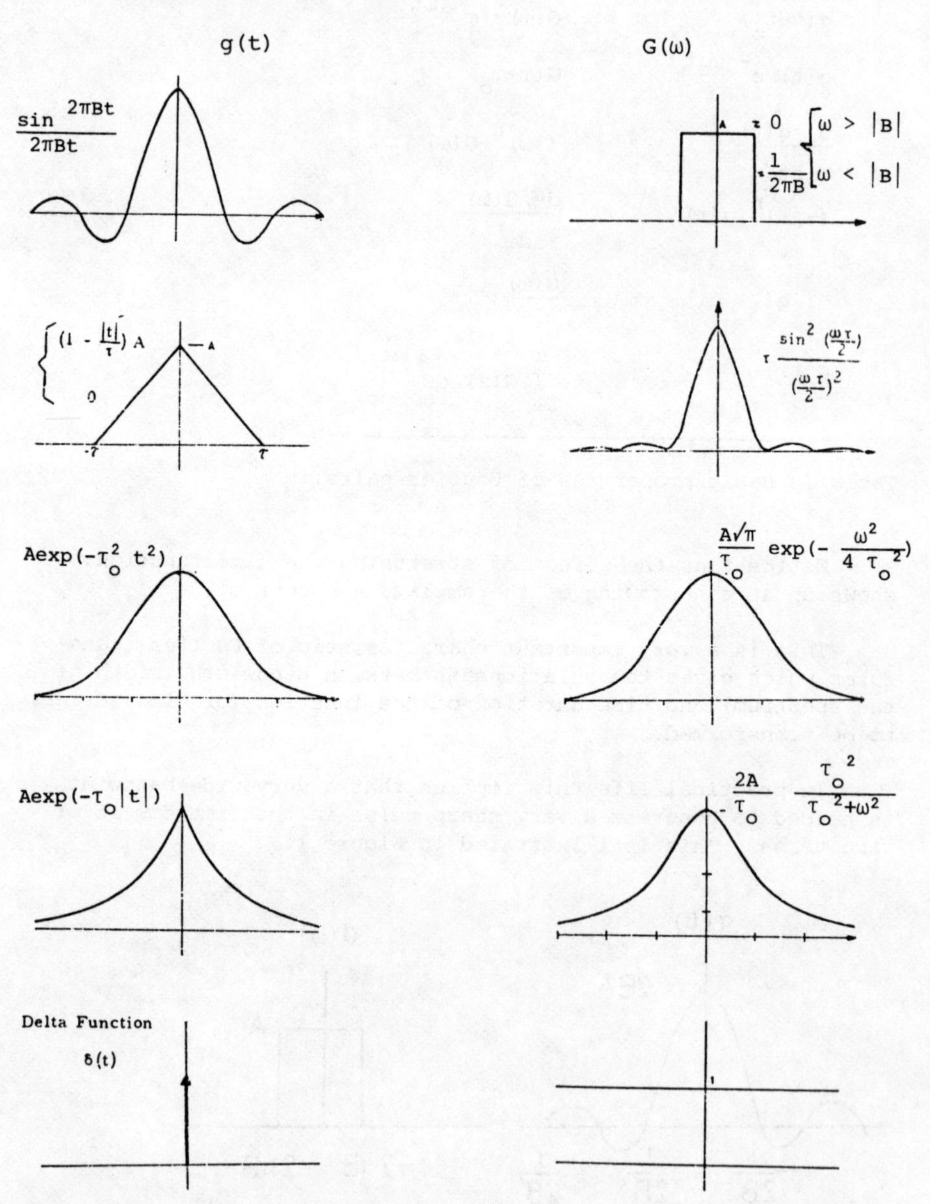

Table 2 Graphical representation of some transform pairs.

4. CORRELATION FUNCTIONS - POWER SPECTRA

Most of the processes of interest, in particular within atmospheric sciences, are random processes.

A very useful approach to give a mathematical description of such processes is to use the correlation function and the corresponding power spectrum.

4.1. Correlation functions

The crosscorrelation between two random functions f(t) and g(t) is

$$R_{fg}(\tau) = \lim_{T\to\infty} \frac{1}{2T} \int_{-T}^{T} f(t+\tau)\, g(t)\, d\tau$$

which in the case of an aperiodic function can be reduced to

$$R_{fg}(\tau) = \int_{-\infty}^{\infty} f(t+\tau)\, g(t)\, d\tau \qquad (4.1)$$

If f(t) = g(t) we have the autocorrelation function

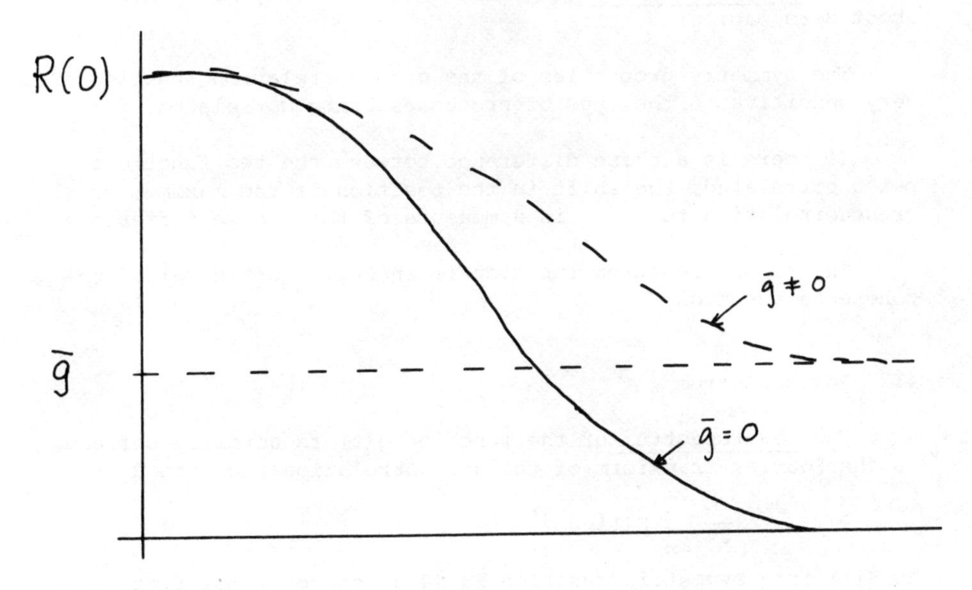

Figure 2 Autocorrelation a signal with and without zero mean.

$$R_g(\tau) = \lim_{T\to\infty} \frac{1}{2T} \int_{-T}^{T} g(t)\, g(t+\tau)\, d\tau \tag{4.2}$$

We have assumed that

$$\overline{g(t)} = \overline{f(t)} = 0$$

If the mean value is not zero, we must include a second term consisting of the square of the mean value. The effect of not subtracting the mean value before doing the autocorrelation is shown in Figure 2.

The autocorrelation at $\tau = 0$ is the variance of the process.

Generally we have that

$$R(-\tau) = R^*(\tau)$$

but if g(t) is real, which normally is the case, we find that the autocorrelation is symmetric about τ equals zero

$$R(\tau) = R(-\tau)$$

The crosscorrelation function is necessarily not symmetric about zero lag.

The symmetry properties of the crosscorrelation function are very sensitive to the type of processes being correlated.

If there is a phase difference between the two functions being correlated, the shift in the position of the maximum of the crosscorrelation function is a measure of this phase difference.

The autocorrelation function is in optics often called the coherence function.

4.2 Powerspectrum

The powerspectrum of the function g(t) is normally defined as the Fourier transform of the autocorrelation function $R_g(\tau)$

$$P(\omega) = \frac{1}{2\pi} \int_{-\infty}^{\infty} R(\tau)\, e^{-j\omega\tau}\, d\tau \tag{4.3}$$

As $R(\tau)$ is a symmetric function Eq (4.3) can be simplified to

$$P(\omega) = \frac{1}{2\pi} \int_{-\infty}^{\infty} R(\tau) \cos \omega\tau\, d\tau \tag{4.4}$$

The autocorrelation function $R(\tau)$ and the powerspectrum $P(\omega)$ are therefore a Fourier pair.

The cross powerspectrum can similary be obtained by transforming the crosscorrelation function.

$$C(\omega) = \frac{1}{2\P} \int_{-\infty}^{\infty} R_{fg}(\tau)\, e^{-j\omega\tau}\, d\tau \qquad (4.5)$$

Another way of obtaining the powerspectrum which is much more efficient, (using Fast Fourier Transform techniques) is to first take the transform of the function $g(t)$ to obtain the amplitude spectrum. The energy spectrum $E(\omega)$ is then defined as

$$E(\omega) = G(\omega)\, G^{*}(\omega) \qquad (4.6)$$

which can be considered as a purely mathematical definition. To ensure that the signal $g(t)$ has finite energy it is useful to construct a new signal $g_T(t)$ from $g(t)$ by truncation as shown in Eq (2.9).

The powerspectrum can now be defined as

$$P(\omega) = \lim_{T\to\infty} \frac{1}{2T} G_T(\omega)\, G_T^{*}(\omega) \qquad (4.7)$$

Similary we obtain the cross powerspectrum as

$$C(\omega) = \lim_{T\to\infty} \frac{1}{2T} G_T(\omega)\, F_T^{*}(\omega) \qquad (4.8)$$

The powerspectrum is real and symmetric and it has no information about the phase only the power of the process.

The cross powerspectrum on the other hand may be complex

$$C(\omega) = C_o(\omega) + j\, Q(\omega) \qquad (4.9)$$

where $C_o(\omega)$ is the real part and is called the cospectrum.

$Q(\omega)$ is the imaginary part and is called the quadrature spectrum. The quadrature spectrum contains information about the phase difference between the functions being correlated.

If the crosscorrelation function is symmetric, the quadrature component will be zero and hence

$$C(\omega) = C_o(\omega) \qquad (4.10)$$

The phase angle ϕ may also be expressed as

$$\tan \phi(\omega) = \frac{Q(\omega)}{C_o(\omega)} \tag{4.11}$$

Applications of the phase angle in different aspects will be discussed later.

4.3 Coherence

The coherence function is here defined as a crosscorrelation function in frequency

$$\mathrm{coh}(\omega) = \frac{|C(\omega)|^2}{P_f(\omega)\, P_g(\omega)} = \frac{C_o^{\,2}(\omega) + Q^2(\omega)}{(P_f(\omega) P_g(\omega))} \tag{4.12}$$

The coherence function has no information about phase. It only gives the correlation coefficient between different frequency components in the two functions $g(t)$ and $f(t)$.

4.4 Structure function

In our definitions of autocorrelation, crosscorrelation and their corresponding Fourier transforms we have assumed stationarity.

In many cases we have signals or functions which have mean values which are variables in time.

In such a case it is more convenient to use what we call a structure function instead of the usual autocorrelation function.

Our starting point is to replace $g(t)$ by

$$g_s(t) = g(t+T) - g(t)$$

if T is not too large, the low frequency component of $g(t)$ does not affect this difference.

If this difference $g_s(t)$ is a stationary function of t, $g(t)$ is usually called a random function with stationary increments.

The definition of structure function is then

$$D_g(\tau) = \overline{\left(\{g(t+\tau) - g(t)\} - \overline{\{g(t+\tau) - g(t)\}}\right)^2} \tag{4.13}$$

where overbar denotes time average. For a stationary process we have a simple relationship between the autocorrelation function and the structure function

$$D_g(\tau) = 2\left(R_g(0) - R_g(\tau)\right) \tag{4.14}$$

or

$$R_g(\tau) = \tfrac{1}{2} D_g(\infty) - \tfrac{1}{2} D_g(\tau)$$

This relationship is schematically shown in Figure 3.

If we have some doubt about the stationarity of a process, it is better to construct its <u>structure function</u> and not the correlation function. In practice, the <u>structure function</u> is more reliable, since $D_g(\tau)$ is not affected by errors in the mean value.

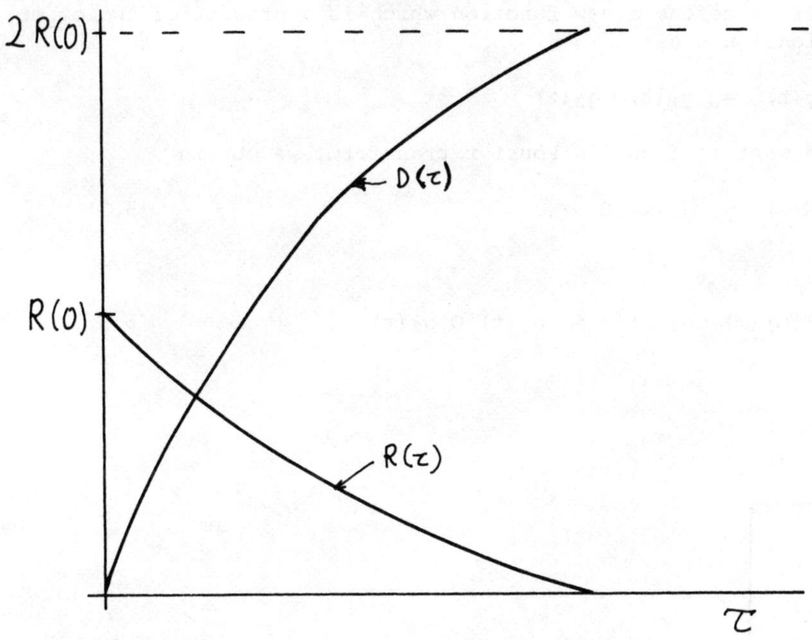

Figure 3 Relation between autocorrelation function and structure function.

5. CONVOLUTION

Convolution is one of the most important and most used theorems within the whole Fourier analysis.

The convolution is defined as

$$f(t) \otimes g(t) = \int_{-\infty}^{\infty} f(t_1)\, g(t-t_1)\, dt_1 \qquad (5.1)$$

where $\otimes$ denotes the convolution operator.

Convoluting two functions is a way of combining them together. The effect is often abroadening of the one by the other.

An example of graphical representation of a convolution is shown in Figure 4.

If we define a new function which is a product of two other functions such as

$$g(t) = g_1(t) \cdot g_2(t)$$

and we want to find its Fourier transform, we obtain

$$G(\omega) = G_1(\omega) \otimes G_2(\omega) \qquad (5.2)$$

or

$$F\{G_1(\omega) \cdot G_2(\omega)\} = g_1(t) \otimes g_2(t) \qquad (5.3)$$

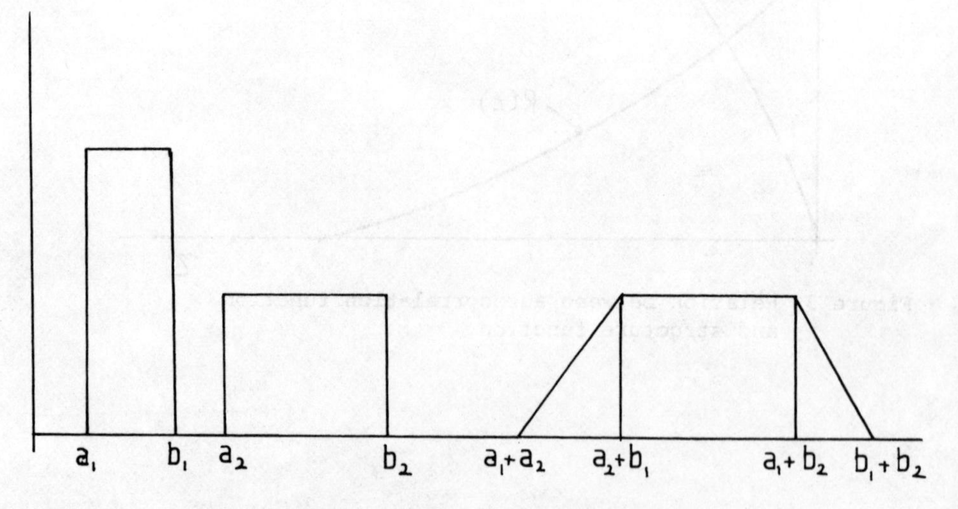

Figure 4 Example of graphical representation a convolution.

In other words, multiplication of two functions in one domain and convolution in the other domain is accordingly Fourier pair.

6. FOURIER PAIRS INVOLVING CORRELATION AND CONVOLUTION AND SOME IMPORTANT FOURIER TRANSFORM FORMULAS

Table 3 is summarized some important relationships between correlation and convolution functions.

Table 4 gives a review of some important Fourier transform formulas.

Autocorrelation in time	:	$\int_{-\infty}^{\infty} g(t_1)g^*(t_1-t)dt_1 \leftrightarrow G(\omega)G^*(\omega)$
Autocorrelation in frequency	:	$g(t)g^*(t) \leftrightarrow \int_{-} G(\omega_1)G^*(\omega_1 \cdot \omega)d\omega_1$
Crosscorrelation in time	:	$\int_{-\infty}^{\infty} g_1(t_1)g^*(t_1-t) \leftrightarrow G_1(\omega)G_2^*(\omega)$
Crosscorrelation in frequency	:	$g_1(t)g_2^*(t) \leftrightarrow \int_{-\infty}^{\infty} G_1(\omega_1)G_2^*(\omega_1-\omega)d\omega_1$
Convolution in time	:	$g_1(t) \otimes g_2(t) \leftrightarrow G_1(\omega)G_2(\omega)$
Convolution in frequency	:	$g_1(t)g_2(t) \leftrightarrow G_1(\omega) \otimes G_2(\omega)$

Table 3. Summary of Fourier pairs involving correlation and convolution.

7. APPLICATIONS

Correlation techniques and Fourier transform are so widely used that we here due to lack of space have to limited the discussion to those applications which in our connection may be the most useful ones.

As Fourier transforms more or less have their origin in communication theory we will discuss some application related to that field. These may also be useful in other sciences.

Secondly we will study some applications in atmospheric science.

TYPE OF FUNCTION	APERIODIC	RANDOM FUNCTION
Function	$g(t) = \frac{1}{2\pi} \int_{-\infty}^{\infty} G(\omega) e^{j\omega t} \, d\omega$	not defined
Amplitude spectrum	$G(\omega) = \int_{-\infty}^{\infty} g(t) e^{-j\omega t} \, d\omega$	not defined
Auto-correlation	$R_g(\tau) = \int_{-\infty}^{\infty} g(t) g(t+\tau) dt = \frac{1}{2\pi} \int_{-\infty}^{\infty} E(\omega) \cos\omega\tau d\omega$	$R_g(\tau) = \lim_{T\to\infty} \frac{1}{2T} \int_{-T}^{T} g(t) g(t+\tau) dt = \frac{1}{\pi} \int_{-o}^{\infty} P(\omega) \cos\omega\tau d\omega$
Cross-correlation	$R_{gt}(\tau) = \int_{-\infty}^{\infty} g(t) f(t+\tau) \, dt$ $= \frac{1}{2\pi} \int_{-\infty}^{\infty} E_{fg}(\omega) \, e^{j\omega t} \, d\omega$	$R_{gt}(\tau) = \lim_{T\to\infty} \frac{1}{2T} \int_{-T}^{T} g(t) f(t+\tau) dt$ $= \frac{1}{\pi} \int_{o}^{\infty} C(\omega) \, e^{j\omega\tau} \, d\omega$
Energy spectrum	$E(\omega) = \lvert G(\omega) \rvert^2 = 2 \int_{o}^{\infty} R_g(\tau) \cos \omega\tau \, d\omega$	not defined
Cross energy Spectrum	$E_{fg}(\omega) = F^*(\omega) \, G(\omega)$ $= \int_{-\infty}^{\infty} R(\tau) e^{-j\omega\tau} \, d\tau$	not defined
Power spectrum	not defined	$P(\omega) = \lim_{T\to\infty} \frac{1}{2T} \lvert G_T(\omega) \rvert^2 = \lim_{T\to\infty} \frac{1}{2T} \lvert \int_{-T}^{T} g(t) e^{-j\omega t} \, dt \rvert^2$ $= 2 \int_{o}^{\infty} R_g(\tau) \cos \omega\tau \, d\tau$
Cross power spectrum	not defined	$C(\omega) = \lim_{T\to\infty} \frac{1}{2T} F^*(\omega) \, G(\omega) = \int_{-\infty}^{\infty} R_{ft}(\tau) \, e^{j\omega\tau} \, d\tau$

Table 4 Some important Fourier transform formulas.

7.1 Linear systems - impulse response

A linear system may be considered as a "black box" that converts an input function g(t) into an output y(t). The output is also called the response of the system

g(t) [system] ⟶ y(t)

With a shift-invariant linear system the output is related to the input through the impulse response function h(τ) as

$$y(t) = \int_{-\infty}^{\infty} g(\tau)\, h(t-\tau)\, d\tau \qquad (7.1)$$

(The definition of the impulse response function is strictly speaking related to the response of a delta function input).

Remembering in the definition of convolution in chapter 5 we see that Eq (7.1) is identical to Eq (5.1) and hence

$$y(t) = g(t) \otimes h(t) \qquad (7.2)$$

Since convolution and multiplication was a transform pair, the Fourier transform of the output is

$$Y(\omega) = G(\omega)\, H(\omega) \qquad (7.3)$$

where H(ω), which is the transform of the impulse response function h(t), is called the energy transfer function of the system

$$h(t) \leftrightarrow H(\omega) \qquad (7.4)$$

If the input is a random signal (not a simple waveform) with finite energy the output powerspectrum $P_y(\omega)$ is given by

$$P_y(\omega) = H(\omega)\, H^*(\omega)\, P_y(\omega) = |H(\omega)|^2\, P_g(\omega) \qquad (7.5)$$

where $P_g(\omega)$ is the powerspectrum of the input signal.

If the input signal is white noise with a powerspectrum $P_g(\omega) = A$ (constant) we have

$$P_y(\omega) = A|H(\omega)|^2 \qquad (7.6)$$

and the autocorrelation function is

$$R_y(\tau) = A \int_{-\infty}^{\infty} |H(\omega)|^2\, e^{j\omega\tau}\, d\omega \qquad (7.7)$$

7.2 Correlator

One of the most useful ways of detecting a known input signal in noise or to find the impulse response of a system is to use correlation techniques.

With an input function $g_T(t)$ (as defined in Eq (2.9)) the output amplitude spectrum $Y(\omega)$ is

$$Y_T(\omega) = G_T(\omega)\ H(\omega) \tag{7.8}$$

or

$$Y_T(\omega)\ G_T^*(\omega) = G_T^*(\omega)\ G_T(\omega)\ H(\omega) \tag{7.9}$$

Using the definition of the power spectrum we obtain

$$P_{gy}(\omega) = P_g(\omega)\ H(\omega) \tag{7.10}$$

or

$$R_{gy}(\tau) = R_g(\tau) \otimes h(t)$$

In the case of white noise input we find that

$$P_{gy}(\omega) = A\ H(\omega)$$

or

$$R_{gy}(\tau) = A\ h(t)$$

which means that the impulse response function of a system can be found by forcing white noise into the system and measuring the crosscorrelation between the output and input of the system.

If the input signal is coloured the energy transfer function is, according to Eq (7.10)

$$H(\omega) = \frac{P_{gy}(\omega)}{P_g(\omega)} \tag{7.11}$$

Recalling the definition of coherence, the transfer function $H(\omega)$ can be written as

$$H(\omega) = \mathrm{coh}_{gy}(\omega)\ \sqrt{\frac{P_y(\omega)}{P_g(\omega)}} \tag{7.12}$$

7.3 Signals in noise

A problem we often have to deal with is extraction of information from a noisy signal.

The signal g(t) may simply consist of an input signal s(t) plus noise n(t) or the signal can be blurred, by the effects of an echo which may be considered as a convolution.

In such cases we want to construct a filter which removes the noise. The optimum characteristics of such a filter is given by Eq (7.11).

If we have a signal given as

$$g(t) = s(t) + n(t) \tag{7.13}$$

and take the autocorrelation, we obtain

$$R_g(\tau) = R_s(\tau) + R_n(\tau) + R_{sn}(\tau) + R_{ns}(\tau) \tag{7.14}$$

If the input s(t) and the noise n(t) are uncorrelated we have

$$R_g(\tau) = R_s(\tau) + R_n(\tau)$$

The crosscorrelation between g(t) and s(t) is

$$R_{gs}(\tau) = R_s(\tau) + R_{ns}(\tau) = R_s(\tau) \tag{7.15}$$

assuming again uncorrelated input signal and noise.

The autocorrelation of the signal g(t) can then be expressed in terms of crosscorrelation as

$$R_g(\tau) = R_{gs}(\tau) + R_n(\tau) \tag{7.16}$$

The optimum characteristic of the filter is then given as

$$H(\omega) = \frac{P_{gs}(\omega)}{P_g(\omega)} = \frac{P_s(\omega)}{P_s(\omega) + P_n(\omega)} \tag{7.17}$$

Another important case is when the signal is blurred by reverberation. As mentioned earlier this may be interpreted as a convolution

$$g(t) = s(t) \otimes m(t)$$

where m(t) represents the reverberation.

Again using our definition of the powerspectrum in Eq (4.8) we may write the cross powerspectrum between g(t) and s(t) as

$$P_{gs}(\omega) = \lim_{T\to\infty} \frac{1}{2T} \{S^*(\omega)\ S(\omega)\ M^*(\omega)\}$$

$$= P_s(\omega)\ M^*(\omega) \tag{7.18}$$

Using the same definition the powerspectrum of g(t) is

$$P_g(\omega) = P_s(\omega)\ M(\omega)\ M^*(\omega)$$

and the optimum characteristic of a filter in the case of a blurred signal is

$$H(\omega) = \frac{P_s(\omega)\ M^*(\omega)}{P_s(\omega)\ M(\omega)\ M^*(\omega)} = \frac{1}{M(\omega)} \tag{7.19}$$

The criterion which is satisfied by Eq (7.19) is that the mean square of the difference between the actual output a(t) and the desired output S(t) is a minimum.

If

$$\varepsilon(t) = a(t) - s(t) = g(t) \otimes h(t) - s(t) \tag{7.20}$$

and again introducing truncated signals as defined in Eq (2.9), using our definition of power spectrum and Eq (7.11) we obtain the power spectrum of $\varepsilon(t)$ as

$$P_\varepsilon(\omega) = P_s(\omega) - \frac{|P_{gs}(\omega)|^2}{P_g(\omega)} \tag{7.21}$$

Using our definition of the coherence

$$P_\varepsilon(\omega) = P_s(\omega)\ \{1 - \mathrm{coh}_{gs}^2(\omega)\}$$

If it is a complete correlation between actual and desired output i.e. $\mathrm{coh}(\omega) = 1$ we see that the powerspectrum of the error is zero as expected.

7.4 Applications within atmospheric science

The few scattered applications discussed on the last pages have of course also bearing on atmospheric science, as some type of sensors and measuring system always are involved when doing measurements of atmospheric parameters.

After having taken into account the different limitations of a measuring equipment or eliminated any noise via different filter-techniques, we still have to retrieve information about the atmospheric parameters of interest from one or more time series with fluctuating signals.

In this chapter we will discuss one application of Fourier transforms in atmospheric research, namely obtaining information about velocity from spatially separated sensors using cross-correlation and cross-spectrum techniques.

7.1 Detection velocity from spatially separated recording

In this application we consider two (or more) separated recordings as shown in Figure 5.

The sensors placed in A and B may be pressure sensors to detect gravity waves, wind sensors or any other sensors.

The signals at A and B might be generated from a steady drift across the sensor by a spatially varying medium perturbated by turbulence, or by a wave motion. The phase lag between the two recordings can be found via the crosscorrelation function as shown in Figure 6. [1]

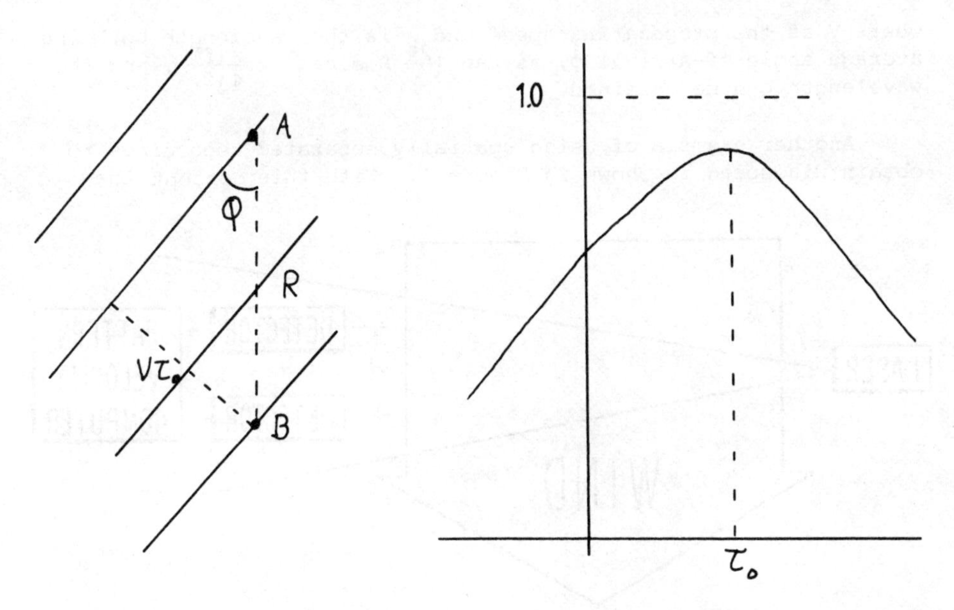

Figure 5 An array for measuring drifting speed

Figure 6 Corresponding autocorrelation

If the medium is frozen, the crosscorrelation will be 1. at a time lag corresponding to τ_o, but normally this is not the case.

The crosscorrelation gives only the average travel time from A to B along the connection line through A and B.

If the medium is dispersive, the cross spectrum must be used. According to Eqs (2.5) and (2.6) the real and imaginary part of the co-spectrum can be written as

$$C_o(\omega) = C(\omega) \cos \omega\tau_o$$

$$Q(\omega) = C(\omega) \sin \omega\tau_o \qquad (7.22)$$

and the phase lag

$$\omega\tau_o = \tan^{-1} \frac{Q(\omega)}{C_o(\omega)} \qquad (7.23)$$

If the medium is propagating at right angle to the connection line between A and B, the crosscorrelation will be symmetric and the cross powerspectrum will be real $(Q(\omega) = 0)$.

Using Taylor's hypothesis,

$$\omega = \frac{2\P}{\lambda} V$$

where V is the propagating speed and λ is the wavelength both the average angle-of-arrival ϕ, as $\tan\left(\frac{2\P}{\lambda} R \sin \phi\right) = \frac{\phi(f)}{\phi(f)}$, and the wavelength can be obtained.

Another example of using spatially separated recordings to obtain windspeed is shown in Figure 7. With this concept the

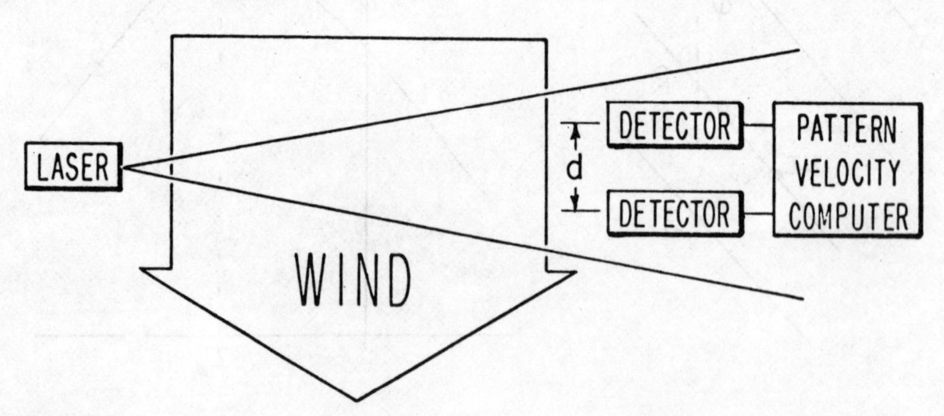

Figure 7 Laser cross wind system. [2]

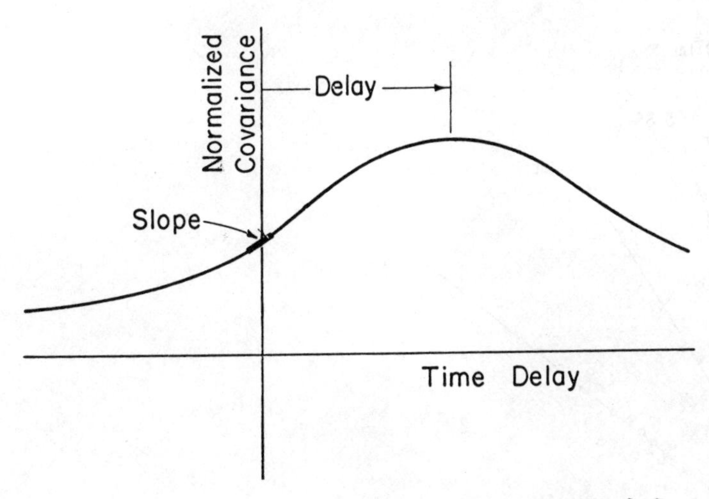

Figure 8 Using slope to obtain wind speed. [2]

average component of the wind across a laser beam can be determined by performing the crosscorrelation between the signals at the two detectors. In this case it is more suitable to use the slope at zero time lag because of the rapid decay of the medium.

This is schematically shown in Figure 8.

Using temperature irregularities as traces the mean wind speed can be obtained from cross-spectrum analysis of signals from temperature sensors spaced along the mean wind direction.

The wind speed is found by plotting the phase $\theta(f)$ of the cross-spectrum against frequency $(\theta(f) = \tan^{-1} \frac{Q}{C_o})$. [3]

The slope of the plot is then related to the apparent mean wind speed of the irrgularities as

$$V = \frac{2\pi f d}{\theta}$$

where d is the separation of the pair of sensor.

In Figure 9 is shown such a plot for four spacing.

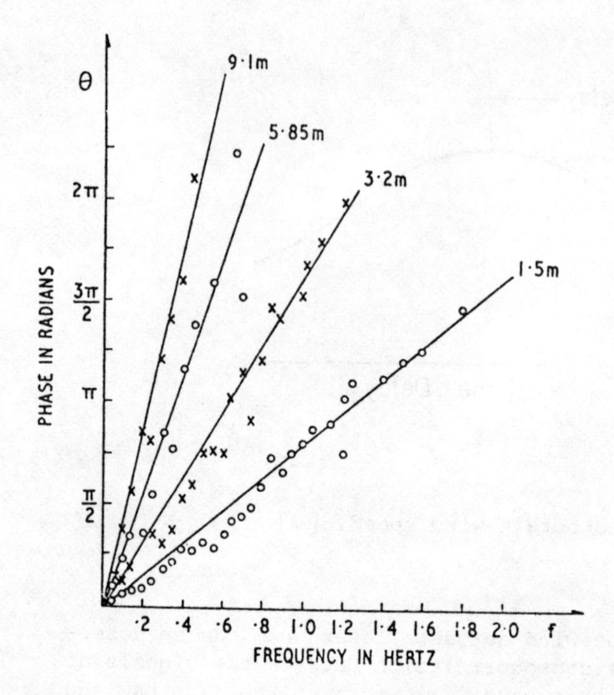

Figure 9 Phase of cross-spectrum against frequency for different spacing of temperature measurements.

BIBLIOGRAPHY

1. Thomas, J.B, An Introduction to Statistical Communication Theory. John Wiley & Son, 1969.
2. Champeney, D.C., Fourier Transforms and their physical Applications. Academic Press, 1973.
3. Papoulis, A., Systems and Transforms with Applications in Optics. McGraw-Hill, 1968.
4. Tretter, S.A., Introduction to discrete time signal processing. John Wiley & Son, 1976.

REFERENCES

1. Gossard, E.E. & W.H. Hooke, Waves in the atmosphere. Elsevier Scientific Publishing Company 1975.
2. Lawrence, R.S, G.R. Ochs and S.F. Clifford, The use of scintillation to measure average wind across a light beam. Appl. Opt., 11, 239-243, 1972.
3. Gjessing, Dag T., A.G. Kjelaas and E. Golton, Small-scale atmospheric structure deduced from measurements of temperature, humidity and refractive index. Bound-Layer Meteors, 4, 475-492, 1973.

PATTERN RECOGNITION - WITH SPECIAL EMPHASIS ON IMAGE PROCESSING

T. Orhaug

National Defence Research Institute, Box 1165,
S-581 11 LINKÖPING, SWEDEN

ABSTRACT. This paper reviews some of the more important methods in computerized pattern recognition and image processing including supervised and nonsupervised techniques, parametric and nonparametric methods as well as transformation of multidimensional data vectors to lower dimensional space for data presentation. The importance of preprocessing and feature extraction is stressed.

1. INTRODUCTION

Pattern recognition is a collective term for various methods which may be useful for studying, analyzing and/or classifying complex data of which various degrees of *a priori* knowledge is available. The field is not a unified field; it has a broad spectrum ranging from principal questions dealt with in bionics and artificial intelligence to practical questions of technical implementation. Also, the development in the field has been dependent upon, and has gained from, a large number of fields like statistical decision theory, automata theory, control theory, signal and communication theory to name just a few.

It could also be commented that the methods available in the field do *not* offer general solutions to complex data interpretation problems. Today, one has come to realize the complexity, efficiency and flexibility of the human brain for pattern interpretation and problem solving and the expectations one might have had some years ago to automate the complexity of human perception has not been fulfilled. A more realistic attitude to the field of pattern recognition is to

T. Lund (ed.), Surveillance of Environmental Pollution and Resources by Electromagnetic Waves, 353–379.

say that it offers some interesting and sometimes even efficient tools for studying and analyzing complex data.

A vast amount of literature in the field of pattern recognition and image processing is available. A limited number of recent textbooks and also some conference proceedings are listed in the Reference section, (1) - (13).

2. PROBLEM FORMULATION

The various methods that come into use in pattern recognition may be grouped into the following two main sets:

- statistical (or geometrical) approach
- structural (linguistic or syntactic) approach

These two approaches can be characterized as follows. We consider the pattern as represented by a *sampled* version of the pattern and each set of sampling values can be considered as a vector in multidimensional space (whose number of dimensions is equal to the number of sampling values). Alternatively, we may construct the vector, not by using all the sampled values but some characteristic measurements (*features*) extracted and/or calculated from the pattern. In the first approach, the problem is to partition either the pattern or the feature space. The reason for using a feature description of the pattern is both that the features are supposed to be more *invariant* with respect to various distortions and also that this description is supposed to be less *redundant*. The recognition problem can now be formulated as follows:

- to extract the "good" or efficient features from the pattern
- to make the decision as to what *class* of patterns a new input pattern should be assigned based upon the partition of pattern or feature space.

A simple illustration visualizing the principle of the formalism above is shown in Figure 1 for the case of an imaging system. The scene can be described in a large number of ways and the inherent dimensionality is very large. The imaging process proper constitutes an important abstraction since the dimensionality of the observable signals (resulting from the interaction of the illumination with the scene properties) is much less than the inherent dimensionality. The resulting imagery data may either be subject to visual interpretation (after suitable "filtering" (processing)) *or* to machine (computer) analysis. The principle of this process is indicated in Figure 2 where the sampling or the feature extraction process may be considered as a further abstraction of the scene information.

The basic idea of the statistical approach is that if the "efficient" features are extracted from the pattern, then patterns coming from the same kind of pattern group or class would be clumped (or clustered) close together in the multidimensional representation. For this reason this approach is also called *geometrical*; geometrical distances or other metrics are used for representing and expressing "closeness". Another quality of this method is that the variability of the samples representing the same pattern class is due to statistical effects.

The *strength* of this approach is that a large number of algorithms which are both well documented and well analyzed are available. The *weakness*, on the other hand, is the question of the good features. A general theory indicating what features to use is *not* available.

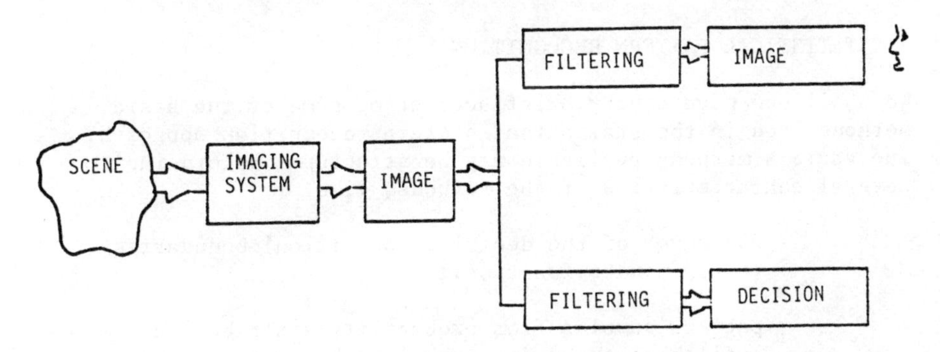

Fig. 1. The imaging and image interpretation process

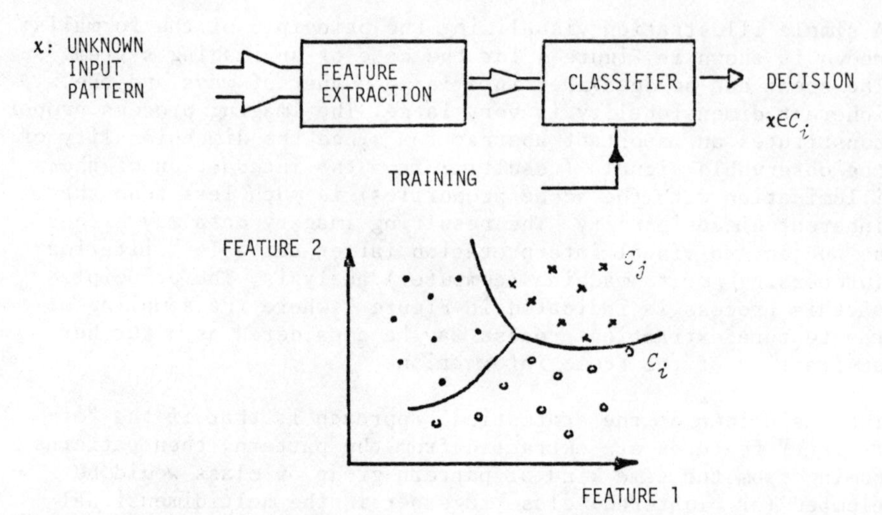

Fig. 2. Illustrating the problem of pattern recognition (statistical approach)

This latter point is also the reason for the development of the second approach, i.e. the structural approach to pattern recognition; the idea underlying this approach is that a complex pattern may be described in terms of simpler pattern elements ("primitives") and subelements. In this manner also the structural information in the pattern is used and the recognition process is not limited to classification but may also include *pattern description*.

3. STATISTICAL PATTERN RECOGNITION

We shall now give a very brief account of some of the basic methods used in the statistical pattern recognition approach. The various methods available may be structured accordning to several characteristics of the methods, like

- the *shape* of the decision (partition) boundaries (*discriminant* function);
- pattern samples from probability distribution are available; these distributions have known/unknown parameters;
- pattern samples of known classbelonging are/are *not* available (*labelled/nonlabelled samples*).

In the last case one may also talk about *supervised/nonsupervised* methods.

3.1 Supervised methods

We introduce the sampled pattern or pattern vector

$$x = (x_1, x_2, \ldots\ldots, x_N)^T \tag{1}$$

(where $(\ )^T$ denotes matrix transposition) and a set of K pattern classes

$$C: C_1, C_2, \ldots . C_K. \tag{2}$$

We shall first consider the case when each class is represented by a *mean* or a *prototype* pattern vector $x_{C_i} \equiv x_i$. The "classical" method of detecting a signal whose properties are known *a priori* is to compute the correlation ρ between the (unknown) signal and candidate signals x_i:

$$\rho = x_i^T x = |x_i| \cdot |x| \cos \alpha \tag{3}$$

A measure of the "closeness" between x and x_i is the (multidimensional) angle α between the two vectors. If we accept x as belonging to C_i for $\alpha_i - \Delta\alpha \leq \alpha \leq \alpha_i + \Delta\alpha$, where α_i is the direction of x_i, then the corresponding decision boundaries are planes through the origin (or straight lines in the two-dimensional illustration in Figure 3).

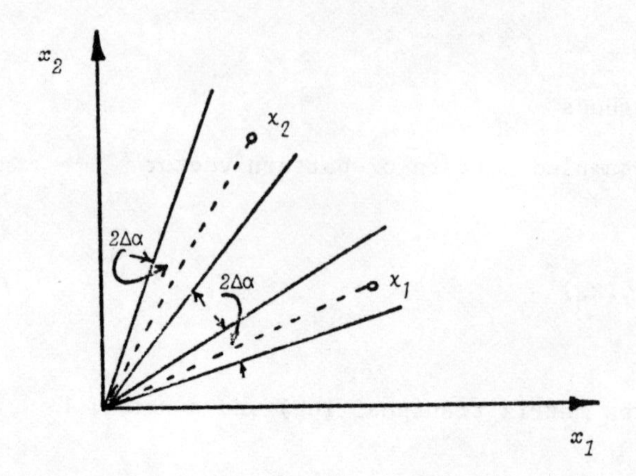

Fig. 3. Decision surface for correlation detection. (Signals are assumed to contain the same energy).

A more general (does not necessarily go through the origin) but still *linear* decision surface is obtained by weighting the *augmented* pattern vector Y,

$$Y^T = [X^T, 1] \tag{4}$$

with a weighting function as in

$$D_i(X) = w^T Y. \tag{5}$$

See references (1-3) for a more detailed discussion on linear discriminant functions.

If a number of reference or prototype vectors are available ($X_1, X_2, \ldots, X_K$), each defining a particular class, one may calculate the distance between the input (unknown) vector and the various reference vectors and use this distance as the classification criterion. The corresponding discriminant function, the so called *minimum distance* classifier, is also for this case *linear* (see Figure 4).

A still more complex discriminant function is obtained if one computes the distance between the input vector and all neighbouring vectors (having class labels). The input vector is either assigned to the class of the nearest vector (*nearest neighbour* rule) or to the class represented by the majority of the K-nearest vectors (*K-nearest neighbour* rule). In this manner, the associated decision surface is *piecewise linear* (see Figure 5). For a more detailed discussion the reader is referred to (1-3).

Up till now, we have taken the actual *distribution* of the labelled vectors into account in only a crude way; either the cluster centers or the cluster edges have been used for defining the decision surfaces In the *parametric* approach the feature vectors of each class are assumed to originate from multivariate probability distributions. The conditional probability density for class C_i, $(p(x|C_i))$, is assumed known. Sometimes, also the *a priori* probabilities $P(C_i)$ as well as the cost functions (*i.e.* the penalty of assigning a vector to class C_i if the correct class is C_j) may be known: The optimum decision may be defined as the one which minimizes the mean cost (*maximum likelihood classifier*).

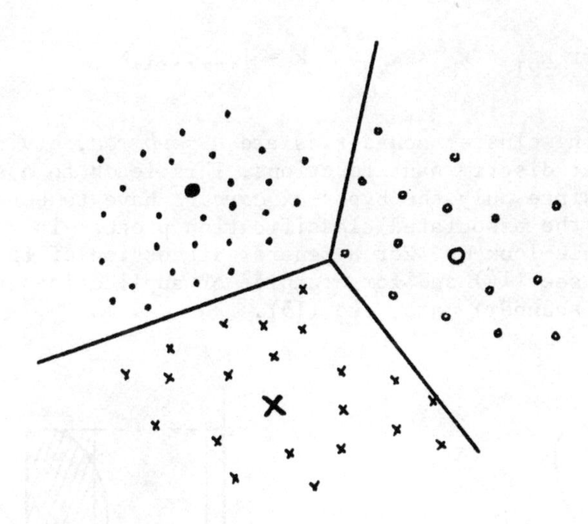

Fig. 4. Illustrating decision boundaries for the minimum distance classifier.

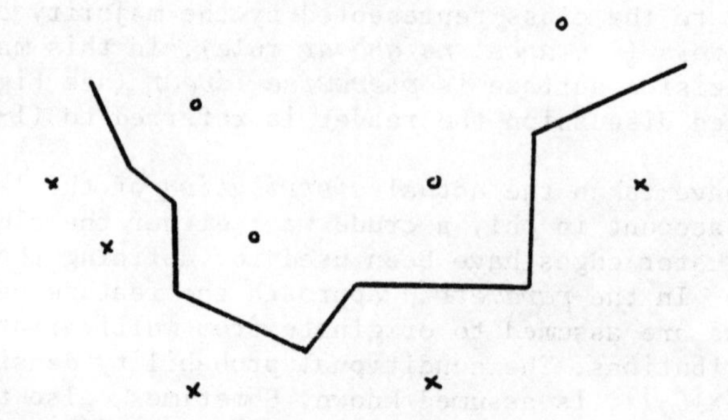

Fig. 5. Depicting a piecewise linear classifier.

The simplest principle assumes *uniform* probability density and *independent* vector components:

$$P(x|C_i) = P_o \text{ for } x_{k1} < x_k < x_{k2}, \quad k = 1,2,\ldots,N$$

The corresponding cluster boundaries are *hyperboxes*, giving piecewise linear discriminant functions. This leads to a simple decision rule since only the hyperbox *corners* have to be stored as a table and the associated classification process is executed by a simple table-look-up. For a general discussion of this type of classifier, see (14) and for examples of applications to MSS (multispectral scanner) data, see (15).

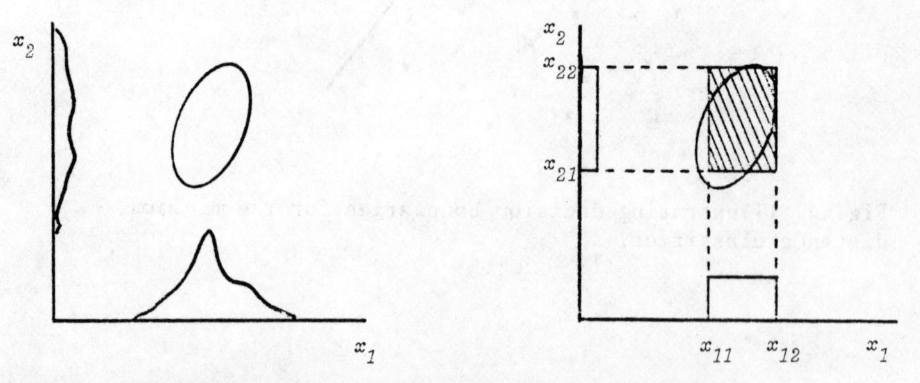

Fig. 6. The table-look-up classifier (uniform probability density and independent vector components).

In the more general case, $p(x|C)$ is assumed to be multivariate Gaussian, leading to the following discriminant function:

$$D(x) = \eta_o - \frac{1}{2}(x - m_i)^T A_i^{-1}(x - m_i) \qquad (6)$$

where η_o: threshold function (dependent upon *a priori* probabilities and cost functions)

m_i: mean value (centroid) of class i

A_i: autovariance function of class i

These decision boundaries are generally surfaces of 2:nd order (quadratic surfaces). For several special cases (diagonal covariance matrices among the classes) the decision boundaries become linear planes (see Figure 7). For diagonal A, the discriminant functions becomes $x^T x$, or similar to the minimum distance classier or correlation detector. For a more detailed treatment of the normal case, see (1), (2), (3) and (14).

In practical applications of supervised recognition methods the procedure consists of two main stages: training and recognition (classification). In the training phase, pattern samples with known class belonging are used to estimate the parameters of the probability distribution (assuming a parametric method like the Gaussian maximum likelihood classifier). First, one must assure that the training samples are homogeneous, the test of which can be performed by testing the probability distribution using "vector" histograms. If a homogeneity condition is *not* fulfilled the samples which distort the distribution must be omitted. This procedure may often be carried out iteratively. Another way to obtain homogeneous sample sets is to use a clustering algorithm to find pixels representing homogeneous parameters (see Section 3.2). The homogeneous samples are used to estimate the unknown parameters (A_i and m_i) for each class. From this knowledge, the discriminant function is also known and new sample vectors may then be feed to the classifier for class assignment.

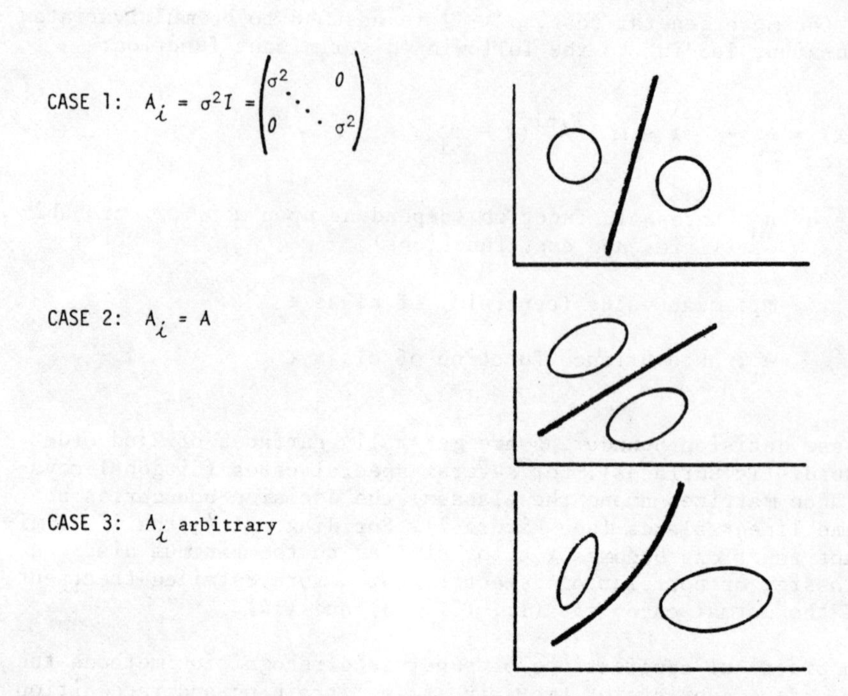

Fig. 7. Illustrating decision boundaries for Gaussian probability distributions

Finally in this section, we shall summarize the various supervised methods by referring to Figure 8. The data are two-dimensional vectors (representing green and red densities measured on color film) and this feature space is partitioned using various pattern discrimination methods (16).

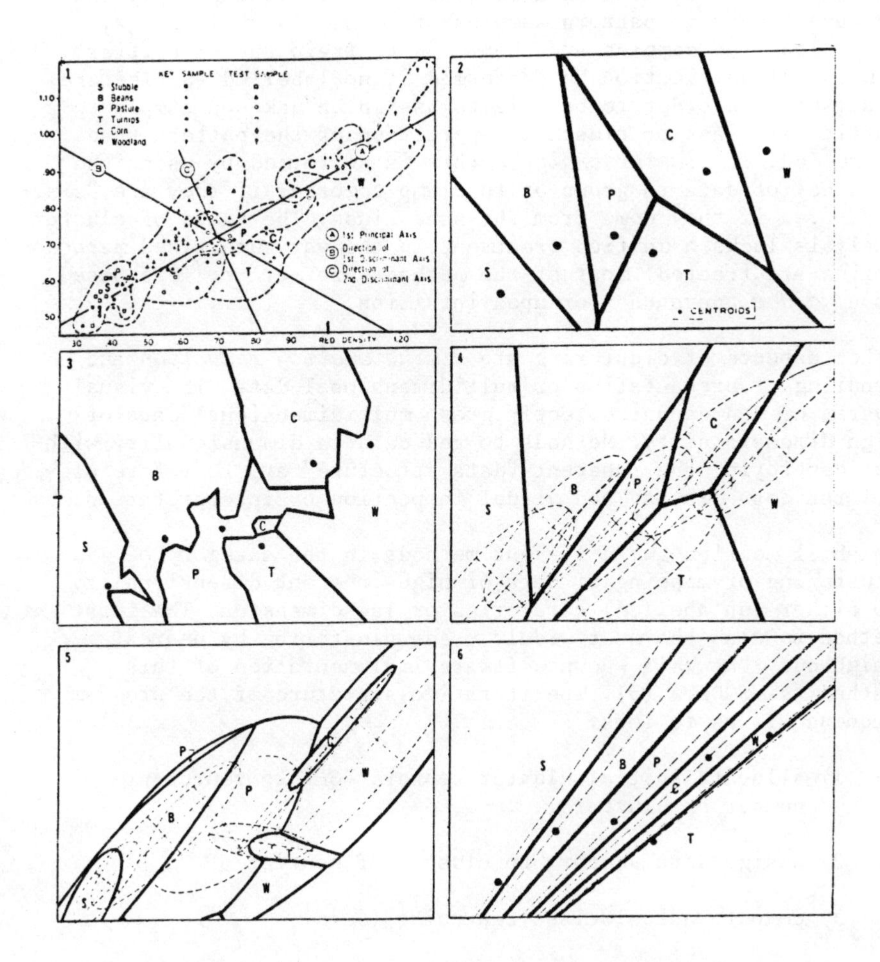

Fig. 8. Partition of two-dimensional feature space.
1: Scatter of input data and partition of data according to ground truth. Axis A: direction of 1:st principal component (using overall distribution). Axis B and axis C: 1:st and 2:nd discriminant axis.2: Linear boundaries based on distance to centroids (minimum distance rule), 3: Piecewise linear boundaries based on the maximum likelihood method and normal distributions assuming equal covariance matrices. 5: Quadratic boundaries, maximum likelihood, normal distributions, covariance matrices not identical. 6: Linear boundaries based on angular proximity (correlation detection).After (14).

3.2 Unsupervised methods

The supervised methods we have just been discussing imply that we have access to pattern samples of *known* classbelongings; these labelled samples are then used to train the classifier. How can classification be performed if no labelled samples are available? In order to be able to assign an unknown sample to a particular class or cluster, a partition of the pattern space is needed. In *cluster analysis* this is perfomed by using the tendency of data to group or to clump together if they are "similar" *i.e.* if they come from the same class. The field of cluster analysis lacks a unified treatment in the way supervised methodologies are treated; most of the methods are more or less based upon *ad hoc* approaches or upon intuition.

Other aspects of clustering are *dimensionality reduction* and handling or presentation of multidimensional data. The visual system of man cannot directly grasp multidimensional data of high dimensionality. Methods to reduce data dimensionality without destroying the inherent "data structure" are therefore useful for data display and visual inspection or interpretation.

We shall mention two different methods in the category of clustering or mapping of data of high inherent dimensionality to either (unlabelled) categories or low dimension. The first method is also the most widely used, clustering by nearest neighbour. One well-known software implementation of this method is ISODATA (5). The iterative structure of the program sequence is as follows:

1) allocate initial cluster centers *(SP)* to represent number of clusters,

2) assign data vectors to cluster of nearest *SP*

3) compute actual class centroids *(CC)*

4) use *CC:s* as new *SP:s*

5) repeat until a convergence or threshold criterion is fulfilled

6) discard clusters having $N < N_1$ (N = number of members within a specific cluster)

7) split clusters having $N > N_2$

8) repeat if necessary.

For actual data, the rate of convergence is often rapid as indicated in Figure 9 where the two seedpoints (SP_1 and SP_2) in a matter of a few iterations will converge to the "good" cluster centers (CC_1 and CC_2)

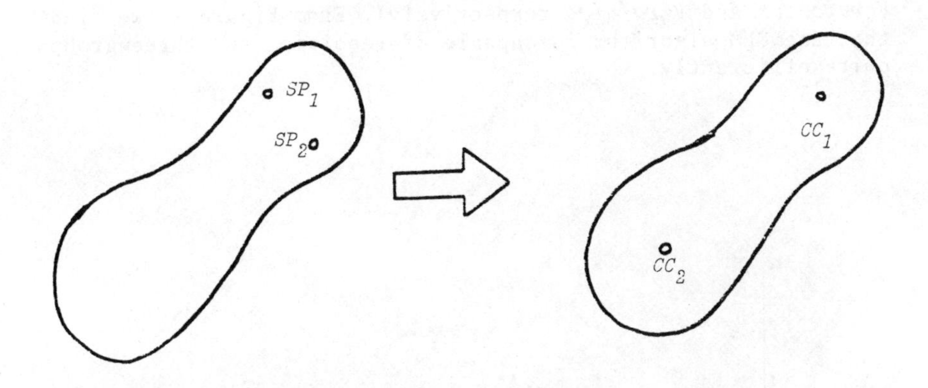

Fig. 9. Illustrating convergence of clustering sequence

The other clustering method is more to be considered as a *mapping* method by which high-dimensional data can be mapped to a low-dimensional space. The procedure of finding a low-dimensional configuration of points whose interpoint distances correspond to dissimilarities between points in the original high-dimensional space is named *multidimensional scaling* (1). This mapping process is *nonlinear*. A special version of nonlinear mapping has been developed by Sammon (17) whereby the interpoint distances are chosen in such a way that they are as equal as possible (in a mean sense) to the interpoint distances of the original data vectors. The purpose of this method is to preserve the eventual clustering tendency of the data. If the "image" space is of dimension 3 or 2, the data may easily be displayed and therefore visually inspected. The method which may start with image data randomly distributed, is iterative and roughly 30 iterations are sufficient to obtain a good match (18). Mapping in steps ($N_o \rightarrow N_1 \rightarrow N_2$) where N_o is the number of dimensions of the original space and $N_2 = 2$) has also been investigated (18). This method seems to give a better preservation of clustering properties.

In order to indicate some properties of the various methods to be discussed we shall exemplify the methods using the same data set. The examples are due to Chien (19) who used the well-known *Iris* data consisting of 150 samples of fourdimensional data. The variables are size measurements (sepal length and width and pedal length and width) of three different species of *Iris* (*Setosa*, *Versicolor* and *Virginica* respectively). From Figure 10 we find that the NLM algorithm is capable of resolving the three groups quite efficiently.

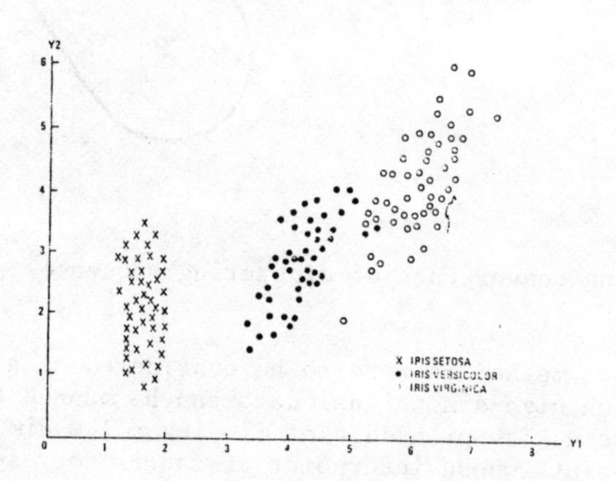

Fig. 10. Result from non-linear mapping of four-dimensional *Iris* data to two dimensions (after Chien (19)).

Also *linear transformations* of data come into frequent use, both as a tool for data presentation and data analysis. The two most commonly used methods are *discriminant analysis* and *factor analysis*. Such transformations can be considered as rotation and scaling of the axes of the original space. In discriminant analysis, the *separation* of classes is emphasized. For K classes, a new space of $K-1$ axes is constructed (discriminant vectors). For a two-dimensional representation the first two axes (corresponding to the two largest eigenvalues) may be chosen (as in Figure 11). For $K = 2$, the data are projected onto a line (*Fisher's linear discriminant*). In Figure 11, two of the three *Iris* classes are presented in a two-dimensional space (using the second vector orthogonal to Fisher's discriminant).

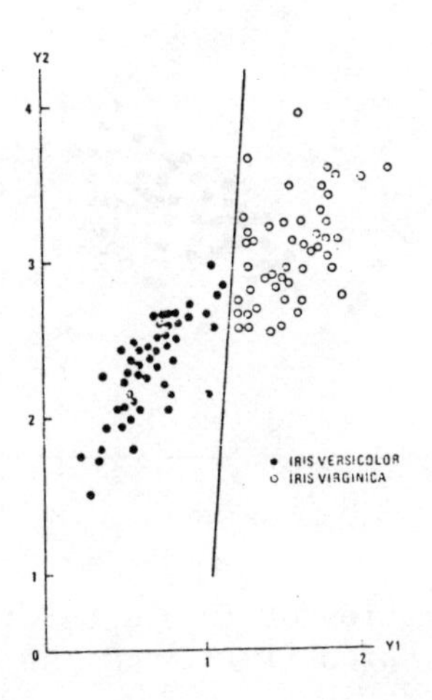

Fig. 11. Two classes of *Iris* data displayed in a two-dimensional space using discriminant analysis (after Chien (19)).

The purpose of *factor analys* is to find a linear mapping composed of the eigenvectors calculated from the scatter matrix (factor analysis) or the covariance matrix (principal component analysis or *Karhunen-Loeve* transformation). An important property of this transformation is that it preserves the *spatial* relationship of the data. The various principal components are also uncorrelated. In Figure 12 is shown a two-dimensional display of the three *Iris* groups using the first two principal components. The *Setosa* class is well separated from the two other classes which in turn are *not* separated from each other. The ability of this method for clustering, cannot be measured by any general method.

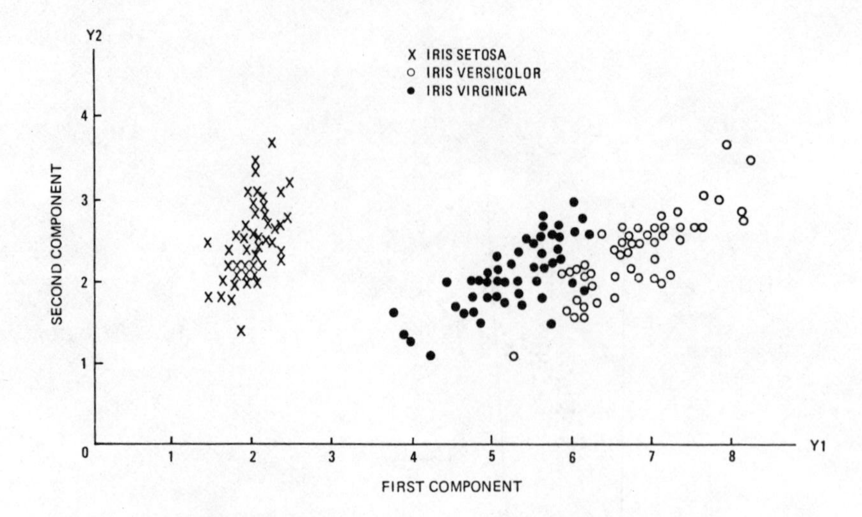

Fig. 12. Linear transformation of *Iris* data using the principal component method (after Chien (19)).

The last method for data presentation which we shall mention is the *functional display* of highdimensional data. The purpose of this method is to transform the multivariate data into functions of single variables. This may be achieved by assigning amplitude values of the various harmonics to the various components of the multidimensional data as follows:

$$f(t) = x_1 + \begin{cases} x_i sin(\frac{i}{2}t), & i \text{ even} \\ x_i cos(\frac{i-1}{2}t), & i \text{ odd} \end{cases} \quad i = 1,2,\ldots\ldots N$$

and by plotting $f(t)$ as a function of t $(\pi \leq t \leq \pi)$ for a number of data vectors.The data distribution will cause a modulation of the harmonic curves (see Figure 13).

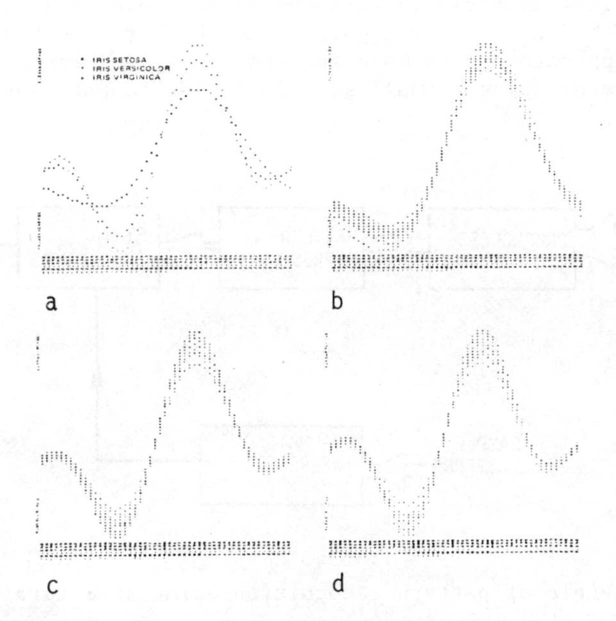

Fig. 13. Functional display of *Iris* data (after (19)).
a: Mean vectors of three *Iris* classes.
b: Functional cluster of 10 samples from the *Setosa* class.
c: Functional cluster of 10 samples from the *Versicolor* class.
d: Functional cluster of 10 samples from the *Virginica* class.

4. STRUCTURAL PATTERN RECOGNITION

In many cases, the *structural* information in the pattern is of importance; this is particularely true for many pictorial patterns. Also, if the pattern is complex having large degree of variability, the number of features necessary to describe the pattern is also large. In such cases, the ordinary methods of pattern recognition which we have described in section 3 are not necessarely efficient for tackling the pattern recognition problem. In the structural approach to pattern recognition, the pattern is considered composed of simple subpatterns which in turn are composed of simpler elements etc. in hierachical structure. In order to represent the structural information in each pattern, the relationship within the pattern element hierarchy may be described as a *graph e.g.* a *tree structure*. For this task, the *linguistic* approach is often used, based on the analogy between the tree-like structure and the syntax of languages. The idea behind this approcah is that the pattern elements or *primitives* should be easier to recognize than the pattern themselves. Also, this approach can be used for describing a large set of complex patterns using a small set of simpler subpatterns.

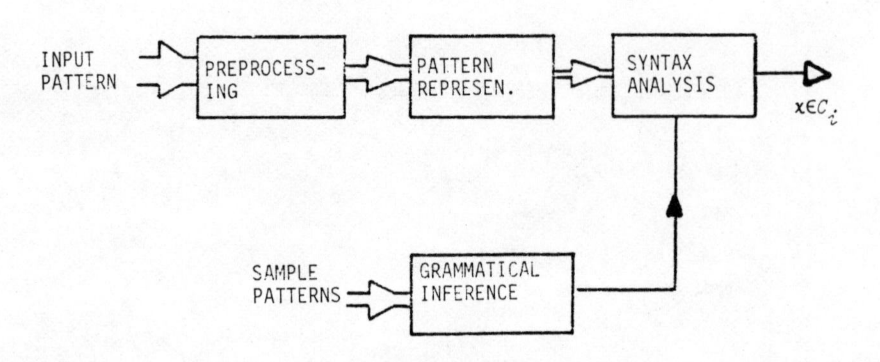

Fig.14. Principle of pattern recognition using structural method (after (20)).

The principle recognition system using the structural approach is shown in Figure 14. Like the systems using statistical methods (Figure 2), the linguistic system also incorporates a preprocessing part, a learning part and a recognition part. The preprocessing may include sampling, encoding, filtering, cleaning, restoration etc. The preprocessing is followed by a representation part leading to a (language-like) structural representation. The operations achieving this is often a *pattern segmentation* followed by *feature (primitive) extraction*. The former usually incorporates *region identification* whereby regions having uniform brightness and/or texture are detected and identified.

The decision process is carried out by the "syntax analyzer" which determines if the pattern belongs to a particular class. The *learning phase* is used to obtain a structural description of sample patterns of known class belongings. This description, which is in the form of a grammer, is used for the syntax analysis.

The structural method is applied to a wide variety of problems like character recognition, spoken digits, bubble-chamber images, chromosome images, fingerprint, aerial photographs etc. An example of a hierachical structural description of the objects and content in an aerial photograph is shown in Figure 15. For a more detailed discussion of the structural approach the reader is referred to (20).

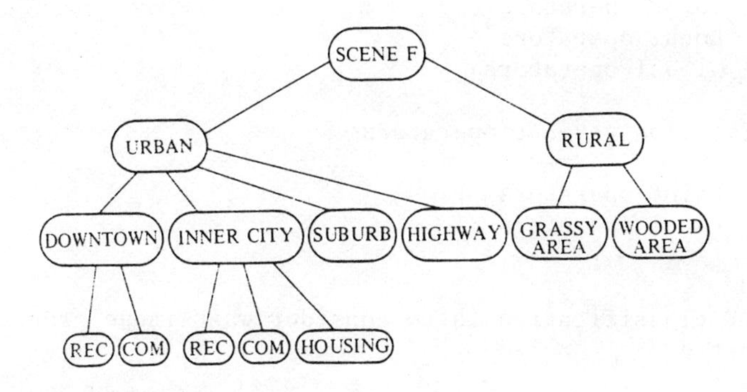

Fig. 15. Example of hierachical description of the information in an aerial photograph (after (18)).

5. IMAGE PROCESSING

5.1 Introduction

The field of *image* processing has evolved into a very large as well as a very diversified field. To discuss the various methods being used is not an easy task. The reason is partly that we may consider the various methods from so many different viewpoints. We may, as an example, consider the operators with respect to the properties they have in the spatial or other domains or we may consider the operators with relation to the purpose of their applications. Some operators are, however, more general than others being more problemorientated. Also, a simple and unequivocal structure is difficult since the same operators may be used for completely different purposes.

5.2 Classification of image operators

In order to help structure the more important operations applied in image processing we shall consider some different ways of operator classification.

OPERATOR CHARACTERISTICS

- Linear/nonlinear

- Position invariant operators

  Point operators
  Local operators
  Global operators

- Position variant operators

  Point operators
  Local operators

Another way of classification is to consider what image properties are effected:

- Radiometric transformation

  Enhancement
  Preprocessing
  Correction

- Geometric transformations

  Image rectification
  Transformation into predescribed coordination system

Also, one may distinguish the way the image itself is effected by considering the input/output images as follows:

- Image to image transformation for visual presentation

- Image to image transformation for further machine processing

- Image to decision transformation

- Image to description transformation

Still another way of distinguishing the various operators is the following listing which addresses the purpose of operation and/or the applications:

IMAGE CODING

Approximation
Sampling
Quantization
Series expansion or transformation

IMAGE ENHANCEMENT

Density slicing
Contrast stretching
Image combination
Addition, substraction,ratio etc
Image smoothing
Image sharpening

IMAGE RESTORATIONS

Geometrical restoration
Radiometric restoration
Desmoothing

IMAGE CLASSIFICATION

Object detection and classification
Region classification

IMAGE SEGMENTATION

Preprocessing and thresholding
Tracking
Region growing

IMAGE DESCRIPTION

Description using numerical features
Symbolic (high level) description

A more detailed discussion of the various operators, their characteristics and their applications can be found in the literature (e.g. (10)).

Many operators are used for image to image transformation with the purpose of generating images for visual presentation. This problem area shall not be considered here. Also, a broad class of operators are used for processing aiming at *scene analysis* (*i.e.* analysis of 3-D scenes viewed or imaged at relatively close distances).For further discussion on these points, the reader is referred to the literature.

5.3 Linear transformations

An important class of image operations are those which are based upon *linear transformations*. The "classical" transformation of this kind is the Fourier transformation and the 2-D application of this tool may be considered as an extension of the classical filtering procedures used in 1-D signal processing to the 2-D (image) signals. One reason why the Fourier transform is of importance is the fact that many image degradation processes are linear and therefore can be analyzed and desribed in terms of Fourier transforms in the spectral domain.

The transformations we consider belong to a special class of linear transforms, the *unitary* transforms, having the property that the linear operator is invertible and also satisfies certain orthogonality conditions (see (21)). An image transformation may be interpreted in several ways. One is to consider the image data as being decomposed into a 2-D spectrum. Another view is to consider the image as being synthesized with the use of 2-D basis functions having a special (complex) amplitude distribution. The various linear transforms are utilizing different basis functions and the more important unitary transforms which have been applied in image analysis are the following:

1) Fourier transform

2) Cosine transform

3) Sine transform

4) Hadamard-Walsh transform

5) Haar transform

6) Slant transform

7) Karhunen-Loeve transform

8) Singular value decomposition

The transforms *1*, *2* and *3* all use the fimiliar trigonometric functions for synthesizing the original image. These functions may be considered as "corrugation" in the image plane or as standing (non oscillating) waves. In every image point the weighted sum of the various basic wave functions add to the image function value in that particular point. Transform *2* and *3* are special cases of *1* and use *cosine*- and *sine*-terms respectively for synthesizing the image. This is brought about in *2* by generating a symmetric image and in *3* by introducing a special image autocorrelation function for transform *7*. Transform *4* is based upon *Hadamard* matrices and may be viewed as a synthesis of the image using *square* waves (instead of sinusoidals). The name *BIFORE (B*inary *Fourier*) is also used for this transform. The *Haar*-transform 5 also is of squarewave nature; this transform emphasizes the *differential* energy in the original images. The transform in *6* is similar to the Hadamard transform and possesses a square wave nature.

The transforms *7* and *8* are different from the previous since the basis functions are image *dependent*. In the case of *7*, the basis functions are the eigenfunctions to the covariance function of the image and therefore this transform is dependent upon some *statistical* property of the image. The transform *8*, on the other hand is directly dependent upon the image itself (and not upon a *class* of images as in *7*). For more detailed discussion of the properties of the various transforms, see (21).

Another point which should be mentioned is that of computational requirements. For an image of NxN pixels the number of operations (= multiplication (division) *and* addition (substraction) needed for computing the 2-D transform values is in general N^4. For the transformations *1-6* fast computation is possible (by expressing the transform matrix as a product of sparse matrices) resulting into $N^2 \log_2 N$ operations only. This procedure makes 2-D image transformations practical, even on smaller computers.

For *separable* transforms (the transform kernel can be written as a product of two 1-D kernels) the transformation can be carried out in two (one-dimensional) steps. For the second step the image matrix must be transposed. Fast and efficient ways of matrix transposition have been described which enables transposition to be carried out when only small core memory is available (22).

Unitary transforms find their application in many different areas like image coding, compression, restoration etc. As an example of application in pattern recognition or object detection we refer to Figure 16 showing the result of detecting handwritten characters (the numbers 0 to 9) sampled in a 12x16 = 192 point raster (3). A minimum distance classifier with respect to class mean was used and Figure 16 shows the percentage of correct classifications using spectral (or transform) features from transforms 1, 4, 5 in addition to pattern space itself. It is evident that using Fourier or Hadamard-Walsh features improve classification accuracy considerably compared to pattern space. The best classification results are, however, obtained with the Karhunen-Loeve transform.

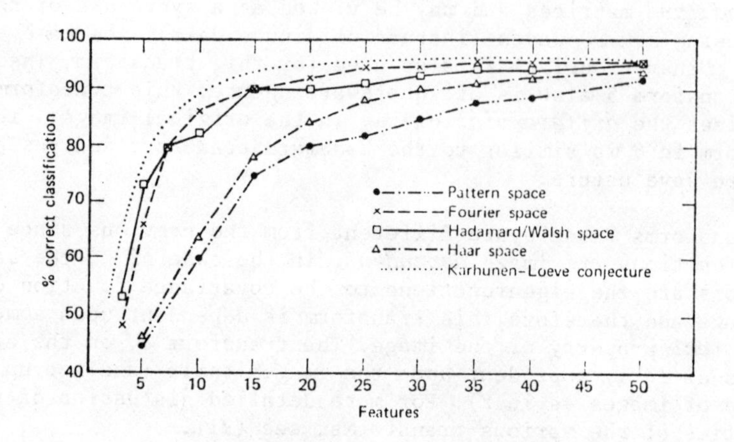

Fig. 16. Result of classification of numbers using pattern space and features from three linear transformations (after (3)).

6. APPLICATIONS IN REMOTE SENSING

Applications of image processing and pattern recognition techniques in remote sensing cover many different areas and many different kinds of problem. As far as computeraided image analysis or classification is concerned, one can not strongly enough emphasize the importance of the MSS-technique. Using this method, *spectral* information or features are used and based upon the spectral signature and its characteristics, objects may be detected and identified provided the spectral signature is unique. This means that one, in a way, is substituting the *spatial* properties, which are so important for visual scene and image interpration, with *spectral* properties. In particular in the area of vegetation and land use, the spectral features are quite efficient for gross classification. In its simplest form, pixelwise classification (each pixel(= picture element)is treated independent on its neighbours), this technique has been operational for many years, in particular from a methodological point of view. The pattern recognition schemes which are applied are in particular various forms of the statistical methods. Of the supervised techniques, simple table-look-up is often applied on small computers while maximum likelihood (normal distribution) is normally applied when large computers when large memory and computing capacity is available. Examples of applications may be found in the rich literature available in this field, in particular from conferences ((23) and (24)).

One problem applying this technique is the question of the homogeneity of the data (and their statistics). One way to find regions exhibiting homogeneity is to use unsupervised methods (hierachical clustering) to find such regions whereafter supervised methods are applied.

In order to decrease computation load, attempts have been made to find regions having the same signature. Then classification methods can be applied to the whole region. This technique (region classification) requires fairly large homogeneous regions having smooth boundaries. Also, linguistic approaches have been tried in remote sensing applications (see e.g. (25)). In this way, the *spatial* distribution of the various categories (found by spectral classification)is utilized.

Another way of incorporating spatial information is to use the image texture. Even if image resolution is too low for object resolution, (as in the case for LANDSAT images) spatial information may still be available as texture information. Also, for types of classes of interest whithin earth resources, individual objects cannot easily be defined. Various methods for extraction textural information are discussed in (14) which also contains many other interesting aspects of image processing in remote sensing. Since for many cases, the use of spectral data alone for classification purposes either gives too low accuracy and/or is subject to large variability, there is an urgent need for incorporating spatial data using for instance texture.

7. ACKNOWLEDGEMENTS

The author is grateful for many comments and suggestions made by Eklundh, Hauska and Åkersten during the outline of this paper and to Ms Lantz for manuscript preparation and typing.

REFERENCES

1. R. Duda, P. Hart: Pattern Classification and Scene Analysis, J. Wiley and Sons, New York, 1973.
2. T. Young, T. Calvert: Classification, Estimation and Pattern Recognition, Elsevier, New York, 1974.
3. H.C. Andrews: Introduction to Mathematical Techniques in Pattern Recognition, Wiley-Interscience, New York, 1972.
4. K.S. Fu (editor): Digital Pattern Recognition, Springer, Berlin, 1976.
5. M. Anderberg: Cluster Analysis for Applications, Academic Press, New York, 1973.
6. T. Pavlidis: Structural Pattern Recognition, Springer, Berlin, 1977.
7. The Third International Joint Conference on Pattern Recognition, IEEE Catalogue No. 76CH1140-3C
8. K.S. Fu, A.B. Whinston (editors): Pattern Recognition Theory and Application, NATO Advanced Study Institute Series, No. 22, Noordhoff, Leyden, 1977.
9. A. Agrawala (editor): Machine Recognition of Patterns, IEEE Press, 1977.
10. A. Rosenfeld, A. Kak: Digital Picture Processing, Academic Press, New York, 1976.
11. A. Rosenfeld (editor): Digital Picture Analysis, Topics in Applied Physics, Springer, Berlin, 1976.
12. J.C. Simon, A. Rosenfeld (editors): Digital Image Processing and Analysis, NATO Advanced Study Institute Series No. 20, Noordhoff, Leyden, 1977.
13. J.K. Aggarwal, R.O. Duda, A. Rosenfeld (editors): Computer Methods in Image Analysis, IEEE Press, 1977.
14. R.M. Haralick: Automated Remote Sensor Image Processing in (11).
15. M. Cassel: A Computer Program for Rectangular Table Look-Up Classification of Multispectral Data, FOA report D 30065-E5, March 1977.
16. D.S.Steiner et.al.: Automatic Processing and Classification of Remote Sensing Data, Chapter 3 in R.F. Thomlinson (editor): Geographic Data Handling, IGU Commission on Geographic Data Sensing and Processing, also FOA reprints B 30015-E1, 1975/76: 30.
17. J.W. Sammon, Jr.: A Nonlinear Mapping of Data Structure Analysis, IEEE Trans. Comp., C-18, 401-409, May 1969.
18. T. Orhaug, K. Severinson-Eklundh: Clustering by Nonlinear Mapping, Proceedings from The Third International Joint Conference on Pattern Recognition, Nov. 8-11, Corona, Calif., pp 386-393, IEEE Cat.No. 76CH1140-3C, FOA reprints B 30018-E1, 1976/77:8.
19. Y.T. Chien: Interactive Pattern Recognition: Techniques and Systems, Computer. vol 9 No. 5, 11-25, May 1976.

20. K.S. Fu: Syntactic (Linguistic) Pattern Recognition, in (4) pp 95-134.
21. W.K. Pratt: Two Dimensional Unitary Transforms, in (12) (pp 1-21).
22. J-O Eklundh: Efficient Matrix Transposition with Limited High Speed Storage, FOA report vol 12., No. 1, B 30021-E1, 1978.
23. Purdue: LARS Symposium Proceedings, Machine Processing of Remotely Sensed Data, June 29 - July 1, 1976, IEEE Cat., No. 76CH1103/MPRSD.
24. Proceedings of the 11th International Symposium on Remote Sensing of Environment, 25-29 April, 1977.
25. J.M. Brayer, K.S. Fu: Application of a Web Grammar Model to an Earth Resources Satellite Picture, pp 405-410 in (7).

DIGITAL ANALYSIS OF MULTI-CHANNEL RADAR DATA AT CCRS

Philippe M. Teillet

Methodology Section, Applications Division
Canada Centre for Remote Sensing
2464 Sheffield Road
Ottawa, Ontario, Canada
K1A 0Y7

Because of the coarse spatial resolution of LANDSAT scanner data, the four or five spectral radiance measurements for each pixel contribute more to successful target identification than does the spatial information. On the other hand, the higher spatial resolutions achieved by synthetic aperture radar (SAR) lead to an increased importance of textural features in pattern classification algorithms. Thus, it is not surprising that automated schemes developed for use on LANDSAT data often work very poorly on radar data. Moreover, the coherent speckle, which is characteristic of SAR imagery, precludes pixel by pixel classification unless the data are smoothed considerably.

The Canada Centre for Remote Sensing (CCRS) has recently developed a rapid, interactive tool for multi-channel image analysis: the CCRS Image Analysis System (CIAS). A description of the hardware and software structure of this system has been described by Goodenough (1977). In designing the CIAS, special emphasis was placed on meeting LANDSAT image analysis requirements. Nevertheless, the system has been successfully used to study multispectral data generated by various airborne systems in use at, or accessible to, CCRS.

Data from the four-channel, simultaneous X-L band SAR of the Environmental Research Institute of Michigan (ERIM) (Rawson and Smith 1974) have been analysed on the CIAS. The chief radar parameters of the ERIM system are listed in Table 1.

T. Lund (ed.), Surveillance of Environmental Pollution and Resources by Electromagnetic Waves, 381–385.

This SAR has just been installed in the CCRS Convair 580 aircraft and is to be flown in support of a variety of pilot projects and research experiments over an 18-month period (Inkster and Kirby 1977). To support these projects, CCRS is developing a software package on the CIAS which is especially suited to the digital analysis of multi-channel radar data.

Both analog and digital output are available from the data recording and processing systems at ERIM. Optical processing yields four strips of 70 mm film, one for each radar channel. The hybrid optical-digital processor generates 800 b.p.i. LARSYS-format magnetic tapes. At CCRS, the CIAS includes a PDS micro-densitometer sub-system which can be used to scan the optically processed image film to produce disk files for subsequent digital analysis on the IMAGE 100 sub-system, or data on magnetic tape from the hybrid optical-digital correlator can be converted to 1600 b.p.i. CCRS/JSC-format tapes by means of a program on the CCRS PDP-10 Time Sharing System. These tapes can then be used as input to the CIAS.

Computer programs have been developed to radiometrically and geometrically correct radar data prior to analysis by converting slant range data to ground range and by removing antenna pattern and range dependent radiometric errors. The geometric distortion can be specified with accurate knowledge of the location of nadir return on the image, the exact aircraft altitude, the geoidal surface elevation, and the sampling interval. The correction involves the careful resampling of radar returns in uniform ground range intervals chosen to reflect either the worst or the best resolution available in the scene under investigation. Another alternative not currently available on the CIAS is to actually use the slant range sampling interval to over-sample in ground range.

Radiometric non-uniformities arise from gain variations in antenna pattern and range losses, both of which are functions of range. The radar range loss due to two-way propagation [$(\text{range})^4$] is moderated by the gain associated with aperture synthesis [$(\text{range})^1$] to yield a net range loss proportional to $(\text{range})^3$. The ERIM radar antenna pattern is reasonably well known and the necessary weighting functions can be calculated for each slant range pixel, given the exact aircraft altitude and the exact size of the slant range pixel in question. It is likely that the system will have to be recalibrated in the Convair 580 configuration. If the terrain under study has uniform reflectivity as a function of angle, an alternative correction method presents itself. The average intensity in each range cell over all azimuth samples is calculated and a weighting factor is determined which would result in all range cells having the same average value.

Once disk files of radiometrically corrected images have been created, the next step depends on whether the four channels will be studied independently or together in various combinations. Automatic classification schemes for multi-channel radar imagery cannot be carried out successfully without prior registration of all channels to be used. Many investigations have considered the X- and L-bands and their respective modes of parallel and crossed polarization separately. Under these circumstances, such as in the study of utility of SAR imagery for ocean wave and ship observation (Lowry et al. 1977), each channel can undergo geometric correction independently without need for inter-channel registration.

The two-dimensional Fast Fourier Transform (2-D FFT) on the CIAS has been modified to provide greater flexibility in the Fourier analysis of radar imagery of the ocean. The ability to perform averaging, resampling, and sliding of 128 x 128 2-D FFTs with the input data were incorporated into the original program.

Vegetation classification of multi-channel ERIM SAR data by computer is feasible and quite promising (Shuchman and Lowry 1977). It is necessary to register the four channels after radiometric correction, but before performing geometric correction, using ground control points in each channel. This step is essential because of variations in film stretching, CRT sweep non-linearities, film drive rate variations, etc. between channels.

To reduce the speckle which would interfere with automatic classification, the data for the vegetation classification were smoothed by means of a 3 x 3 pixel low pass filter. Simple parallelepiped classification was then carried out interactively, making appropriate use of knowledge of radar scattering, textural, and polarization properties of the terrain. Although somewhat broad, the machine-generated classifications showed little overlap and corresponded reasonably well with the limited ground truth at hand.

There remains much to be done in the development of digital analysis techniques specifically suited to SAR data. Presently implemented clustering algorithms place considerable emphasis on Gaussian statistical distributions with variances small compared to means; radar data is characterized by either Rayleigh or exponential distributions with variances comparable to the means. There is thus a need for segmentation routines that can classify areas of like statistics. It would be desirable to use directly the spatial properties of radar imagery for automatic classification purposes. The KANDIDATS spatial feature extraction and clustering programs (Haralick et al. 1976) are currently being implemented on the CIAS together with the results of CCRS research on spatial filtering and segmentation. Spatial features such as gradients, segments, and texture features based on first and second order statistics (Haralick et al. 1973) can be generated and merged into the image files as additional channels. In this way, spatial information can be used to either increase the dimensionality of the feature space used in machine classification or replace some of the spectral features which may be inappropriate for radar data. The merits of such an approach compared to spectral classification, followed by spatial filtering or segmentation, followed by spectral classification are presently under study at CCRS for LANDSAT MSS, airborne MSS, and SAR imagery (Goodenough 1978).

Further research of this kind will be required in the near future if the strengths and weaknesses of SAR data are to be as well understood as are those of LANDSAT and other visible and near-infrared data. Of specific interest to Canada is knowledge of the optimum satellite and airborne sensors as well as processing algorithms for the operational surveillance and inventorying of natural resources. Research into the digital analysis of multi-channel radar data is therefore likely to play a key role in the development of such a remote sensing surveillance program.

Table 1

Parameter	X-Band	L-Band
Centre Frequency	9.450 GHz	1.315 GHz
Resolution	3 x 3 m	3 x 3 m
Transmitter (Peak)	1.2 kw @ 2% DC	6 kw @ 2% DC
Antenna Gain	28 dB	16.5 dB
Antenna Beamwidth	1.1°	7°
Polarization Isolation	23 dB	19 dB
Depression Angle	0 to 90°	0 to 90°
Maximum Range	24 km	24 km
# of Spots/Scan	8,000	4,000

References

Goodenough, D.G., 1977, "The Canada Centre for Remote Sensing's Image Analysis System (CIAS)", Fourth Canadian Symposium on Remote Sensing, Quebec, Quebec, Canada, p. 227.

Goodenough, D.G., M. Goldberg and P.M. Teillet, 1978, "Spectral and Spatial Features for Remote Sensing Classification", in preparation for submission to the Fifth Canadian Symposium on Remote Sensing, Victoria, B.C., Canada.

Haralick, R.M., D.P. Johnson, W.F. Bryant, G.J. Minden, A. Singh and C.A. Paul, 1976, "KANDIDATS Image Processing System", Proceedings of the Purdue Symposium on Machine Processing of Remotely Sensed Data, p. 1A-8.

Haralick, R.M., K. Shunmugam and I. Dinstein, 1973, "Texture Features for Image Classification", IEEE Transactions on Systems, Man, and Cybernetics, 3, p. 610.

Inkster, R. and M. Kirby, 1977, "A Synthetic Aperture Radar (SAR) Program for Environmental and Resource Management in Canada", Fourth Canadian Symposium on Remote Sensing, Quebec, Quebec, Canada, p. 469.

Lowry, R.T., D.G. Goodenough, J.S. Zelenka and R.A. Shuchman, 1977, "On the Analysis of Airborne Synthetic Aperture Radar Imagery of the Ocean", Fourth Canadian Symposium on Remote Sensing, Quebec, Quebec, Canada, p. 480.

Rawson, R. and F. Smith, 1974, "Four Channel Simultaneous X-L Band Imaging SAR Radar", Proceedings of the Ninth International Symposium on Remote Sensing of the Environment, Ann Arbor, MI, USA, p. 251.

Shuchman, R.A. and R.T. Lowry, 1977, "Vegetation Classification with Digital X-Band and L-Band Dual Polarized SAR Imagery", Fourth Canadian Symposium on Remote Sensing, Quebec, Quebec, Canada, p. 444.

EXPERIENCES FROM APPLYING AIRCRAFT AND SATELLITE MSS-DATA TO EARTH RESOURCES INVENTORY PROBLEMS IN SWEDEN

T. Orhaug*, L-E Gustafsson*, L. Wastenson**, S.I. Åkersten*

* National Defence Research Institute, Box 1165, S-581 11 LINKÖPING, SWEDEN

** Department of Physical Geography, University of Stockholm, Box 6801, S-11386 STOCKHOLM, SWEDEN

ABSTRACT. This paper gives a brief report on some of the experiences gained using satellite (Landsat, NOAA) as well as aircraft MSS-data. Projects have been carried out in various application fields like forestry, vegetation and land use. The aim of the projects have been both to introduce to the application community the MSS imagery data as well as the digital computer technique for handling, processing and analyzing such data. In a longer perspective, the aim has also been to investigate possible future operational applications of MSS-data. Forest inventorying management seems to be one promising field for more large scale and operational application of MSS-data.

1. INTRODUCTION

A cooperative research and application oriented project was started in 1972 in order to investigate the potential applications of multispectral scanner (MSS-) imagery data for Swedish conditions. The project, which has been partly sponsored by the Swedish Board for Space Activity, started using Landsat 1 MSS digital data, but has later utilized both analog VHRR data from NOAA satellites and analog MSS data from airborne scanner. The aims of this project have been the following:

T. Lund (ed.), Surveillance of Environmental Pollution and Resources by Electromagnetic Waves, 387–397.

- to introduce to the application managers and to the user communities in Sweden an awareness of the novel data source, MSS imagery data,
- to introduce to these communities the novel means of handling, processing and analyzing imagery data by digital computers and associated techniques.

The way of achieving these aims has been:

- to develop a batch-oriented general purpose handling processing and analyzing software system for MSS-data on a general purpose computer,
- to make this system available through remote terminals,
- to develop handling and processing routines on a dedicated minicomputer,
- to develop software routines for transforming analog MSS data to digital data having the special format needed for the software processing system mentioned above,
- to develop handling and output routines to present color-coded classified data as color ink images,
- to apply these routines to well-defined application oriented studies in various fields of interest,
- to support groups and individuals with a need and a interest for information on remote sensing in general and its data aspects in particular,
- to take part in preparations for and execution of applications projects of varying extent,
- to participate in policyforming committees, advisory groups and long term planning activities.

The application oriented projects have been utilizing the following data sources

- Landsat (1 and 2) digital imagery MSS-data
- NOAA VHRR analog MSS imaging data
- MSS analog data acquired from aircraft as an imaging instrument carrier.

In the sequel the main interest will be focussed upon the first and last of these projects.

2. DATA SOURCES AND DATA CONVERSION

Of the data sources used, the Landsat MSS data present least problems as they are delivered in a well-defined format stored on magnetic tape. Before the Landsat data are fed to the processing software (see later) they are subject to a pre-editing procedure which gives as its output product a (1600 bpi) tape storing the full Landsat (185 km x 185 km) registration in a first order, nearest neighbour, earth rotation corrected form.

The NOAA VHRR analog data have been delivered from Tromsö Telemetry Station. A special program system has been developed which formats the A/D-converted data to a suitable digital tape format. The A/D conversion for NOAA and airborne MSS-data is also carried out using instruments available at FOA (see 1-3 for a description of these routines).

The airborne MSS data were collected during July 1975. The aim of this project was to study various problems with data of better accuracy (both geometrical and spectral) than given by the present Landsat data. Data were collected with a 10 channel analog scanner (Daedalus) on a loan from CNES in France. During 10 days, 17 test areas were registered for a total effective recording time of 106 minutes. These regions were carefully selected to study various applications like vegetation, forest, land use, higher water vegetation, water pollution and water quality. More than 100 persons were engaged in field work in connection with and during the recording period. A brief description of the data collection phase of the project and the various application projects has been given in (4).

3. SOFTWARE SYSTEM

The software system (5) is general in its form and is therefore neither restricted in its use to particular problems nor to the use of a particular data type like Landsat-data. The programs available accept all of the different tape formats offered by NASA and converts the tapes to "line interleaved" format chosen as FOA 355 standard.

Very broadly, the programs constituting the software system can be divided into groups such as data handling, image manipulation and object identification. For a general discussion of digital processing of multispectral data see (6). An instruction oriented introduction to MSS digital processing featuring personal exercises but not requiring access to a computer has also been given in (7).

Programs of the data handling type are the initial editing programs, programs for transfer of data from the edited tapes to computer disk memory for immediate availability in subsequent processing or for the transfer of processed data from disk memory to tapes. Such tapes may be used for transporting data from the GP computer to the dedicated minicomputer PDP 11/40 for display purposes.

An interesting display device used for representing color images is an ink jet drum plotter interfaced with the PDP 11/40. With the use of this plotter, a ready to use 3 color hard copy of 28 cm x 19 cm is made in a couple of minutes. A description of the ink plotter used for displaying multispectral images has been given in (8).

In the group of programs for image manipulation we might refer to programs like transformations of image intensities, scale changes through stepwise sampling or repetition of pixels, generation of composite images (sum - or difference images, ratio images etc), elimination of systematic effects like scan angle effects, six-line striping etc. Also image display using line-printers, symbol coding etc should be mentioned. Programs for geometric rectification and geographical transformations are presently being added to the subroutine library.

In the group of image analysis or object identification three different programs are available, table-look up classification, maximum likelihood classification and non-linear mapping. (For a reference to the first and the third programs see (9-10)). The two first programs belong to the supervised classification type. In this processing mode, the two phases, training and classification respectively, are equally important for a good result using computer-aided analysis.

The careful selection of good type areas is by far the most important aspect of supervised classification and also the most timeconsuming for the investigator. One must first identify the exact pixels representing the type area in question and then study the multispectral characteristics of those pixels. This procedure must usually be repeated several times till a satisfactory result is obtained.

The maximum likelihood classification algorithm is of the special type described by Dye and Chen, (11) using a canonical transformation. In this method a more rapid search for the correct class is possible and thus the number of arithmetic operations is reduced compared to $N(N+1)K$, the number of operations needed per pixel for the usual method. (Here N is the number of spectral channels and K is the number of classes) The computational advantage of the canonical method is specially important for large $N \cdot K$-values.

The classified image may also be subject to a postfiltering or a "cleaning" process in order to produce an improved cartographic product. This image is generated by convolving the classified image with a "majority" filter. If, within a 3 x 3 window, one particular class is found in 5 (or more) of the pixels, the center pixel is also assigned to this class.

4. STUDY OF FOREST INVENTORY

One rationale for the project studying forest inventory potentials is the need to complement the detailed forest maps on the scale 1:100.000 with an updated general map *e.g.* in 1.1.000.000 or 1:250.000 which can be used in planning and controlling forest regeneration. Information of clear cuts and wind thrown stands is of interest for planning immediate replanting. Information about logging residue, ground vegetation (grass - or sub-shrub type), frequency of decidous shrubs in coniferous regeneration areas and density of coniferous plants is of interest in planning clearing of the ground, cleaning of decidous shrub and thinning of the regeneration stands. Using two Landsat scenes (Sept. 4, 1972 (1043-09574) and June 18, 1973 (1330-09523)) an area in Dalsland, Sweden, was investigated. The classification system used is shown in Table 1.

A study of classification accuracy has revealed that regeneration areas with a preponderance of young decidous weeds have *not* been separated from areas dominated by young coniferous stands. Also, the apportionment between regeneration areas *with* and *without* predominance of coniferous saplings is not entirely correct. The latter class contains regeneration areas where the contribution of reflected light from decidous shrubs overpowers that from coniferous plants as a result of their relative frequency of occurrence. This is rather common in those parts of the test area where farm woodlots prevail. In the part of the area where forest service land is in majority, decidous shrubs are only sparsely occurring and the result of computer classification is here satisfactory. For a more detailed discussion of the forest inventory project, see (12) and (13).

TABLE 1. CLASSIFICATION SYSTEM - FOREST INVENTORY - DALSLAND

Land use class	Characteristics of type object
Lake 1	Pure water lake
Lake 2	Pure water lake and shorelines
Old forest > 40 years	
• spruce	> 7/10 spruce
• pine	> 7/10 pine
• mixed coniferous	mixture
Clear cut areas <1 year old	*Clear-cut-72*. Clear cut during the winter 1972. Logging residue still green in September 1972.
	Clear-cut-72. Clear cut during the winter 1973. Logging residue still green in June 1973.
Regeneration area	
• replanted but *not* dominated by coniferous plants	11 years old regeneration area without domination of decidnous shrubs or coniferous plants
• replanted area dominated of young coniferous trees	16 years old regeneration area with domination of coniferous plants
Barren land	
• mire	mire 1, mire 2 dominated by Carex species
• bedrock outcrops	
Agricultural area with crops	Agriculture 1, Agriculture 2 Cultivated land dominated by crop(s). Scattered farm settlements
Rejects	

Using the 10 channel MSS data from the airborne scanner project, 3 different forestry test areas have been studied. Registration altitude for all three was 6000 m and the corresponding ground resolution 15 m. These data are therefore interesting to investigate in anticipation of the use of Landsat D MSS data of 40 m resolution. These areas have been classified using up to 32 classes and the general result seems to be quite accurate. Control of accuracy for discriminating old, growing pine against old, *not* growing pine shows, however, that differences are partly dependent upon ground conditions.

In order to illustrate the spectral signatures for forest inventory, Figure 1 shows two types of such signatures ("simple and "complicated").

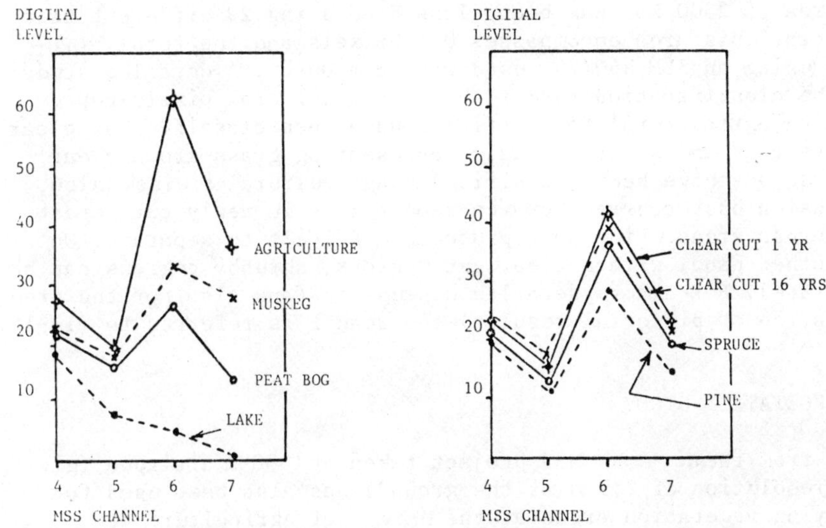

Figure 1. Examples of spectral signatures (from the Dalsland study).

5. LAND USE MAPPING

In connection with the airborne MSS project, the Landsat 2 MSS sensor was opend over one of the test areas (Linköping). This scene (July 3, 1975) has also been used for studies in land use mapping (14-15). The classification system utilized is shown in Table 2.

TABLE 2. CLASSIFICATION SYSTEM - LAND USE - LINKÖPING

1. Urban and Built up land
   - 11. Town, city: commercial services
   - 12. Residential
     - 121 Apartment houses
     - 122 Villa quarter
   - 13. Industrial, Institutional, Communicational land
     - 131 Industrial
     - 132 Communication
     - 133 Institutional
2. Agricultural Land
   - 21. Crop land
   - 22. Pasture and parks
3. Forest Land
   - 31. Coniferous
     - 311 Pine
     - 312 Spruce
   - 32. Mixed
   - 33. Deciduous
   - 34. Regeneration
4. Hydrography
   - 41. Lakes
   - 42. Streams
5. Wet Land
6. Rejects

An area of 2500 km^2 has been classified using 29 different test objects. This area encompasses 0.5 Mpixels and the total CPU-time (using an IBM 360/75) used was 28 minutes. A detailed study of the classification result shows that serveral pixels representing agricultural fallow fields have been classified as clear cut regions and several pixels representing grass-grown clear cut regions have been classified as agricultural fields. Also, confusion has occurred between some pixles in newly cut forest and urban areas. Pine and spruce is difficult to separate. On the other hand, within clear cut regions, shrubby regions can be separated. For a more detailed discussion (and also for the presentation of pictorial results) the reader is referred to (13-15).

6. VEGETATION STUDY

Data from the airborne MSS project taken at 1000 m altitude (giving resolution of 2,5 m at the ground) has also been used for studying vegetation areas at the Univ. of Agriculture at Ultuna (16). In these studies investigation of classification accuracy was also carried out. Generally, classification accuracy, is dependent upon (and therefore gives information on) distribution normality, parameter homogeneity and within cluster scatter (in relation to between cluster distances). A summary of accuracy results is given in Table 3.

Table 3. Classification accuracy (vegetation study)

Object	Accuracy %
Barky 1, grass pras	98 - 100
Road (asphalt)	93
Wheat (fall-type fallow-field	87
Barky 2	73
Oats	64

The reason for the low accuracy obtained for oats, barley 2 and wheat is indicated in Figure 2, the spectral signatures for these crops are very similar.

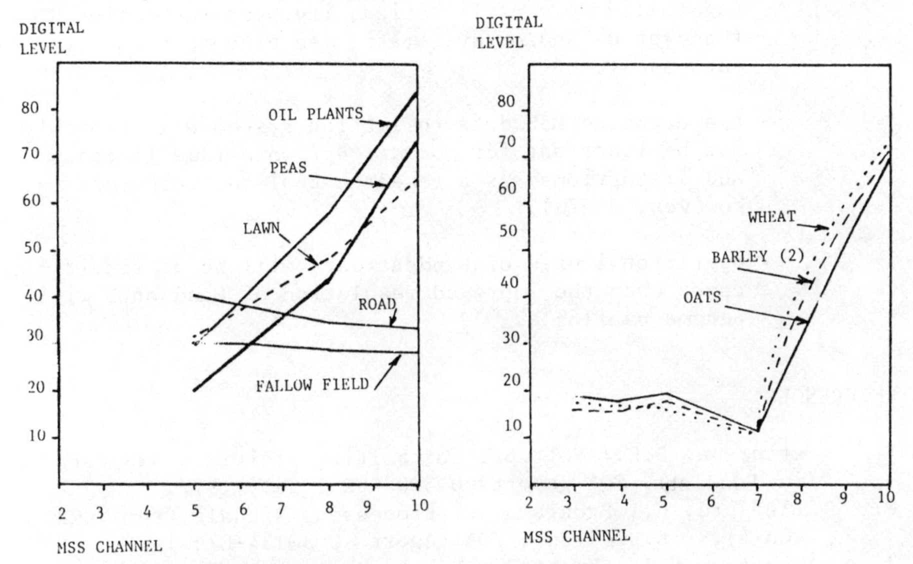

Figure 2. Spectral signatures for vegetation study (airborne MSS-data).

7. CONCLUSIONS

Based upon several years of experience of using MSS-data we may summarize the experience as follows:

- the availability and overall quality of the digital MSS data from the Landsat system has been of significant importance for the introduction of computer techniques to the user community,

- the airborne MSS data has higher cost figures, more severe geometrical (and radiometric) distortions,

- as a result of higher distortions, classification accuracy for the airborne data decreases drastically towards the edges (scan-angle effect) unless adjusted,

- in the central region, the accuracy of the classified airborne MSS data is significantly higher than for Landsat, due to the increased geometrical resolution,

- for Swedish inventory problems - forest inventory seems to be a promising problem area; classification accuracy must, however be improved (either by applying multidate classification, improved resolution (Landsat D) and/or systematic use of complementary databases),

- the airborne MSS data collection system will probably not be important for *operational* work (due to costs and distortions) as a research tool the technique is, however, useful,

- operational uses of Landsat data will be of significance when the improved resolution of Landsat D will become available.

REFERENCES

1. Petterson, S.E.: Software for Sorting Digitized MSS Data (in Swedish), FOA report D 30079-E5, July 1977.
2. Erlefjord, G.: Programs for Processing Signals from VHRR (NOA 4), (in Swedish), FOA report C 30134-E1, 1978.
3. Åkersten, S.I., Engström, B., Lundgren, H., Petterson, S-E.: Digitizing and Handling of Analogregistered MSS Data, FOA report, 1978, (under preparation).
4. Wastenson, L., Borg, C-G,: MSS-75, A Swedish Study Using Airborne Scanning Radiometer (in Swedish), Svensk Lantmäteritidskrift, 1977:3, pp 193-197, FOA report B 30019-E1, 1977/78:1.
5. Gustafsson, L-E, Åkersten, S.I.: Program Library for Handling and Processing of Remotely Sensed Multispectral Data, FOA report, D 30055-E1, October 1976.
6. Orhaug, T., Åkersten, I.: Digital Processing of Multispectral Data, FOA report C 30075-E1, July, 1976a.
7. Orhaug, T., Åkersten I.: A Workshop Introduction to Digital Image Processing, FOA report, D 30053-E1, September 1976b.
8. Hertz, H., Orhaug, T.: The Ink Jet Plotter: A Computer Peripheral for Producing Hard Copy Color Imagery; Computer Graphics Image Processing, Vol 5, No. 11, pp 1-12, January, 1975.
9. Cassel, M.: A Computer Program for Rectangular Table Look-Up Classification of Multispectral Data, FOA report, D 30065-E5, March 1977.
10. Orhaug, T., Severinson-Eklundh, K.: Clustering by Nonlinear Mapping, Proc. the Third Int. Joint Conf. on Pattern Recognition, Nov. 8-11, 1976, Corona, Calif., IEEE Cat No. 76CH1140-3C; FOA report B 30018-E1, 1976/78:8

11. Dye, R.M. Chen, C.S.: Divergence Analysis of Bendix Feature Extraction and Classification System, Proc. Symp. on Machine Processing of Remotely Sensed Data, IEEE Catalogue No. 75 CM 1009-0-C, pp IA-23--IA-27, 1975.
12. Orhaug, T., Wastenson, L., Åkersten, I.: Forest Inventory Using Landsat Digital Imagery, FOA report A 30008-E1, October 1976.
13. Orhaug, T., Wastenson, L., Åkersten, I.: Processing of Digital Landsat-Data - methodology and Application studies for Forest Inventory (in Swedish), Svensk Lantmäteritidskrift, 1977:3, pp 178-185, FOA B 30019-E1, 1977/78:1.
14. Wastenson, L., Boberg, A, Gustafsson, L-E., Ihse, M.: Digital Landsat Data for Land Use Inventory (in Swedish), Svensk Lantmäteritidskrift, 1977:3, pp 186-192, FOA report B 30019-E1, 1977-78:1.
15. Orhaug, T., Wastenson, L., Åkersten, I.: Forest Inventory and Land Use Mapping by Automatic Classification of Digital MSS-data from Satellite and Aircraft, Paper presented at 4th Canadian Symposium on Remote Sensing, Québeck, May 16-18, 1977, FOA B report, 1978.
16. Bertilsson, B., Gustafsson, L-E, Wastenson, L., Åkersten, I.: Detection of Crops Using Airborne Scanner Data (in Swedish), Svensk Lantmäteritidskrift, 1977:3, pp 108-204, FOA report B 30019-E1, 1977/78:1.

WORKSHOP REPORT

WORKING GROUP: SIGNAL PROCESSING, PATTERN RECOGNITION AND IMAGE PROCESSING

PARTICIPANTS: Allen, Alparslan, Ataman, Attema, Bingen, Haydn, Holt, Offner, Röyset, Schmalfuss, Thompson, Wenstøp, Wooton

CHAIRMAN: T Orhaug

1. INTRODUCTION

Pattern recognition and image processing have proven to be both a valuable and necessary technique in many remote sensing applications. The importance of the Landsat data for introducing and promoting computer applications for handling, processing and analyzing digital imagery data can hardly be exaggerated. The experience gained by applying image processing and pattern recognition to remotely sensed data was also recognized as an important background for the discussions during the working group sessions. The discussions were partly focussed upon the interaction between the data processing techniques and the optical and microwave sensors being studied during this Advanced Study Institute.

2. DIGITAL VS. ANALOG TECHNIQUES

Signal and image processing methods may be implemented by analog or digital techniques. Analog implementation may be either electrical or optical. The latter has played an important role for certain applications, in particular when data volumes and/or speed requirements have been large. An example of a data-processing problem which is still commonly solved by optical techniques is the processing of SAR- (Synthetic Aperture Radar) data. With the important developments in digital techniques and computer systems it seems that digital techniques will be used also for problems which hitherto have been restrained to optical processing. The limitation of optical system in terms of I/O-interfaces and processing operators is also of significance for

T. Lund (ed.), Surveillance of Environmental Pollution and Resources by Electromagnetic Waves, 399–402.

this evolution.

The rapid development of the digital methodology and digital systems and the associated applications for pattern recognition and image processing is *not* however, problemfree. The rapid changes of systems and system parts makes systematic development difficult. Also, standardization, both for programs and data would facilitate a better interchange of ideas, methods, data and results. This would easen an improved interaction between research and application groups, a highly desirable goal.

3. INVERSION PROBLEMS

Often the observable quantity is given as a (weighted) linear combination of wanted quantities. Such onerous facts are often met both in one-dimensional problems (e.g. laser profiling) and in two-dimensional problems (limited geometrical resolution). The calculation of the wanted quantity then leads to mathematical inversion problems. A number of methods for solving such problems exist, and a number of proven algorithms are also available.

4. APPLICATION OF PATTERN RECOGNITION METHODOLOGY

The application of the various pattern recognition methods available will be dependent upon the characteristics of the problem at hand. In the field of *optical* spectroscopy as an example, the *a priori* knowledge concerning spectral location and shape is often so large that the more simple detection techniques like matched filtering or correlation detection are sufficient.

Unsupervised recognition techniques like clustering and various kinds of linear (and non-linear) transformation techniques are likely to be of importance in various phases of sensor system development and application. In these *exploratory* phases such techniques can be utilized for investigating clustering tendencies and "sufficient" discrimination features of the collected data. Examples of such investigations were given at this NATO Institute sessions by Schanda (microwave radiometry) and by Attema (microwave imaging radar). In the *application* orientated phases, similar techniques may be applied to study homogeneity in the image (or in the data). This is of importance for selecting homogeneous training areas. When the various sensors which are now in the exploratory phase are to be used in the applications phase, the exploration of the various pattern recognition (supervised and unsupervised) techniques will come into more frequent use to explore the characteristics of the data being collected by the sensors.

5. SPECTRAL AND SPATIAL RESOLUTION

Investigations of the classification accuracy for vegetation and its dependence upon improved spectral and spatial resolution for passive MSS sensors have shown that the accuracy does not improve significantly when spatial resolution is increased with a factor of 2 - 4 compared with the present Landsat system. The highest improvement in performance is obtained when doubling the number of spectral channels from 4 to 7; increase in accuracy goes from approximately 85 % to 90 %. It will be of great interest to see the results from application orientated projects using microwave imaging radars to compare the accuracy with that obtained using passive MSS techniques. The weather independence of the former sensor makes it especially valuable in operational systems.

Another point of great interest is the application of active optical systems like lidars to crop and land use classification. For such sensors, the spectroscopic *microstructure* may be utilized. Imaging optical systems of this kind should be feasible in the future, at least when carried by aircraft platforms. Since for most applications as indicated above, the inherent spatial resolution of the lidar is far higher than needed, a coarse spatial sampling can be used so as to avoid increasing data-volume and data rate beyond manageable proportions.

6. DATA REDUCTION

The increased use of various data collection systems in remote sensing, particularly of imaging types, amounts to a significant growth in data rate and data volume as compared to previous experiences. Even when taking into account the important developments in data and image processing which are taking place, there is an increased need for efficient data volume reduction. Various methods may be used to facilitate this:

- data encoding and compression
- "on-board" processing
- dedicated or "alarm" systems

The problems of data encoding are being studied at various laboratories and so are means and methods of data processing to be carried out in connection with the data collection. The easiest way to facilitate this is to use an "all purpose" instrument and to transmit (or store) for specific uses only those data (e.g. selected spectral channels for MSS-sensor) that are of significance for the specific problem at hand. Another way to ease the data problem is to design special purpose or "alarm" systems which function for specific and well defined purposes. The MTI radar system is a "classical" example of this method. Another example was given at this Institute session by Gjessing suggesting the use of multidimensional matched filtering techniques. If in such systems, the signature variability can be taken into account, they will constitute very efficient ways of data volume reduction.The consideration of such systems in parallel with more multipurpose instruments is considered to be of vital importance in the future.

7. DEVELOPMENT OF OPERATIONAL TECHNIQUES

The development of operational systems often goes through the following phases: sensor development, signature studies, application oriented studies (often including application oriented signature studies in which phase the application or user community studies the signature problem and in particular the dependence of signatures upon various parameters) and operationally oriented studies. An efficient use of imagery data requires geometrical rectification and transformation software or programmable special processors so that multidate images from one sensor can be integrated and/or images from different sensors can be integrated. In the operational phase, it is often necessary to integrate imagery data (in raw or processed form) with other kinds of data. The kind of work now being carried out or initiated at various laboratories using Landsat data in conjunction with data base systems gives an indication of what will be important also in the field of various microwave imaging systems.

Also, the integration of microwave imageries with optical ones, thus extending - and broadening the spectral range, will be of importance for the near future. In this way a more efficient combination of spectral bands may be feasible. Geometrical transformation is a prerequisite for such an integration.